T0329167

Nutrition Economics

Nutrition Economics

Principles and Policy Applications

Suresh C. Babu
International Food Policy Research Institute (IFPRI), Washington, DC, United States

Shailendra N. Gajanan
University of Pittsburgh, Bradford, PA, United States

J. Arne Hallam
Iowa State University, Ames, IA, United States

AMSTERDAM • BOSTON • HEIDELBERG • LONDON
NEW YORK • OXFORD • PARIS • SAN DIEGO
SAN FRANCISCO • SINGAPORE • SYDNEY • TOKYO
Academic Press is an imprint of Elsevier

Academic Press is an imprint of Elsevier
125 London Wall, London EC2Y 5AS, United Kingdom
525 B Street, Suite 1800, San Diego, CA 92101-4495, United States
50 Hampshire Street, 5th Floor, Cambridge, MA 02139, United States
The Boulevard, Langford Lane, Kidlington, Oxford OX5 1GB, United Kingdom

British Library Cataloguing-in-Publication Data
A catalogue record for this book is available from the British Library

Library of Congress Cataloging-in-Publication Data
A catalog record for this book is available from the Library of Congress

ISBN: 978-0-12-800878-2

For Information on all Academic Press publications
visit our website at https://www.elsevier.com

Publisher: Nikki Levy
Acquisition Editor: Nancy Maragioglio
Editorial Project Manager: Billie Jean Fernandez
Production Project Manager: Julie-Ann Stansfield
Designer: Maria Inês Cruz

Typeset by MPS Limited, Chennai, India

To
Susan Hallam
Rekha Gajanan
Chitra Jayachandran

Contents

Preface

The motivation for this volume comes from the conviction that the hunger and malnutrition challenges facing humanity today require a multidisciplinary and multisectoral approach. The multidisciplinary and multisectoral capacity to tackle nutritional problems remains a major stumbling block in setting the nutrition policy agenda, analyzing policy options, designing intervention programs, implementing interventions, monitoring and evaluation of the programs' costs and benefits, and feeding back the evidence on the impact of these programs to the policy process. While there exists a large body of literature on the economic and public policy analysis related to nutritional challenges, this information is not synthesized, and is certainly not accessible in a single collection. This book is a modest attempt to fill this gap.

This volume aims to bring together the disciplines of human nutrition and applied economics and, in the process, addresses a longstanding void and critical knowledge gap that decision-makers face in designing nutrition and health policies and programs. The book attempts to present the state of knowledge in the emerging field of the economics of nutrition.

This book is divided into sections that contain 17 topics and could be used as a semester long textbook for an advanced undergraduate course, or a course in a first year post-graduate program in the faculties of colleges of home economics, nutrition, public policy, and applied economics. Each chapter of this book introduces current theories, models, and conceptual frameworks, and their application to the study of nutrition issues and challenges facing policy makers.

The content of this book covers the interests of a wide ranging audience. What kind of audience will benefit from this book? This book is written primarily for teaching a one-semester public policy course at senior undergraduate level, or at first year graduate level in nutrition, agriculture, economics, development studies, and public policy faculties. It will help a multidisciplinary set of students coming from diverse backgrounds, yet interested in nutrition issues, to understand how economic principles and econometric methods are applied to study nutrition problems and evaluate public policies and programs.

This book will also be of interest to general readers and policy makers willing to explore and update their perspectives about the thematic issues related to the economics of nutrition. This book is intended to help government policy analysts by bringing them up to speed on nutrition policy analysis and program evaluation issues. Nutritionists in the public sector, private sector, and civil society organizations can update their knowledge on the policy and program issues confronting them, and can quickly be on the same page with their economics and other social science counterparts. Program officers in the international agencies and the NGOs working in nutrition issues can have a better understanding of the nutrition policy literature and applied practical tools in designing, monitoring, and evaluating program interventions.

Academic policy researchers can apply policy analysis tools demonstrated in each of the chapters to build their research and analytical capacity toward real-world problem solving. Professionals not familiar with econometric methods, but who are still interested in nutrition policy issues and the current results from the literature on various issues debated and discussed in the nutritional policy arena, can gain from the book as well. For these readers, the book offers the current state of global, regional, and national issues in nutrition public policy making. They can skip the theoretical chapters and data analysis part of various chapters and can still benefit from the rest of the content. Finally, this book is a good reference for those dealing with nutrition policy issues on a regular or occasional basis, including international and national nutrition and development consultants.

The authors of this book have been researching and teaching the issues covered in the chapters for more than 25 years. Although the initial concept for this book was conceived only recently in the present form, the impetus for bringing together the development challenges facing the policy and nutrition community through a unified theme of nutrition economics came from initial discussions we have had with colleagues at Iowa State University, Cornell University, IFPRI, the World Bank, and the University of Pittsburgh.

After teaching various elements covered in this book in several developed and developing country's universities in North America, Europe, Asia, Africa, and Latin America, it has become clear to us that there is a dire need to fill the gaps in the current nutrition policy curriculum. Certainly, this will help educators to successfully train future nutrition policy researchers and policy makers to address the problem of malnutrition in the context of economic policy making. More importantly, there is another compelling reason for writing this book. There is a general frustration among policy makers that the current knowledge on nutrition policy is not available in a single volume. Unfortunately, there is very little incentive for the researchers and faculties to write a multidisciplinary and synthesis oriented volume, as more emphasis is placed on working on original research papers which have a higher marginal value to a researcher. Yet, much of the nutrition policy literature published in high quality journals remains inaccessible to the researchers and policy makers, particularly in developing countries for whom policy making on nutrition matters most.

Policy impact and reduction in nutrition challenges will require the preparation of a new generation of policy researchers and analysts who can solve their own problems without waiting for an external team of researchers to implement research studies and generate evidence for them. In this context, Professor Christopher Udry (1997) notes that a major failing in development studies is the lack of publicly available data along with the relevant computer codes that can help and inform students and new researchers. This book is a contribution to address this capacity challenge in a small way. Over the years, however, data along with relevant computer codes are gradually becoming available, as in the very helpful source by Broussard (2012), which we have liberally adopted in Chapter 13, Economics of School Nutrition: An Application of Regression Discontinuity, on nutritional implications of social protection. Likewise, we have used several simple example-data sets to illustrate the workings of the STATA code and interpretation of the outputs. The data and examples are meant to illustrate the workings of the underlying theory, policy questions, and debates. Thus, this book illustrates a possible direction in a multidisciplinary approach to teaching development problem solving.

While this book can be taught in a standalone semester-long course, it could be taught as a sequel to another preparatory course which can help students gain skills with basic statistical methods. A book covering this basic material, also published by Elsevier/Academic Press, entitled *Food Security, Poverty, and Nutrition Policy Analysis* (Babu et al., 2014) presents content for such a full preparatory course at the undergraduate or first year postgraduate level. These two courses together can form an integral part of a curriculum offering "nutrition policy" as a specialization major or minor at the undergraduate, honors, or postgraduate levels.

The content and their organization in the present book form largely benefited from interactions with a large number of colleagues in several institutions over the years. Without implicating them, we were most influenced and benefited from the research conducted by the following colleagues, reviewing the course contents and reading materials, in some cases teaching jointly with them, listening to their presentations on topics covered in this book, as well as interacting with them on the contents of the book. They and their contribution to the field of nutrition economics have all been a source of inspiration and motivation at various stages of the development of the book.

Akhter Ahmed, IFPRI; Harold Alderman, IFPRI; Abhijit Banerjee, Economics, MIT; Christopher Barrett, Cornell University; Jere Behrman, University of Pennsylvania; Alok Bhargawa, University of Maryland; Katherine Cason, Clemson University; Kenneth A. Dahlberg, Western Michigan University; Timothy Dalton, Kansas State University; Cynthia Donovan, Michigan State University; Paul Dorosh, IFPRI; Esther Duflo, Economics, MIT; Shenggan Fan, IFPRI; Constance Gewa, George Mason University; Stuart Gillespie, IFPRI; Dan Gilligan, IFPRI; Craig Gundersen, University of Illinois at Urbana-Champaign; Jean Pierre Habicht, Cornell University; Lawrence Haddad, IFPRI; Sudhanshu Handa, University of North Carolina; Derek Heady, IFPRI; Sheryl Henriks, University of Pretoria; John Hoddinott, Cornell University; Sue Horton, University of Waterloo; Helen H. Jensen, Iowa State University; Raghbendra Jha, Australian National University; Rolf Klemm and Keith West, Johns Hopkins University; Jane Kolodinsky, University of Vermont; Jef Leroy, IFPRI; Jim Levinsohn, Tufts University; Emily Levitt Ruppert, Agriculture – Nutrition Community of Practice; William Masters, Tufts University; Dan Maxwell, Tufts University; Ellen Messer, Boston University; Bruce Meyer, University of Chicago; Sendhil Mullainathan, Harvard University; Rosamond L. Naylor, Stanford University; Marion Nestle, New York University; Christine Olson, Cornell University; Robert Paarlberg, Harvard Kennedy School; Rajul Pandya-Lorch, IFPRI; Prabhu Pingali, Cornell University; Per Pinstrup-Andersen, Cornell University; Barry Popkins, University of North Carolina; Agnes Quisumbing, IFPRI; Susan M. Randolph, University of Connecticut; Jonathan Robinson, University of California; Marie-Claire Robitaille-Blanchet, University of Nottingham; Beatrice Rogers, Tufts University; Marie Ruel, IFPRI; David Sahn, Cornell University; Prabuddha Sanyal, Sandia National Laboratories; Meera Shekar, The World Bank; Prabhakar Tamboli, University of Maryland; James Tillotson, Tufts University; Peter Timmer, Center for Global Development; Maximo Torero, IFPRI; Francis Vella, Georgetown University; Joachim von Braun, University of Bonn; and Partick Webb, Tufts University.

This book's content has also benefited from the comments on initial drafts of various chapters from several colleagues in universities worldwide. The participants of the courses that we regularly offer on food security, nutrition, and poverty-related themes have helped us shape the pedagogy and the presentation to a common reader interested in the subject matter. Over the years, we have benefited specifically from offering selected contents of the course in the following universities; American University, Washington, DC; University of Pretoria; University of Malawi; Eduardo Mondlane University, Mozambique; Indian Agricultural Research Institute, New Delhi; Tamil Nadu Agricultural University, Coimbatore; Tashkent State Agrarian University, Uzbekistan; University of Nairobi; University of Kwazulu-Natal, South Africa; University of Maryland, College Park; Indira Gandhi National Open University, New Delhi; Royal Veterinary University, Copenhagen; National Agrarian University, La Molina, Lima, Peru; Tufts University, Medford; University of Hohenheim, Stuttgart, Germany; University of Development Studies, Tamale, Ghana; University of Agricultural Sciences, Uppsala, Sweden; University of Zimbabwe, Harare; Iowa State University, Ames; and the University of Pittsburg, Bradford.

Finally, our editors at Elsevier/Academic press, Nancy Maragioglio and Billie-Jean Fernandez, have constantly kept us on track and have provided a high level support to finish the project in time. We are grateful to them. Needless to say, authors alone are responsible for the contents of this book.

Suresh C. Babu, Shailendra N. Gajanan and J. Arne Hallam

PART

INTRODUCTION

CHAPTER

WHY STUDY THE ECONOMICS OF NUTRITION? 1

Investing in early childhood nutrition is a surefire strategy. The returns are incredibly high.
Anne M. Mulcahy

Although advances have been made in science and technology, food production has increased in many developing countries and globalization has made the transportation of food easier, the challenge of malnutrition remains a major development concern. The recent "Global Nutrition Report" (IFPRI-Global Nutrition Report, 2015) summarizes the scale of malnutrition from various sources: about 794 million people are undernourished in terms of calorie deficiency; 161 million children under 5 years of age are stunted—an indicator of chronic malnutrition; 51 million children under 5 years of age are wasted—an indicator of acute malnutrition in the community; micronutrient deficiency, also called hidden hunger, a collective term used for deficiencies in the intake of Vitamin A, iodine, iron, and zinc, affects about 2 billion people; and overweight or obesity affects 1.9 billion adults. Malnutrition thus remains a global development challenge (IFPRI-Global Nutrition Report, 2015).

Malnutrition affects sustainable development of nations in different ways. At the individual level, several manifestations of malnutrition, such as under-nutrition, over-nutrition, unbalanced dietary intake, and hidden hunger in the form of micronutrient deficiencies, have high costs in both the short run and the long run. In the short run, reduced productivity of the work force results in low levels of output produced, and in the long run it increases the cost of health care due to the disease burden associated with poor nutrition. Also in the long run, malnutrition results in the loss of returns to human capital resulting from the reduced ability to fight disease and the reduced human potential of growing children (Hoddinott et al., 2013; Hoddinott, 2016). Finally, the continued high levels of malnutrition in different forms result in high costs, and the recent technical brief of the Global Panel on Agriculture and Food Systems for Nutrition estimates that malnutrition costs up to US$3.5 trillion per year to the global economy (GLOPAN, 2016).

Studying the economics of nutrition cannot just focus on the economics of food intake and its nutrient content. Nutrition is an outcome of a complex process that includes effective use of food and nutrients along with clean water, sanitation, care, health services, and safe food sources. The achievement of good nutrition through these factors involves adequate levels of these elements, which in turn depend on several other determinants including the income levels of individuals, households, and communities. In general, the availability of food and nonfood ingredients to achieve good nutrition depends on the socioeconomic and geographic factors affecting individuals, households, communities, and nations (Smith and Haddad, 2015).

Studying nutrition economics from a policy perspective requires understanding of, among other things, the changes in food and nutrition intake patterns in various societies. The pattern of food

Nutrition Economics. DOI: http://dx.doi.org/10.1016/B978-0-12-800878-2.00001-3

and nutrient intake changes as the countries transform their economies in general, and their rural and agricultural sector in particular. Further, as the income of households increases, the intake of high quality foods increases, along with foods that contain high levels of saturated and trans fats, sugar, and salt contributing to overweight and obesity, and to the development of noncommunicable diseases. Moreover, members of households with higher income levels and urban households tend to have sedentary lifestyles and eat processed foods. Households at the lower strata of income also add more oil and sugar to their already highly calorific diet when incomes increase (Popkin et al., 2012; Hawkes et al., 2007). Thus, studying changes in the pattern of food and nutrient intake is key for designing policies and programs that can help attain optimal nutrition and health for any society (Ruel and Alderman, 2013; Reardon et al., 2012).

A broad set of factors affect nutritional outcomes, although the causes of malnutrition could be wide ranging and complex to tie down. Among the major contributors are food security, clean water, health, sanitation, care, gender relations, and the availability of nutritional and health interventions (Smith and Haddad, 2015). Several technological, environmental, political, cultural, and socioeconomic factors affect determinants of nutrition intake in the community. Consequently, the factors that affect these determinants will have a second round effect on nutritional status. For example, to the extent that climate change affects food security and environmental conditions, it will have a profound effect on the nutritional status of the affected population (Springmann et al., 2016). Similarly, food safety could influence nutritional status by affecting the quality of food and health-related outcomes. Analyzing the determinants of nutritional outcomes can help generate evidence for designing policy and program interventions.

Malnutrition challenges result in economic losses at the individual, household, community, national, and global levels (Horton and Steckel, 2013). Various forms of interventions that address specific nutritional challenges at different stages of life do not come without costs to the countries (Bhutta et al., 2013; Shekar et al., 2015; Rollins et al., 2016). While on the one hand, countries are struggling to reduce under-nutrition, on the other hand, the increasing problems of over-nutrition and obesity are posing emerging threats to the already limited resources countries can invest in health and nutrition (Black et al., 2013; Shekar et al., 2016).

The global community has set the following nutritional targets to achieve by 2025: reducing stunting among children under five years of age by 40%; reducing anemia among women of reproductive age by 50%; increasing exclusive breast feeding to 50%; reducing wasting of children to 5%; reducing low birth weight 30%; and arresting the increase in the childhood obesity (WHO, 2014). Achieving these targets will require concerted efforts at global, national, and community levels. It is estimated that in order to reach the stunting, anemia, breast feeding, and wasting targets an additional annual investment of $7 billion is needed over the next 10 years (Shekar et al., 2016). Even if such levels of resources are mobilized by governments of the countries and the global donor community, reaching these targets will require capacity at local levels, which is grossly lacking, to effectively use these investments and translate them into nutritional outcomes.

There has been renewed and increasing global interest in addressing the nutritional challenges that countries face (WHO, 2014; UNICEF, WHO, and World Bank 2015; Shekar et al., 2016). Malnutrition, both under- and over-nutrition, has been recognized as a key goal in attaining sustainable global development (UN, 2015). The international community has set targets for reducing malnutrition as part of the Millennium Development Goals, and now the recently agreed upon Sustainable Development Goals. Developing countries have devolved national strategies to address malnutrition over the past two decades. Food security, a necessary condition, but not a sufficient condition, has been recognized as a fundamental human right. International declarations and manifestos have called for various measures to address the problem of malnutrition: matching the global nutrition assistance with national goals and interests; increasing the sustainability and resilience of food systems to provide balanced diets to communities; designing and implementing holistic public policies and institutions; developing food systems that provide balanced diets that are safe; and promoting food marketing and distribution systems that minimize waste (FAO and WHO, 2014). Yet implementation of such strategies remains a challenge at all levels because there is a lack of evidence on how well these policy and program interventions contribute to achieving the above-mentioned global development goals. The study of the economics of nutrition from the policy analysis perspective can help in generating such evidence.

What does it take to effectively tackle the problem of malnutrition at the individual, household, community, national, regional, and global levels? How can the vulnerable groups that are prone to malnutrition problems be identified? What programs and policies need to be designed and implemented in order to eliminate malnutrition? How will the multiple sectors that contribute to and are affected by the nutritional challenges come together to address this common malaise? What kinds of food production, marketing, trade, and distribution systems are needed to enhance nutritional outcomes at the national, regional, and global levels? How can the challenges related to uncertain weather patterns and volatile food prices be managed through increasing the resilience of the food systems? And how can common measures and standards be established to deliver safe and nutritious diets to the populations? These are some of the common questions debated frequently in the global, regional, and national forums.

Answering these questions requires evidence that is context specific. Generating timely and credible evidence requires capacity of the professionals in the countries who advise the policy makers on various programs and policies. Unfortunately, such capacity remains weak, and as a consequence many of the nutritional challenges continue to be addressed under the veil of a poor information base. In addition, the capacity to collect, process, analyze, and interpret the data on indicators of food security and nutrition has been a chronic challenge in developing countries (Babu and Pinstrup-Andersen, 1994; Babu, 2015). This book aims to fill this gap by providing content and methods that could help in developing the analytical skills of multidisciplinary policy researchers and analysts to generate evidence for designing and implementing nutrition policies and program interventions.

ORGANIZATION OF THE BOOK CHAPTERS

The chapters of this book are divided into eight parts. The remaining chapters in Part A set the stage for the rest of the book. Chapter 2, Global Nutritional Challenges and Targets: A Development and Policy Perspective, presents the global nutrition challenges and the trend in nutritional indicators. It also provides definitions for the nutritional indicators and clarifies the age-old confusion of terminologies such as hunger, food insecurity, malnutrition, over-nutrition, under-nutrition, and hidden hunger. A common set of definitions can help readers form various disciplines to speak the same language on nutrition challenges and solutions.

Chapter 3, A Conceptual Framework for Investing in Nutrition: Issues, Challenges and Analytical Approaches, uses a widely accepted conceptual framework for studying the determinants of nutrition security that is applied throughout the book, and shows how analytical approaches described in the chapters of the book are interconnected in addressing malnutrition as a multidisciplinary and complex problem that requires a multisectoral approach for developing interventions and their implementation.

Part B introduces basic micro- and macroeconomic principles in the context of nutrition policy analysis. Chapter 4, Microeconomics of Nutrition Policy, presents the microeconomic aspects of nutrition in the context of demand for food and nutrients. It also extends microeconomic principles in the context of intra-household allocation of nutrients among family members, and further considers optimal nutrition as an outcome of a production process involving inputs such as food intake, health status, care, and water and sanitation. Macroeconomic aspects of nutrition policies and the implications of macro policies on nutritional outcomes are presented in Chapter 5, Macroeconomic Aspects of Nutrition Policy.

Part C of the book deals with the economic analysis and policy applications of nutrient intake. Chapter 6, Consumer Theory and Estimation of Demand for Food, deals with methods of estimating the demand for food and nutrients. These estimates can help in understanding the implications of changes in food prices and income that consumers face with regard to their nutrient intake. Policies that change prices through taxes and subsidies could be analyzed with better estimates of nutrient prices and income elasticities. Chapter 7, Demand for Nutrients and Policy Implications, specifically addresses the issues of food subsidies and food taxes, and how such policies could be analyzed in different nutritional contexts of developing and developed economies.

Nutritional status of the population, however, is affected by several factors other than food prices and income. Part D of the book is dedicated to analyzing these factors. These factors go beyond the economic determinants to social, environmental, health, and sanitation issues. Chapter 8, Socioeconomic Determinants of Nutrition: Application of Quantile Regression, begins with an analysis of the socioeconomic determinants of nutrition, and shows how understanding the contribution of various socioeconomic variables can help in the process of developing program and policy interventions at national, local, and community levels. Chapter 9, Intra-household Allocation and Gender Bias in Nutrition: Application of Heckman Two-Step Procedure, is concerned with how nutrients are allocated among the members of a household. It also addresses the issue of gender bias as a determinant of the nutritional status of women and female children.

Chapter 10, Economics of Child Care, Water, Sanitation, Hygiene, and Health: The Application of the Blinder–Oaxaca Decomposition Method, takes a specific look at the key nonprice determinants of nutrition security such as care, water, sanitation, hygiene, and health, also called WASH factors. Health status is an input to nutritional status, and also an outcome of good nutritional status. Understanding this interconnectedness and the cyclical nature of their relationship can help in identifying causal factors in various contexts of under- and over-nutrition problems.

Nutrition program interventions can help in achieving optimal nutrition. Yet increasing the effectiveness and efficiency, as well as the accountability, of nutrition programs requires evaluation of their intended objectives. Part E of the book is dedicated to nutrition program evaluation. The methods of program evaluation are described in Chapter 11, Methods of Program Evaluation: An Analytical Review and Implementation Strategies, starting with a literature review and how program evaluation methods have evolved over the years and been applied to refining nutrition intervention programs. Chapter 12, Nutritional Implications of Social Protection: Application of Panel Data Method, reviews the literature on the analysis of the specific set of interventions called social safety net programs. The issues of conditional and nonconditional cash transfers are addressed, and the methods of analysis are demonstrated. Chapter 13, Economics of School Nutrition: An Application of Regression Discontinuity, explores another set of nutritional interventions through school feeding programs. School feeding programs help not only to reduce hunger among school children, but also increase enrollment in schools and improve educational outcomes. Such multiobjective programs require bringing nutrition and education communities together to develop and implement nutritional interventions.

Part F of the book is concerned with the economics of the "Triple Burden": under-nutrition, over-nutrition, and micronutrient deficiencies all occurring in the same community, and sometimes in the same household. Chapter 14, Economic Analysis of Obesity and Impact on Quality of Life: Application of Nonparametric Methods, addresses over-nutrition in the context of overweight and obesity resulting from over-consumption of fat, sugar, and salt, and from unbalanced diets. It specifically looks at policies and programs that have been put in place to keep a check on the obesity problems, including the controversial tax on certain foods and beverages in order to regulate their consumption.

Part G of the book addresses special topics related to agriculture and food systems that can contribute to solutions to malnutrition and optimal nutrition planning. Chapter 15, Agriculture, Nutrition, Health: How to Bring Multiple Sectors to Work on Nutritional Goals, analyzes policies related to agriculture-nutrition linkages that encourage optimal nutrition transition. Countries going through economic transformation also face distorted nutritional transition in moving from under-nutrition to optimal nutrition. The quality of food and its contribution to the optimal nutrition status is analyzed in the context of developing policies and programs that can help enhance nutritional contributions of the local food and agriculture systems and food value chains. Chapter 16, Designing a Decentralized Food System to Meet Nutrition Needs: An Optimization Approach, deals with the optimal diet problem that has implications for the acceptance of new foods and taste for food and nutrients. The use of linear programming to guide nutrition policies and program interventions is demonstrated.

Finally, Part H and the final Chapter 17, Future Directions for Nutrition Policy Making and Implementation, deals with understanding the nutrition policy process in which multiple actors and players operate from global to local levels, and come from several related sectors. This chapter highlights the challenges of getting the evidence into the hands of policy makers, and the process of policy making, adoption, implementation, and revision. It synthesizes the nutritional challenges of global communities in the context of applications of the methods demonstrated in the chapters of the book, and also highlights the challenges of communicating nutritional policy and programs.

CHAPTER

2 GLOBAL NUTRITIONAL CHALLENGES AND TARGETS: A DEVELOPMENT AND POLICY PERSPECTIVE

If a mother can feed her infant at workplace her baby would be healthy and the mother will also be tension-free and more sincere to her work.

Sheikh Hasina, Prime Minister, Bangladesh, and Member of the SUN Movement Lead Group (Scaling Up Nutrition News, August 12, 2015)

The nutritional status of the population is both an input into the process of economic development of a nation, as well as an outcome of that process (Hoddinott, 2016). A well-nourished individual is likely to be more productive and can better contribute to the solving of individual, communal, and societal challenges. Thus, maintaining a population that is healthy and well-nourished becomes an economic investment in the future of the country. As in any economic activity the objective of studying nutrition economics is to understand how best to allocate scarce resources in order to maximize the nutritional benefits to the society. Since a well-nourished society forms the foundation of healthy and productive human capital of a country, the study of the economics of nutrition broadly relates to applications of economic principles and analytical methods to nutrition problem-solving in order to maximize the nutritional outcomes at the individual, household, community, national, and global levels.

In this chapter, we review the nature and magnitude of the nutritional challenges that the development community faces to provide the background to the thematic issues addressed in the rest of the chapters. The global community has also developed a set of nutritional targets to achieve in the next 15–20 years to address these nutritional challenges. We review them in order to highlight the task ahead for the global nutrition community. We briefly review various organizations and recent initiatives that are currently engaged in achieving the targets at the global, regional, and country levels, and highlight the lessons learned in implementing nutrition policy and program interventions.

GLOBAL NUTRITION CHALLENGES

The study of the economics of nutrition can help address nutritional challenges at various levels. However, it is important to have a basic understanding of the nutrition challenges. Box 2.1 presents

Nutrition Economics. DOI: http://dx.doi.org/10.1016/B978-0-12-800878-2.00002-5

BOX 2.1 BASIC DEFINITIONS OF FOOD SECURITY AND NUTRITION INDICATORS

Food security exists when all people, at all times, have physical, social, and economic access to sufficient, safe, and nutritious food that meets their dietary needs and food preferences for an active and healthy life.

Under-nutrition results from a national status that is less than optimal. Micronutrient and macronutrient deficiencies, low weight for age, shorter height for age, and lower weight for height are all indications of under-nutrition in children.

Malnutrition is a term used to indicate a wide range of nutrition problems and results from the deprivation of one or more factors contributing to good nutrition. This term includes both over- and under-nutrition, and is caused by a number of factors including food and nonfood factors. Any imbalance in optimal nutrition can result in malnutrition.

Stunting is defined as height-for-age less than 2 standard deviations below the WHO child growth standard median for children aged under 5 years. Stunting becomes a public health problem when more than 20% of the population is affected. Stunting reflects long-term and chronic malnutrition.

Wasting is a short-term measure of malnutrition caused by an acute shortage of nutrition. A child is classified as wasted if he or she has weight-for-height less than −2 standard deviations below the WHO child growth standard median for children aged under 5 years. Wasting becomes a public health problem when it affects more than 5% of the population.

Underweight: Measures for children under 5 years of age. Weight for age less than −2 standard deviations below the median value of the healthy population defined by the WHO growth standard. It can result from short- or long-term under-nutrition.

Overweight: A child under 5 years of age is characterized as overweight if the child has weight too high for his or her height—more than +2 standard deviations above the median in a healthy population defined by the WHO growth standard.

Body mass index (BMI): Generally measured for individuals more than five years of age, and expressed as their weight/height (kg/m^2). The normal range for BMI is 18.5–24.99. Below 18.5 the individual is considered thin. Above 24.99 the individual is considered as overweight. A person with BMI above 29.99 is considered obese.

Micronutrient deficiency: When measured for all individuals it is defined as a functional lack of one or more essential vitamins and minerals such as iron, vitamin A, iron, or zinc.

Sources: Based on FAO, 2006. Food Security. Policy Brief Issue 2, June 2006. Available from: <http://www.fao.org> (accessed October 2007). United Nations Food and Agriculture Organization (FAO), Rome; Webb, P., Block, S., 2012. Support for agriculture during economic transformation: impacts on poverty and undernutrition. Proceedings of the National Academy of Sciences of the United States of America. 109 (31), 12309–12314; IFPRI Global Nutrition Report 2014: Actions and Accountability to Accelerate the World's Progress on Nutrition. Washington, DC: International Food Policy Research Institute WHO (2013).

a quick review of the definitions of the food security and nutrition indicators. The nutritional challenges facing the development community can be summarized as follows (FAO 2015; IFPRI-Global Nutrition Report, 2015; UNICEF/WHO/World Bank, 2015):

- about 800 million people eat a calorie-deficient diet every day;
- about two billion people face hidden hunger in the form of micronutrient deficiencies;
- among children under 5 years of age, 161 million are stunted for their age, 51 million are wasted, and 42 million are overweight; and
- about 2 billion people are overweight or obese.

CALORIE-DEFICIENT DIET

Let us compare these recent figures with the figures at the beginning of this century. The level of food insecurity in the developing and developed world is based on the latest data available, published by the Food and Agriculture Organization of the United Nations. Food insecurity at a global level affected 800 million people in 2015. This number was also around 800 million 15 years ago. South Asia and Sub-Saharan African countries still have the largest number of food insecure people. Considerable progress has been made to reduce food insecurity through increasing the supply of food, yet due to the increase in population and poor distribution policies, there are still a large number of people to whom food is not available at the national level. The number of poor people in world has also been a challenging factor contributing to food insecurity. Nutrition poverty is measured in terms of the daily availability of income to meet the nutritional requirements to live a healthy life. In 2015, the number of people who lived on less than US$1.25 a day was 1.1 billion (World Bank, 2015).

While food insecurity at the national level depicts the level of the hunger problem (at least partly, as hunger levels at the individual level may vary), the nutritional status of the population is expressed using several indicators. We discuss them below.

MICRONUTRIENT DEFICIENCIES

The challenge of malnutrition as a global development problem has also been evolving. In the 1980s researchers and the development community were concerned with under-nutrition arising from low food intake and as indicated by nutritional status indicators, such as stunted and underweight children. In the 1990s a new phenomenon was observed, although it probably existed for a long time. The households which had underweight children also had overweight members (usually adults). This phenomenon came to be known as "Double Burden," and posed a different set of challenges to the development community, as such a combination requires a holistic approach to nutrition at the household level (Gillespie and Haddad, 2000).

In the 1990s and 2000s, the serious challenges of micronutrient malnutrition, also called "Hidden Hunger," became prominent as a development challenge and was included in the millennium development goals (MDGs). Added to the under-nutrition and the over-nutrition that exists in the same household, and micronutrient malnutrition, the development community is facing a "Triple Burden" of malnutrition.

Major micronutrient deficiencies that have developmental implications include: vitamin A deficiency, iron deficiency, and iodine deficiency. Anemia, caused mainly by iron deficiency, has been a major micronutrient challenge for development because for pregnant women it goes beyond their own health to the health and nutritional status of children born to anemic women. By far, South Central Asia is the most affected by anemia, where about 70% of women continue to suffer from iron deficiency. Low food intake and intake of food with a low iron content, and poor absorptive capacity contribute to iron deficiency, among other things. Diets focused on a rich iron content, and increasing iron nutrition through the childhood and adolescent periods of girls is important to produce healthy mothers giving birth to healthy children who grow up to fulfill their physical and intellectual potentials.

Iodine deficiency continues to be a major micronutrient challenge, although progress has been made in most of the developing countries. Once again, South East Asia leads as the worst affected region in terms of iodine deficiency. Although there have been improvements through interventions such as the iodization of salt, the challenge continues to reduce iodine deficiency through this approach because of poor regulatory systems that monitor salt iodization and quality control. Lack of iodine is associated with low cognitive ability and mental growth among children, and this can reduce their learning and growth capability.

The micronutrient that has been most limiting in many low income societies is vitamin A. Over the years, supplementary capsule distribution has been a key intervention to address vitamin A deficiency. South Asia and Sub-Saharan African regions continue to lag behind other regions in addressing vitamin A deficiency. Dietary diversity and biofortification of staple foods eaten by poor people have been suggested as sustainable interventions to reduce vitamin A deficiencies. In addition, several other micronutrients are limiting in poor people's diets, causing short- and long-term health and economic losses.

Around the turn of the century, about 2 billion people suffered from anemia. Micronutrient malnutrition has not shown any remarkable decrease in the last 15 years. For example, anemia continues to affect millions of people in the developing world. Anemia, defined as the nutrition challenge when the hemoglobin level goes below 120 g/L of blood, affects women of reproductive age and results from iron deficiency. Lim et al. (2012) attribute about 120,000 deaths in a year to anemia. Poor health resulting from iron deficiency in pregnant women results in low birth weight (LBW) of children, which has further consequences for child malnutrition and health. In the past countries have resorted to interventions such as iron supplementation, and folic acid for adult and pregnant women. For the past two decades, food fortification of grain flours and processed foods has been a popular method of increasing the iron content of foods which are available mostly to only urban populations. Recently, biofortification of selected food crops such as cassava, beans, pearl millet, and wheat has become possible on a limited scale to reach the rural population, as they can grow these crops and consume them without relying on markets. Continued efforts to increase the iron content of the foods consumed by vulnerable groups such as adolescent girls, pregnant women, and lactating mothers is key to reducing iron deficiency anemia.

STUNTING

At the beginning of the century, stunting affected 165 million children and has been a difficult challenge to overcome. Children who are stunted face lifelong challenges and develop into less productive adults due to their reduced cognitive ability, which affects their income earning potential when they are adults. The results of stunting on children are largely irreversible. Due to a weaker immune system resulting from this form of chronic under-nutrition, these children are highly vulnerable to health challenges and are susceptible to death. Under-nutrition is a major causal factor contributing to 45% of all child deaths in developing countries (Hoddinott et al., 2008; Martorell et al., 2010; Black et al., 2013).

Stunting, an indicator of long-term malnutrition among children, is measured by low height for age of children under five years of age. Reflecting a chronic under-nutrition situation, stunting has been related to economic growth in the long run. According to the Global Nutrition Report, at the global level 25% of all the children under five years are stunted, and stunting is most concentrated in the few most deprived countries; about 80% of stunted children live in just 14 countries (IFPRI, 2014). Stunting is a consequence of prolonged underdevelopment in the countries.

In the life cycle, stunting begins with the poor nutritional status of pregnant women. The quality of the diets pregnant women consume affects the growth of the child in utero, and can affect the child's development during pregnancy, and their growth after childbirth. Continued poor diet and growing environment affects the child's growth as well. Stunted children face increased vulnerability to diseases and a high risk of mortality. As multiple factors affect child growth even during normal periods, without any emergencies or natural disasters, tackling stunting as a under-nutrition problem has been extremely challenging (Black et al., 2013; De Onis et al., 2013). It is increasingly recognized that while food quantity and quality diet alone is not enough to reduce the level of stunting, additional investment in providing nonfood inputs such as hygiene, clean water, health services, and child care are all important. This calls for a multisectoral approach at the national, community, and household levels.

WASTING

It is estimated that every year two million children die from acute malnutrition measured in terms of wasting (Shekar et al., 2016). Compared to healthy children, those who are severely wasted are 11 times more likely to die. The short-term nature of the nutritional problem measured by wasting is calculated as a ratio of the weight of the children compared to their height. The problem is most common during food emergencies and in areas affected by natural and manmade disasters such as armed conflicts, although it is increasingly becoming a challenge even in economies which are growing normally. While the solution for wasting calls for emergency responses, in some societies the level of wasting could be high even in normal years, mainly indicating a high level of poverty and the vulnerability of certain segments of the population.

OVERWEIGHT AND OBESITY

Over-nutrition is becoming a major nutrition and health concern; as countries increase their income they also go through transitions in terms of diets and nutritional status (Webb and Block, 2012). Overweight and obesity increase with economic growth and can contribute to close to 3.4 million deaths annually through various pathways (Ng et al., 2014). Among the regions of the world, Latin America and the Caribbean are badly affected by obesity—severe under-nutrition is combined with obesity in the same countries. For example, in Egypt, stunting, wasting, and overweight have all increased in recent years. Some households have people with both nutritional problems. Multiple nutritional disorders are also an individual phenomenon. Women could be overweight or obese, and still be micronutrient deficient such as in iron.

Children who are undernourished in the early part of their life are likely to be vulnerable to overweight and obesity at a later stage of their life, and can become highly susceptible to noncommunicable diseases such as hypertension and diabetes. When children weigh more in proportion to their height, the problem becomes overweight and obese children. Measured in terms of children under five years of age who weigh more than two standard deviations compared to the WHO standards (WHO, 2006), the challenge of overweight and obesity is increasing in both developed and developing countries. Low and middle income countries contribute to about 70% of the overweight children in the world (UNICEF, 2013). As income increases, the lifestyle changes and dietary pattern changes cause a shift toward more sugar and fat-oriented diets of children, resulting in overweight children.

EXCLUSIVE BREAST FEEDING

Exclusive breast feeding of infants up to the first 6 months of their life is an intervention implemented to ensure the nutritional and health demands of a child are met. Increasing the prevalence of exclusive breast feeding to 50% is the current goal of the international nutrition community. Yet, monitoring and measuring this indicator faces data collection challenges and is based on sample surveys. Nutrition education to mothers on the benefits of exclusive breast feeding, hygiene, and weaning foods after six months of a child's life, as well as the interventions that help in providing child care, and storage and use of breast milk in the mother's absence, needs to be promoted (Shekar et al., 2016).

LOWER BIRTH WEIGHT

LBW has major nutritional and health implications from an economic development perspective. It is related to prenatal mortality, and morbidity of children, and LBW children face a higher risk of noncommunicable diseases as adults (Shekar et al., 2016). Any child born with a weight below 2.5 kg is considered a LBW child. LBW is a result of poor nutrition and the health status of the mother during her pregnancy, her health status during adolescent years, and a possible result is the premature birth of her child (WHO/UNICEF, 2005). LBW further contributes to child stunting as the child grows up due to the reduced growth at the time of birth. Several factors contribute to LBW including dietary intake, the work and lifestyle of the pregnant women, and the prenatal availability of heath care for the pregnant women. Globally, LBW continues to be a major cause of child malnutrition. Maternal underweight during pregnancy and lactation add to the problem of child malnutrition.

MALNUTRITION FROM A DEVELOPMENT AND POLICY PERSPECTIVE

At the global level, malnutrition—both under- and over-nutrition—continue to be a major development challenge. While developed countries have overcome the under-nutrition problem, many of them are sliding into the problem of over-nutrition such as overweight, obesity, and the associated health disorders. There is a huge wedge between advanced countries and developing countries on the status of nutrition. However, it should be noted that the focus of preschool malnutrition within developing countries is shifting from Asia to Africa, although a majority of malnourished children in absolute numbers still live in Asian countries. Studying such disparities among the regions of the world is useful for directing the nutrition investments needed at the global level (Shekar et al., 2016).

While malnutrition is in itself a result of society's lack of investment in its population, it has implications for the current and future health status of society. For example, poor nutrition can cause health disorders such as obesity and diet-related noncommunicable diseases. Such health problems account for about 46% of the global burden of disease, and result in about 60% of total global deaths. It is worth noting that overweight coexists in countries where both child and maternal under-nutrition is present, leading to the double burden of malnutrition (Gillespie and Haddad, 2002). The obesity rate among mothers is alarming, with Middle Eastern and North African countries having the highest overweight rates, followed by Latin American and Caribbean countries. The trends in overweight among children in developing countries are also on the rise, particularly among the middle and high income classes.

What costs do the above global malnutrition levels result in to the individual nations and to the global development community? What are the implications of such levels of malnutrition to the health status of the population, the quality of human capital, and hence the productivity of the societies? What amount of investment will it take to address the malnutrition problems of the magnitude described above? Answering these and other related questions requires a thorough understanding of the economic concepts and the implications behind the malnutrition figures.

In the rest of the chapter, we look at how these major indicators are addressed at the global level to understand the nature and extent of the problems and relate them to national and local contexts. Such exposition helps in relating the problem to the global sustainable development goals (SDGs) and identifying context-specific solutions at various levels of analysis. While some progress has been made in developing countries, under-nutrition continues to be a major challenge. Further, increased levels of overweight and obesity are posing a new set of challenges, and have implications for the health of individuals and the increasing level of noncommunicable diseases such as hypertension, diabetes, and heart related illness. In understanding the causes of malnutrition, researchers from different disciplines use different approaches. Economists view malnutrition as a problem of low income and slow economic growth, while nutritionists see the problem arising from having not enough calories or protein, poor nutritional knowledge, micronutrient deficiencies, and infection (Ruel and Alderman, 2013).

HOW CAN THE GLOBAL COMMUNITY ORGANIZE TO ACHIEVE THE GLOBAL NUTRITION GOALS?

Around the turn of the century, the global community set several development goals to guide countries toward progress on various development indicators. These came to be known as MDGs; there were eight specific targets that were set and they were not fully accomplished by 2015, the target year. The development community again geared up to define a new set of goals, calling them SDGs (see Box 2.2). But let us look at how the nutritional trends described earlier in the chapter

BOX 2.2 SUSTAINABLE DEVELOPMENT GOALS

Goal 1: End poverty in all its forms everywhere.
Goal 2: End hunger, achieve food security and improved nutrition, and promote sustainable agriculture.
Goal 3: Ensure healthy lives and promote well-being for all at all ages.
Goal 4: Ensure inclusive and equitable quality education and promote lifelong learning opportunities for all.
Goal 5: Achieve gender equality and empower all women and girls.
Goal 6: Ensure availability and sustainable management of water and sanitation for all.
Goal 7: Ensure access to affordable, reliable, sustainable, and modern energy for all.
Goal 8: Promote sustained, inclusive and sustainable economic growth, full and productive employment, and decent work for all.
Goal 9: Build a resilient infrastructure, promote inclusive and sustainable industrialization, and foster innovation.
Goal 10: Reduce inequality within and among countries.
Goal 11: Make cities and human settlements inclusive, safe, resilient, and sustainable.
Goal 12: Ensure sustainable consumption and production patterns.
Goal 13: Take urgent action to combat climate change and its impacts.
Goal 14: Conserve and sustainably use the oceans, seas, and marine resources for sustainable development.
Goal 15: Protect, restore, and promote sustainable use of terrestrial ecosystems, sustainably manage forests, combat desertification, halt and reverse land degradation, and halt biodiversity loss.
Goal 16: Promote peaceful and inclusive societies for sustainable development, provide access to justice for all and build effective, accountable, and inclusive institutions at all levels.
Goal 17: Strengthen the means of implementation and revitalize the global partnership for sustainable development.

Source: IFPRI Global Nutrition Report (2015), UN (2015), Wage et al. (2015).

are connected to the recently set nutritional goals, and see how important they are as we address the nutritional challenges in order to achieve the larger development goals, and vice versa.

GLOBAL NUTRITION GOALS FOR 2025

Recognizing the serious implications of the continued high levels of malnutrition in all forms, the global nutrition community, led by the United Nations, has set various targets to reach by the year 2025. The international donor community has shown a high level of interest in investing in nutrition, and the level of such investments has been going up. Governments of countries have come together and signed the international declaration during the Second International Conference on Nutrition (FAO and WHO, 2014). More specific commitments have been made as well. For example, more than 57 countries have signed on to the global nutrition initiatives, such as Scaling Up Nutrition (SUN, 2010). All these commitments are aimed at multiple nutrition targets set by the World Health Assembly (WHO, 2012) and are summarized as follows:

- reduction in stunting of children under five by 40% compared to the 2010 global estimates;
- reduction in anemia among women of reproductive age;
- reduction in LBW by 30%;
- prevent any more increase in childhood overweight;

- increase the coverage of exclusive breast feeding in the first 6 months up to at least 50%; and
- reduce childhood wasting to less than 5% and maintain it at that level.

The SDGs set in 2015 have 17 nutrition related goals. Wage et al. (2015) analyzed these goals in the context of health nutrition. The goals are listed in Box 2.2 for easy reference. Although only one of the goals explicitly addresses health and wellbeing, the first three goals contribute directly or indirectly to the nutritional objectives. The linkages and synergies among these goals need to be better understood in the context of designing programs and policies addressing malnutrition.

There is much to be gained by reaching these global nutrition goals. For example, Hoddinott (2016) shows, for an illustrative set of 15 African countries, that meeting just one of the World Health Assembly goals for stunting will add US$83 billion to their national income. The recently announced Foundation for Nutrition Initiative (Shekar et al., 2016) calls for the urgent action needed to address the challenges of malnutrition in all forms. Although the world currently spends about US$3.9 billion annually toward addressing nutritional challenges, this report estimates that an additional annual investment of US$7 billion will be needed to reach the nutritional targets. There is a need for investment in scaling up cost-effective programs by the governments and the development partners which will require an additional US$10 billion annually over the next 10 years. Such an investment, it is estimated, can save 2.2 million lives and result in decreased stunting by 50 million cases in 2025 compared to the 2015 levels (Shekar et al., 2016).

These targets cover a wide range of malnutrition challenges and require multisector interventions, as the causes of these nutritional challenges affect different groups of populations, and in different ways. Collectively they address the challenges of under-nutrition such as stunting, wasting, LBW of children, and low intake of micronutrients causing problems such as anemia, overweight, and obesity among children and adults.

WHAT HAVE WE LEARNED AND HOW DO WE MOVE FORWARD?

Nutrition and policy communities have identified a wide range of intervention programs involving multiple sectors. The food and agriculture systems have been investing in increasing the productivity of food production and increasing the nutrient content of food, with an objective to increase nutrient availability, accessibility, and affordability. In addition, high value agriculture has been growing fast in developing countries to meet the increasing demand for nutrient dense commodities such as milk, livestock, poultry, and fish. The development of value chains and their role in improving the access to these commodities have also been recognized. The role of food safety is also increasingly recognized. Interventions that address vulnerable populations, such as school feeding programs and other forms of social safety nets, have been implemented to increase nutritional accessibility for households. However, the challenge of how to contextualize and to scale up the programs that are already successful remains (IFPRI-GNR, 2015).

In moving forward toward achieving the nutritional targets, several questions remain to be answered.

- How to be focused on nutrition at the national level when there are several other competing resource needs, and multiple sectors have nutrition as part of their mandate but do not give due attention to the nutritional implications of their strategies and policies?

- How do we really develop programs and policies that are nutrition driven and place nutrition at the core of the development agenda?
- What prevents actors and players coming from various disciplinary backgrounds to work together?
- Why is it difficult to develop the needed partnerships across sectors?
- What funding challenges exist in addressing nutrition challenges at the country level?
- How do donors coordinate their investments in nutrition at the country level?
- How to improve the accountability of various partners in delivering on their promises?
- How to improve the data collection, monitoring, and evaluation of the performance of the nutrition interventions to learn and refine them to accelerate the progress toward nutritional targets?

These are some of the common questions we hear at various nutrition conventions and conferences. Yet the challenges of fully understanding the magnitude of the problem, in terms of country and capacity context in designing appropriate interventions, investing resources to back up the policies and programs, effective implementation of the programs and monitoring them for the outputs, outcomes, sustainability, and impact remains.

INCREASING INVESTMENTS IN NUTRITION

Several recent nutritional conventions and summits call for speeding up the process of addressing the nutrition challenges. The 2013 London Nutrition Summit brought together more than 40 organizations to address the slow progress made in the reduction of malnutrition in all forms, and to review the lessons learned in the implementation of the nutrition interventions, the challenges in the scaling up of nutrition programs, and the poor accountability of the actors and players at country, regional, and international levels. The International Conference on Nutrition held in Rome in 2014 revisited the commitment made by the countries during the first International Nutrition Conference held two decades ago. The recent commitment from the Foundation for Nutrition Development highlights the need for coordinated investments from international, national, and private sectors (Shekar et al., 2016).

THE ROLE OF CONTINUED DOCUMENTATION OF EVIDENCE

The importance of good nutrition for sustainable development has been documented well. Recently, issues of the *Lancet* (2013) journal brought together a series of papers that send some key messages to the development community. First, about 45% of child death is contributed to by poor nutrition. Improving the nutritional status of children is thus a key to prevent child death. Second, nutrition investments can give good economic returns. Every dollar invested in nutrition improvement can bring back $16 in return. Third, scaling up of the nutrition program to about 90% coverage can reduce stunting by 20%. This knowledge base makes an important contribution to the understanding of the nutrition challenges countries face as an economic problem that goes beyond human rights, and ethical and moral imperatives that have been used to convince the policy makers to invest in nutrition development (Ruel and Alderman, 2013).

INCREASING THE COMMITMENT TO NUTRITION

Malnutrition affects every country in the world. Rich countries as well as poor countries have to deal with malnutrition in all forms. Given the increases in the overweight and obesity in both developed and developing countries, this double burden requires nutrition policies and intervention programs to be more strategic and context specific. In addition, the micronutrient malnutrition in several societies elevates the challenge to a triple burden. This three-pronged challenge affects countries differently, depending on their status in the economic development process. However, they also bring opportunities to design interventions more strategically. Increased commitments to research, development, and policy in the community is key to exploring the opportunities in this new environment (Gillespie et al., 2012). Preparing the policy makers, as well as other players in the nutrition policy process, to face such a new environment and embrace the new set of challenges to develop innovative interventions by bringing multiple sectors together will require strengthening capacities at all levels (Haddad et al., 2014).

MONITORING AND TRACKING NUTRITION PROGRESS

While the nutrition targets are set at the global level, monitoring how countries progress on these indicators will require country level data systems that routinely monitor the indicators at national and subnational levels. This will help not only in correcting the interventions at the country level, but also will help the global community to track the World Health Assembly indicators. The methods of data collection, processing the data in a timely manner, analyzing them for designing local interventions, and sharing them in open access mode requires countries to strengthen their food security and nutrition monitoring systems. At the global level, such information will be helpful to track how many countries are on target in one or more nutrition indicators and why (Babu, 2015).

INVESTING IN OPEN DATA SYSTEMS

Designing appropriate nutrition interventions and refining them to meet the targets requires quality data on various indicators and causal factors on nutrition. Yet the data systems in developing countries are of low quality, and continue to be externally driven. Many developing countries do not have the data needed for good nutrition planning and policy making. They do not have the capacity to track the changes in nutritional status which is crucial for making appropriate investments and holding the policy makers accountable. In countries where data has been collected on nutritional indicators, they are not frequently updated regularly. This is partly due to lack of evidence that collecting and using reliable and updated data can save resources and result in a better performance of the program implemented (IFPRI-GNR, 2015).

Policy makers often see data collection as a wasteful exercise, and do not invest in data because they are costly. This is partly due to the fact that the data that has been collected is not fully processed, much of what is processed is not analyzed, and what little is analyzed is not reported for policy debate and dialogues. Soon there will be a demand for another round of data collection.

How to improve the use of data and gain the confidence of policy makers to invest in data collection systems must be addressed. This requires capacity at all levels: in data collection, data processing and management and analysis, interpretation, and communicating the results all require capacity at local levels. Until we develop the capacity that will use the evidence coming from the data, and demonstrate the benefit of this, asking countries to invest in data collection is probably not realistic (Babu 1997; IFPRI-GNR, 2015).

RECENT GLOBAL INITIATIVES

At the global level, a number of networks contribute nutrition objectives in order to coordinate their action, resource mobilization, and sharing of information. By mobilizing donor support and identifying priority issues to address at the country level, these learning and operational platforms have been successful in putting nutrition on the top of the development agenda.

For example, Secure Nutrition is one of six of the World Bank's Knowledge Platforms, all of which aim to contribute to the shift toward open development: open data, open knowledge, and open solutions. Secure Nutrition aims to bridge the operational knowledge gap between agriculture, food security, and nutrition. It serves as a source of information on the current challenges and potential solutions, and helps to bring operational knowledge for policy and program interventions.

Another initiative—SUN—began in 2010 to address nutritional challenges at the national level in a focused manner. A major impetus for the initiative came from the uncoordinated efforts of the donor at the country level in supporting the nutrition agenda, and the slow action at various levels toward following up on the commitments made through the First International Conference on Nutrition. The food crisis of 2007–08 further escalated the pressure on the global nutrition community to act fast, and gave the political opportunity to bring nutrition to the top of the development agenda in developing countries. The recently finished evaluation of the SUN movement shows that considerable progress has been achieved in mobilizing the resources at the country level, as the initiative is country-focused and country led, and engages a multisectoral approach by bringing together the stakeholders from the government, private sector, civil society actors, and players. It started by identifying a handful of early raiser countries which showed a high level of commitment to solving nutrition problems, and expanded quickly to 20 countries in the first year.

Currently, 57 countries are members of the SUN movement. The countries in the SUN movement have set nutrition as a high level political priority, and the national plans of the countries attempt to make nutrition objectives coherent across the sectors involved in achieving nutrition objectives. The SUN movement brought the actors and players working on nutrition at the country level together to share experiences and knowledge, and yet the results are not fully achieved as envisioned. This is partly because of the need for better accountability of the participants in the movement. The SUN movement remains a core part of the national and international nutrition architecture in setting nutrition priorities and pushing the nutrition agenda at country levels.

POLICY PROCESS AND BEST PRACTICES

Best practices in addressing nutrition challenges can include designing interventions at the country level. Yet, learning from the successful design and implementation of nutrition interventions is still limited. Learning from examples and drivers of successes in nutrition programs at a national level is important for knowledge sharing and resource saving to avoid costly mistakes others have made. Such documentation will help in understanding the nature of the policy process that is conducive for nutrition problem solving, the nature of sustainable partnerships required between the public and private sector, and various roles the private sector can play in a country context to address the nutrition challenges (Babu et al., 2016).

LOCAL OWNERSHIP AND LEADERSHIP

Several countries and regions have shown a high level of ownership and leadership to address nutritional challenges. Yet how can such leadership be sustained at the country level and how can policies and programs initiated by one set of players be sustained when governments change and leadership changes? One set of players in the nutrition community should be able to advocate for nutritionists, the researchers, and analysts in the public sector, academic institutions, and think tanks. Yet, their capacity to be effective as nutrition advocates needs credibility which can only be built based on the their analytical capacity. Relying on results produced by external groups of researchers may not build the local capacity and credibility needed to address nutritional challenges. It is expected that the governments should take control of the nutrition situation, create an enabling environment for multistakeholder dialogues, design interventions that are context specific, and provide leadership for policies and strategies at the national level. Yet, how to achieve such local ownership and control is not fully understood (Nesbitt et al., 2014).

THE ROLE OF CIVIL SOCIETY

The civil society organizations (CSOs) have increasingly played in important role in the nutrition sector, particularly to keep governments accountable. They also play a critical role in developing new partnerships and implementing these partnerships at the national level. Policy environments that have promoted the participation of CSOs benefit from the broad global knowledge they bring to the discussion at the national level. CSOs have also been part of monitoring government spending on nutrition in several developing countries, particularly as part of the SUN movement. This further helps to integrate the budgeting process in nutrition planning and policy making. Yet, a major challenge for both international and local CSOs is their dependency on the external funding without which their efforts will become unsustainable. The capacity of the CSOs will be a determining factor for their sustainability. Poorly equipped CSOs in terms of their ability to raise resources, conduct analytical studies, and promote nutrition goals in a credible manner is key for their survival as they compete among themselves for limited donor resources within the countries (USAID–Multi-Sectoral Nutrition Strategy, 2014).

MULTISECTORAL APPROACH

The nutrition challenges facing the global community demand a multifaceted and yet integrated approach to problem solving. There is no single approach that will fit all situations and conditions.

The global nutrition community recognizes that creating an enabling environment for the various sectors to come together to develop such integrated solutions is necessary. However, nutrition is a challenge that requires intersectoral coordination of actors and players from multiple sets of stakeholder groups. While there is a natural tendency to work together among these sectors in some countries, such an approach does not work in other countries. This is partly due to a lack of multidisciplinary skills among the professionals working in nutrition. Most professionals are understandably comfortable working in their own fields and would not want to do anything with an intersectoral approach to nutrition. Building local capacity for the multisectoral approach to nutrition has just begun in a few countries, and we have a long way to go. Scaling up the capacity, and strengthening efforts at the country level and at the regional levels can help in the development of a new generation of professionals who have the skills and experience for multidisciplinary problem solving is key for addressing nutrition challenges in the next two decades (USAID, 2014; Gillespie et al., 2015).

EMERGING CHALLENGES

New and emerging challenges continue to confront the nutrition community as well. The role of climate change and how it affects agriculture, food systems, and the health of individuals will have serious implications for nutrition outcomes and the way nutrition programs are designed and implemented (Jones and Allison, 2015). The role of various actors and players, and the new approaches to policy making to address food consumption patterns, such as sugar and soda taxes in several countries, will increase the complexity of policy making and program interventions (Fletcher et al., 2010a,b). In addition, the behavioral changes needed to achieve optimal nutrition go beyond the policy domains to education and communication methods for reaching out to vulnerable populations. Food labeling and communication becomes important as more processed foods are consumed even in developing countries (Andersen, 2015). The role and responsibility of the private sector in dealing with nutritional objectives have also come under serious scrutiny (IFPRI-GNR, 2015).

RESEARCH GAPS

Several research gaps exists in addressing nutrition challenges in an interdisciplinary manner. The Interagency Committee on Human Nutrition (2016) in the United States identifies the following priorities, among others. Further research is needed to understand the role of nutrition as a determinant of nutritional status and human health. Food availability, its access, and the ability of households and individuals to utilize them effectively toward nutritional goals will continue to be a major area of research. In this context, how consumers change their eating habits and their diet intake patterns from a policy induced change in the food system is of critical importance. Nutrition interventions for reducing under-nutrition as well as minimizing over-nutrition will require better understanding of the dual role nutrition plays. Research on healthy eating patterns, factors influencing eating choices, and designing and evaluating interventions that promote healthy eating habits become important to design innovative and sustainable food systems. The role of data systems and innovative analytical methods, and the capacity to use them, is critical for effective nutrition interventions and their evaluation.

CONCLUSIONS

Why do studying the methods of analyzing the factors affecting nutrition and evaluating interventions for their process and outcome impact matter? The problem of malnutrition has a wide range of economic implications. For example, stunted children have a lifelong problem, and the problem is not reversible during their lifetime. The loss in human capacity to produce and contribute to society is reduced considerably by the challenges of malnutrition. Thus, going beyond human rights and ethical issues, malnutrition is an economic issue. More importantly, nutrition challenges and interventions involve the full life cycle, and begin at the first 1000 days of human life—from the time of conception to 2 years of age. It involves all members of the household. Pregnant women, lactating mothers, and adults face specific nutritional challenges. Thus, intervention to address their needs vary and involve a wide range of sectors based on the context of the society one deals with. Nutrition specific interventions require multisector approaches. However, we do not have the people trained multisectorally, and as a result well intended policies and programs do not translate into action on the ground, and the process of delivery of nutrition and health services suffers.

A major challenge in nutrition programming is that we expect specialists to work for nutrition and ask them to address other related concerns, such as gender dimensions, when they are often not trained in these areas. Partly due to this, the design of the interventions is faulty and the implementation suffers as well. Recently, the role of food systems and agriculture, and leveraging them to supply dietary diversity, has been recognized and yet agriculture extension workers are not trained in nutrition, and nutrition front line workers are not trained in agriculture. There is a high level of disharmony in the messages and the activities that are implemented at the community level where nutrition intervention finally matters (Babu et al., 2015). Development of the value chains and integrating nutrition throughout the value chain is talked about, and yet few nutritionists are trained in value chain development, and fewer agribusiness experts are trained in nutrition. Developing public–private partnerships for nutrition requires soft skills such as negotiating abilities, coordination of multisector initiatives, studying and strengthening the policy processes—much of these are not in the nutrition curriculum, even at graduate levels. The rest of the chapters of this book are a modest effort to bring different disciplines together, hoping to make the economic analysis of nutrition challenges more accessible to nutritionists and other social scientists interested in incorporating the economic aspects of nutrition in their tool box.

EXERCISES

1. Identify the malnutrition indicators that are not fully covered in this chapter but still are relevant for nutrition policy making. Prepare a list of them and discuss how important they are in addition to the indicators used for global nutrition targets and why.
2. In order to apply the methods learned in this book for developing policies and strategies, the reader is encouraged to select a country of his or her choice and work out the suggested exercises. Based on the discussions in the chapter, develop a nutrition profile for the country of your choice and discuss the policy strategies that are in place to address the nutritional challenges in the country.

3. Develop an inventory of actors and players in the nutrition policy process in the study country and show how the coordination of nutrition intervention efforts could be improved. Discuss the policy, institutional, and capacity gaps that need to be addressed if nutrition should be addressed as a multisectoral challenge.
4. Selected global and regional initiatives have been reviewed in this chapter. Review other major nutrition initiatives and committees of the multilateral and bilateral donors and discuss their relevance for your region and country of interest. Comment on the adaptation and their linkages to policy and program interventions at the country level? How can they be made more effective and accountable for solving the nutrition challenge in your region?

CHAPTER

A CONCEPTUAL FRAMEWORK FOR INVESTING IN NUTRITION: ISSUES, CHALLENGES, AND ANALYTICAL APPROACHES

3

One in seven people in the world do not get enough to eat, and one in three does not get enough of the nutrients they need to live a healthy and productive life. We have got a very serious malnutrition problem on our hands. Part of that problem can be resolved by understanding the causation including how does the food system affect nutrition, and how does it affect human health?

Per Pinstrup-Andersen, 2001 World Food Prize Laureate and Graduate School Professor at Cornell University

The malnutrition challenges facing humanity have been recognized for a long time. Economists have approached this challenge from an economic growth and investment perspective. For example, two broad sets of economic arguments have been put forth in the last 50 years. The first set of literature focused on the efficiency wage hypothesis, which says the productivity and earning potential of individuals depend on their ability to work. Their ability to work in turn depends on their nutritional status. Well-nourished people could accomplish more than malnourished individuals (Majumdar, 1956; Leibenstein, 1957). Further, the wage rates also determine the income spent on food, and hence there needs to be a certain minimum wage level that can meet the nutritional needs of the population for it to be productive and to contribute to economic activities (Mirrless, 1975; Stiglitz, 1976; Dasgupta, 1995).

Another set of nutrition economists were concerned about the contribution of income increases to nutritional consumption (Timmer, 1981; Wolfe and Behrman, 1984; Behrman and Deolalikar, 1987; Deaton, 1988; Subramanian and Deaton, 1996; Bhargawa, 2008). They approached the problem of malnutrition from the demand for food and nutrition perspective. They argued economic growth is essential for increasing household incomes and this increased income can contribute to the improved nutritional status of the households. This set of studies have estimated the demand changes in the intake of food and nutrients due to changes in the incomes of the households.

We review these studies in detail in Chapter 6, Consumer Theory and Estimation of Demand for Food. The estimates of demand for food and nutrients with respect to changes in income and prices have varied depending on the quality of data availability, methods of estimation, and the level of disaggregation of the households (see some of the earlier literature for example, Pitt (1983), for Bangladesh; Behrman and Deolalikar (1987), Babu (1989), Subramanian and Deaton (1996), Dawson and Tiffin (1998), for India; Sahn (1988) for Sri Lanka; Bouis and Haddad (1992)

Nutrition Economics. DOI: http://dx.doi.org/10.1016/B978-0-12-800878-2.00003-7

for the Philippines; Ravillion (1990), for Indonesia; and Alderman (1988) for Pakistan). The hypothesis that income increases will help address malnutrition depend on the estimated income–calorie relationships in a given society. The debate on the nature and magnitude of these estimated relationships and the methods to improve them continues among economists. Yet the nutritional status of the population does not just depend on income alone, although income to access food and nutrition may be of prime importance. Even if income is adequate, translation of the income at the household level into food consumption and nutrition depend on a host of other factors (Smith and Haddad, 2015). In this chapter we present a conceptual framework to identify these factors. The rest of the chapters in this book explore the linkages among various indicators, and causal factors of food security and nutrition at the household and individual levels.

PROBLEMS OF POVERTY, HUNGER, AND MALNUTRITION

In developing nutrition policies and intervention programs, understanding the nature and extent of malnutrition problems at the global, national, and local levels is a first step. Further, as the problem of malnutrition is often expressed in conjunction with other developmental and welfare indicators such as food insecurity, hunger, and poverty, studying their interconnectedness is important (Babu, 2009). The causal relationship among them and the related challenges on one hand, and their influence on the other can provide additional insights into the understanding of the malnutrition challenges themselves. In this chapter we begin with an overview of such interconnectedness in the developing world which is followed by a conceptual framework for understanding the causes of malnutrition in order to identify a broad set of policy and program directions for reducing malnutrition as identified in the literature.

We begin with the interconnectedness of the concepts of poverty, hunger, food insecurity, malnutrition, and health. The approach to study the multiple dimensions of poverty has recognized nutritional status as a poverty indicator for quite some time (Sen, 1981; Babu and Subramanian, 1988; Rhoe et al., 2008). These concepts can be thought of as indicators of the welfare of the individual or group of individuals at the household level, although ensuring them at the household level does not automatically guarantee all the members will achieve them (see chapter: Intra-Household Allocation and Gender Bias in Nutrition: Application of Heckman Two-Step Procedure on intra-household allocation).

Fig. 3.1 shows the interrelationship of various welfare indicators that also bring together various disciplines. Development economists are often concerned with a reduction in poverty as a major objective. Agricultural development programs focus on the availability of macro and micronutrients (FAO, 2015), while the health indicators include nutrition indicators as the starting point. Educationalists look at nutrition issues from the context of cognitive ability and school achievement of children (see chapter: Economics of School Nutrition: An Application of Regression Discontinuity on school nutrition). These welfare indicators are not only interconnected, but often are caused by same set of subset of variables and have a cyclical causal relationships. One can start the cycle anywhere with any of the variables in one of the boxes, and tell the story of individual and household welfare from the perspective of the indicator.

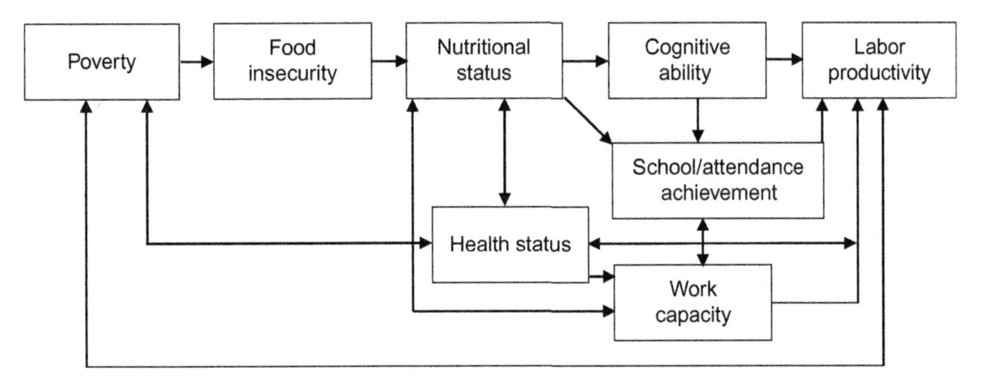

FIGURE 3.1

Interconnectedness of poverty, food security, and nutrition.

Source: *Adapted from Dasgupta, P. (1995). The population problem: theory and evidence. J. Econ. Lit., 33 (4), 1879–1902; Flores (2001); Gillespie et al. (2013); (IFPRI-GNR, 2015).*

In the context of nutritional status, e.g., it can be seen as an end in itself, but also could be treated as an input into good health. Analysts focusing on labor productivity on the other hand will look at health as an input to contributing to work capacity and to labor productivity indirectly and also directly in the short run and in the long run. Increasing health status has been seen as a key welfare indicator in which good nutrition is seen as a key contributing factor along with other variables including lifestyle, taste, preferences, and eating habits, to name a few. The cyclical nature of the linkages can be seen at least from one angle. A highly productive labor force, in an economy that produces adequate employment opportunities, should contribute to reduced poverty which further can result in reduction in hunger and under-nutrition, which can improve the health status and along with other variables can increase labor productivity.

CAUSES OF MALNUTRITION: A CONCEPTUAL FRAMEWORK

In this section we focus on child malnutrition and its causal factors in order to develop a set of policy issues that are confronting the developing community. Fig. 3.2 presents the conceptual linkage of the factors affecting malnutrition at various stages. This time test framework continues to evolve and allows for modification depending on the context of the problem at hand and the emphasis of the variables to be studied. We present the original version with some adaptation to capture the spirit with which the international development community quickly adopted it when it was developed in the early 1990s.

Fig. 3.2 provides a simple but comprehensive framework for understanding the causes of optimal nutrition. The causes of optimal nutrition are separated into three groups of causes. The first group of immediate causes, namely inadequate dietary intake and illness, interact in a negatively synergetic way to contribute to malnutrition. On the next level, household and community deficits

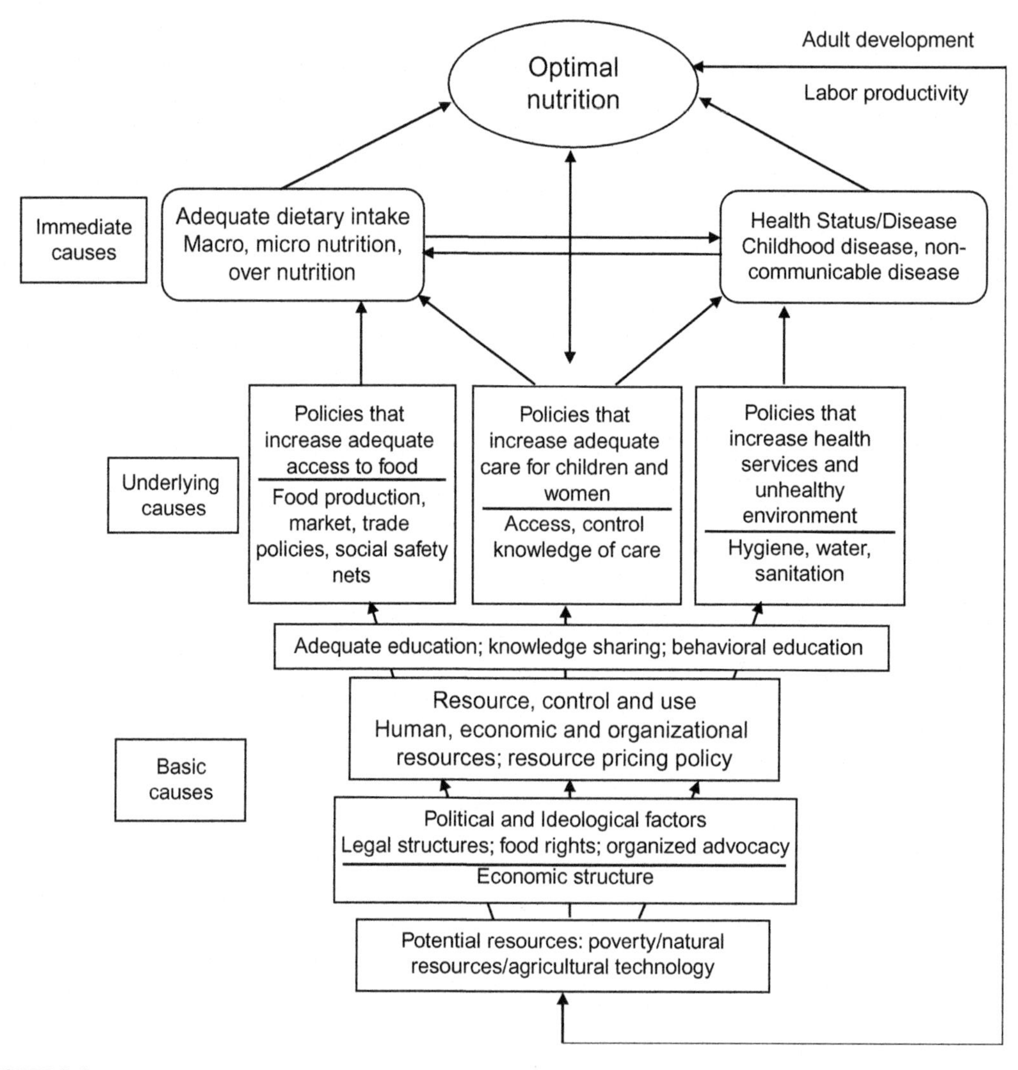

FIGURE 3.2

Determinants of optimal nutrition.

Source: *Based on UNICEF (1991); Smith and Haddad (2000); FAO, 2006.*

in food security, inadequate access to health (reflected by safe water availability, sanitation, and primary health care), and household child care behavior and practices also interact. Summarized as "food security, health, and care" factors, these are the underlying causes of malnutrition. At the more basic level the causes such as control and use of resources, capacity in a society to make use of resources for productive purposes, and the broad set of factors associated with the socioeconomic and political structures of the society determine the status of the underlying causes of malnutrition. Studies have confirmed that the food supplies at the country level are not the most important determinant of nutrition. Other complementary factors such as maternal knowledge, caring practices, access to health services, and water and sanitation are also important determinants of nutritional status (Engle et al., 1999; Smith and Haddad, 2003, 2015).

Much has been learned over the years about addressing the challenges of malnutrition. In the process of attaining nutrition security, achieving secure access to food combined with a sanitary environment, adequate health services, and knowledgeable care to ensure a healthy life for all individuals (World Bank, 2005), several well established facts must be underscored.

Malnutrition is not necessarily just a matter of adequate food intake. Food security does not guarantee nutrition security at the household or individual levels. A nation may have adequate food at the aggregate level for all of its population, but households may not have access to such food, either because of low purchasing power or due to remoteness in physical access to food. Similarly, a household may have adequate food yet individual household members may not have adequate access to food, care, and the health determinants of nutrition. An exception where food intake alone can make some serious dents in the problem of malnutrition is a situation where the population is suffering from prolonged drought and famine conditions (Babu and Pinstrup-Andersen, 1994).

Malnutrition occurs because of poor household purchasing power, even when food is available in the market. Yet, even nonpoor households have malnourished children and individuals, partly due to distorted household priorities in terms of nutrition, which primarily emanates from the lack of information on the benefits of nutrition. Thus, improvements in nutrition do not automatically result from efforts to reduce poverty or broad economic growth (Ruel and Alderman, 2013).

Furthermore, malnutrition does not differentiate between subsistence-oriented households and market-oriented households. Subsistence households can suffer from malnutrition if they have adequate food, but do not have health and appropriate health care practices (Smith and Haddad, 2015). Market-oriented households are also equally vulnerable even if they earn income from the sale of crops, but do not invest that income in improving their nutritional status (Babu and Mthindi, 1994; Babu et al., 2015). Female-headed households are likely to spend their income, even if it is less than male-headed households, on the nutritional objectives of their children (Kennedy and Peters, 1994). Yet, due to a low resource level and poverty that normally strikes female-headed households, they remain vulnerable to malnutrition (Quisumbing and Meizen-Dick, 2012).

Broad-based nutrition strategies are feasible on a massive scale and work in reducing malnutrition (Shekar et al., 2016). Cost-effective programs have been implemented in many countries toward the attainment of better nutrition and health (Bhutta et al., 2013). Malnutrition is a multidimensional problem, measured in multiple facets such as calorie deficit, hunger, micronutrient deficiencies, overweight, and obesity, with varying outcomes ranging from mental impairedness, to blindness, and being a cause of noncommunicable diseases. Thus, addressing malnutrition requires a multidimensional approach (USAID, 2014). Malnutrition has generational implications as well. A low birth weight female grows up to be a stunted girl who is anemic and malnourished, and who as a pregnant woman will give birth to a low birth weight baby, so the cycle of malnutrition continues over the next generation (Gillespie et al., 2015).

Addressing the problems of malnutrition has high economic returns and can enhance the process of economic growth and poverty reduction (World Bank, 2006). Malnutrition does not only affect developing countries. It is also prevalent in middle and high income countries. As countries develop and progress, the nature of the problem changes from one of under-nutrition to that of over-nutrition. The multi-sectoral nature of the nutrition problem involves various sectors such as agriculture, health, and food; ministries have made efforts to jointly address malnutrition, which remains an arduous task in many developing countries (Ruel and Alderman, 2013; USAID, 2014).

Given that malnutrition is an integral part of the human development index, there is a window of opportunity that exists to improve nutritional status, thereby increasing the human development of a country. Recognizing that the Sustainable Development Goals (SDGs) can be achieved through better nutrition helps in increasing the attention of policy makers toward better nutrition investment, since SDGs critically depend on better nutritional outcomes (IFPRI-GNR, 2015; Wage, 2015).

A coordinated effort is needed to focusing on nutritional goals at global and national levels. Small-scale successful interventions never add up to make a large impact on nutrition due to lack of capacity, poor governance, and low priority for nutrition among the decision-makers. The study of economic aspects of nutrition issues, programs, and policies can help in understanding various cost-effective ways and means to improve nutrition. Improving investments at national and global levels and integrating nutrition with a broad set of development plans and programs will require a broader perspective of nutrition (Shekar et al., 2016).

RATIONALE FOR INVESTING IN NUTRITION

The study of the economics of nutrition helps to understand various options and strategies for investing in nutrition. But why should a country, a community, a household, or an individual invest in nutrition? There is broad agreement that improving nutrition increases human productivity. Increased productivity should in turn contribute to increased returns per unit of labor. This helps in increasing the household income if the household is a subsistence household, or when it employs a well-nourished labor force. At the societal level, increased productivity results in efficient provision of both private and public goods. At the national level increased investment in nutrition contributes to a productive labor force, and hence to acceleration of economic growth. The indirect benefits of a better nourished population include reduced health care costs, and less days lost from sick leave and absence from work. Yet, the focus on nutrition as an economic investment in the economy continues to be limited (Hoddinott et al., 2013; WHO, 2013; Shekar et al., 2016).

It is well established that malnutrition leads to direct economic loss (Horton and Ross, 2003; Horton and Stekel, 2013). The strongest linkages between malnutrition and productivity occur through human capital development in the early years of human life. For example, The Cost of Hunger Studies in Africa show that the annual cost of under-nutrition ranges from 3% to 16% of the national income (Hoddinott, 2016). The World Bank (2006) earlier noted that a 1% loss in adult height due to stunting can result in a 1.4% reduction in productivity of a person throughout his or her life. It also noted that eliminating anemia can result in a 5–17% increase in productivity. Such productivity can contribute up to 2% of the GDP where anemia is a serious health issue. It also reported that 60% of the deaths in the world are due to health related challenges that have their origin in poor nutrition (World Bank, 2006).

A lack of investment in improving malnutrition results in high costs in terms of higher budget outlays and eroded national income (World Bank, 2006). A malnourished population will produce less output per worker, will lose opportunities for using their mental capacity effectively due to low cognitive skills in different types of occupations and jobs, and will incur high costs of health related expenses, as they tend to be susceptible to malnutrition-induced illnesses including non-communicable diseases. As children they perform poorly in school and reduce the quality of human capital, which results in significant losses to the economy (Hoddinott et al., 2013). For example, earlier reviews (Behrman et al., 2004; World Bank, 2006) have documented the following specific losses due to malnutrition challenges: low birth weight reduces the IQ of children by 5%; stunting can reduce the IQ by 5–11%; iron deficiency reduces IQ by 10–15%; children who are malnourished early in life have lower scores on the tests of cognitive function, psychomotor function, and fine motor skills; and the reduction of attention span and lower activity levels in school add to the problem of poor performance and a low level of success later in life.

Neglecting investment in addressing nutrition challenges can result in heavy losses to society. However, evaluations of large-scale nutrition intervention programs have shown that nutrition investments in such programs have high returns. Over the years countries have attempted to address the problems of malnutrition through large-scale interventions such as national nutrition programs and early child intervention programs. Experience to date shows that investing in nutrition is a worthwhile economic activity with high benefit cost ratios (Behrman et al., 2004; Hoddinott et al., 2013).

Several earlier estimates have been suggested on the potential benefit of improving nutrition as well. For example, Alderman and Behrman (2004) suggest that preventing low birth weight is worth US$580 per child. Further, the landmark publication (*Repositioning Nutrition as Central to Development*, World Bank, 2006) documents serious evidence on the importance of nutrition investments at country levels as follows: obesity and noncommunicable diseases cost, e.g., China 2% of its national income. In India, the productivity losses from stunting, iodine deficiency, and iron deficiency together result in the loss of 2.95% of its national income. Preventing micronutrient deficiencies in China could increase the GDP between $2.5 and $5 billion, representing 0.2–0.4% of its annual GDP. In India, micronutrient deficiencies continue to cost $2.5 billion per year, which is about 0.4% of its annual GDP. In Sierra Leone, neglect of the problem of anemia among women will lead to an agricultural productivity loss of $94.5 million over the next 5 years (World Bank, 2006).

Nutrition is a sector that will require public intervention, since market forces alone cannot improve the nutritional status of the population. If every individual in a society is guided by the knowledge of the benefits and costs of good nutrition with complete information on how to effectively use food, health, and caring practices to achieve the goal of nutrition, then there would be no need for public intervention. But in many societies, particularly in developing countries, the information that is necessary for educating and motivating individuals and households toward achieving better nutrition is lacking. Thus, identifying such missing elements at the individual, household, and country levels and filling such gaps requires public intervention. The study of the economics of nutrition by understanding individual and household behaviors toward nutritional investment and the associated costs could help in the process of designing and implementing effective policies and programs for reducing malnutrition (Sunny Kim and Phuong Nguyen, 2016). Yet without such strategic interventions, the cost to society of a malnourished population could be substantial.

Malnutrition affects both developed and developing countries. In developing countries undernutrition results in higher health service costs of children due to morbidity, and higher costs due to the additional care needed for malnourished children. In the developed world, over-nutrition affects health care costs as well. For example, in the United States, over-nutrition is a major cause of obesity, which is a major health issue involving an estimated cost of $123 billion (CDC, 2008). However, intervention programs generally have high returns on their investments. Behrman et al. (2004), present the main findings of returns on various nutrition intervention programs. The benefit–cost ratios for nutrition intervention programs are generally quite high. For example, they show that breast feeding promotions in hospitals have a benefit ranging from 5 to 67 times its cost; for integrated child care programs the benefit ranges from 9 to 16 times its cost; iodine supplementation of women ranges between 15 and 520 times cost. The benefits of Vitamin A supplementation for children less than 6 years of age ranges from 4 to 43 times the cost, and that of iron fortification has the highest benefit of all, ranging from 176 to 400 times the cost. Finally, iron supplementation for pregnant women brings a benefit of 6–14 times its cost (Behrman et al., 2004).

Understanding the market failures associated with the provision of nutrition from private individuals and households, and designing policies and programs that can close the information gaps on nutrition through nutrition education can help in designing and implementing public nutrition intervention programs that can result in a well-nourished population—a public good that benefits the society as a whole.

NUTRITION AND OTHER DEVELOPMENT OBJECTIVES

A broad perspective on improving nutrition and its relationship to other development objectives becomes important when the society has competing goals and strategies, all of which require budget resources. How is nutrition improvement related to poverty reduction objectives? While under-nutrition and micronutrient malnutrition are direct indicators of poverty, as mentioned earlier, they can occur in societies where poverty is not a major development problem. Yet, the problem of under-nutrition has strong linkages with income poverty. For example, as the World Bank (2006) noted, the poorest income quantiles have 2–3 times more prevalence of malnutrition than the highest income groups. Furthermore, certain segments of society are more affected with nutritional problems when they live in poverty. For example, in countries where women's nutrition lags behind, improving the nutrition of adolescent girls and women can reduce such inequality among various segments of society (FAO/ADB, 2013).

Finally, poverty and malnutrition reinforce each other through a vicious cycle (Babu and Sanyal, 2008). Poverty is associated with poor diet, an unhealthy environment, physically demanding labor, and high fertility, all of which contribute to malnutrition problems. Chronic malnutrition resulting from poverty reduces human capital, and thus future income prospects of individuals and households. Malnourished women from low income households are likely to give birth to low birth weight babies, thus perpetuating the cycle of poverty. Studying nutrition interventions in the context of poverty reduction strategies and developing poverty reduction programs with a nutritional focus will require a full understanding of the socioeconomic determinants of nutrition, both as a development input and as an outcome of the development process output.

A lot is known about what programs and policies work and what does not in improving nutrition. Much of this knowledge is based on localized and small-scale interventions, although few large-scale interventions have been studied for their nutritional impact (Gillespie et al., 2013). Major knowledge gaps continue to exist when scaling up effective interventions. Several operational challenges should be studied and understood to explain poor scaling up of nutrition interventions. Building a global and national commitment and the necessary capacity to invest in nutrition requires personnel who are knowledgeable about the economic, efficiency, and equity benefits of nutrition interventions. Similar knowledge is needed for mainstreaming nutrition in country development strategies and for reorienting ineffective but potentially beneficial large-scale nutrition intervention programs.

Studying the economics of nutrition can increase capacity for action research on nutrition interventions, and for improving nutrition through learning by doing (Babu 1997a,b, 2011). Such on-the-ground challenges as fine-tuning service delivery mechanisms and strengthening the evidence base for investing in nutrition will require basic training in the evaluation of nutrition programs. Global initiatives continue to face daunting challenges due to inadequate capacity for effective coordination to strengthen high level commitment and funding for nutrition interventions. Focusing on priority countries for investment in nutrition and scaling up of nutrition programs will require trained capacity that has good exposure to develop best practices, mainstreaming nutrition as a development objective. To switch from financing small-scale projects to financing large-scale programs, capacity for program design, implementation, monitoring, and evaluation are also necessary (Babu, 1997c).

Given the key and fundamental role that nutrition plays in the development process, it has been seen as a health and development goal over the years. At the global level as early as 1948 the constitution of the World Health Organization specifically included the improvement of nutrition among its declared functions. Nutrition has since been a major development objective through various international declarations and conventions (WHO, 2005). The declaration of Alma Ata in 1978 recognized the promotion of food and nutrition as one of its essential elements of primary health care (WHO, 1978). The Global Strategy for Health for All formulated in 1981 brought out the importance of monitoring nutrition as an indicator of well-being of populations, and identified nutrition as one of its pillars (WHO, 1981). The World Summit for Children in 1990 identified eight nutrition-related goals for children (UNICEF, 1990). The World Declaration and Plan of Action for Nutrition lists nine goals for nutrition interventions and actions (FAO/WHO, 1992). Health for All in the 21st century also includes malnutrition and micronutrient deficiencies among its targets (WHO, 2000). The World Food Summit in Rome in 1996 identified food and nutrition security for all as its prime goal. The Millennium Development Goals identify directly or indirectly in its six out of eight goals the importance of increasing nutrition (Haddad and Gillespie, 2002). Several of the Sustainable Development Goals either directly or indirectly relate to and depend on achieving the global nutritional targets discussed in the last chapter on "Global Nutrition Challenges" (IFPRI-GNR, 2015; Wage et al., 2015).

Although recognized at global and national levels, nutrition remains a global development challenge and the problems of nutrition along with other food and health related challenges continue to thwart the development process in many developing countries. Identifying nutritional challenges and developing specific nutritional goals and incorporating them into national development policies and strategies needs attention at both the global and national level (Babu, 1999, 2001). Improving food and nutrition security and providing the population with adequate quantities and quality of food continue to be a major challenge. Assessing, analyzing, and monitoring nutrition as an input to the development process, as well as an outcome of development, requires special attention (Babu and Mthindi, 1995a). All this will require a full understanding of the role of nutrition and its economic consequences on society. Designing and implementing nutrition programs and policies requires an understanding of the behavior, incentives, and benefits of nutritional interventions at individual, household, community, national, and global levels (Babu and Mthindi, 1995b).

Various chapters of this book contain analytical tools needed for the study of the economics of nutrition which will enable the reader to provide the basis for understanding complex nutritional challenges and for designing programs and policies that could serve as a major input into nutrition decision-making at various levels.

NUTRITION CHALLENGES AND THEIR ECONOMIC POLICY ANALYSIS

Recent reviews of the nutritional challenges facing development communities highlight several knowledge and capacity gaps in moving forward with evidence based policy making in developing countries. In the last 30 years various factors contributing to nutrition problems have been identified and studied for their policy implications. In the rest of the chapters we try to analyze these determinants of nutrition using different analytical methods. They are worth highlighting here in order to show our choices of policy themes for the chapters and the analytical procedures we demonstrate for readers to gain the needed skills for conducting data-based quantitative analysis and using them in guiding the policy making process in the countries where they work.

Nutrition challenges in a country are inextricably linked to the macroeconomic and microeconomic policy environment in the country. The linkages between the economic growth process, its magnitude and speed, and its effects on poverty and nutrition levels are worth studying in their own right, and to gain better understanding of the inter-sectoral linkages they bring about as they affect several sectors that influence nutritional outcomes. Continued and positive economic growth can bring down poverty, hunger, and malnutrition. Yet as we know, the rate of reduction in poverty is more than the rate of reduction in malnutrition. Why does such a disparity occur? What other missing elements play their role in keeping the malnutrition level stubbornly high in developing countries? Economic transition also brings about dietary transition and the demand for food with higher nutritive value, and the demand for processed food rich in fat and sugar increases (Webb and Block, 2012). The marginal propensity to consume food and nutrient increases as income increases, at least for the poorer segments of the society. Understanding the implications of changes in prices and incomes requires a basic understanding of the economic choices households and individuals make in allocating their resources. The microeconomics of food and nutrition choices are introduced in Chapter 4, Microeconomic Nutrition Policy. The issues related to the implications of macro and trade policies are introduced in Chapter 5, Macroeconomic Aspects of Nutrition Policy.

While nutrition program interventions are increasingly evaluated through randomized control trials, national level policy making on income transfers, food subsidies, and price changes require an understanding of the factors affecting the demand for food and nutrients (Pinstrup-Andersen, 2012). The theory of consumer demand, their assumptions and restrictions as they apply to an empirical estimation of the demand parameters help in understanding the implications of the policy and program interventions on food and nutrient intake (Jensen and Miller, 2012). In addition, sector wide analysis such as multimarket models, and economy wide analysis such as general equilibrium models require estimation of the parameters of demand systems. We introduce the readers to the theory and empirical specifications of demand systems in Chapter 6, Consumer Theory and Estimation of Demand for Food. The nutritional implications of food demand comes from estimating the relationship between price and income changes and nutrient consumption. Applied policy analysis involves estimation of nutrient demand parameters. We introduce this in Chapter 7, Demand for Nutrients and Policy Implications, and show how demand parameters can be used to study nutrition policy.

As discussed in the last two chapters - Chapter 1, Why Study Economics of Nutrition? and Chapter 2, Global Nutrition Challenges - and in this chapter above, malnutrition is a multi-sectoral challenge. This recognition requires the policy analyst to go be beyond the food and agriculture sectors to understand the implications of nonfood factors on nutritional outcomes. Intervention strategies in nutrition have to be coordinated with key ministries such as water, sanitation, gender, education, social protection, and food and agriculture (Smith and Haddad, 2015). However, such coordination requires understanding of the contributions of the factors related to primary health care, immunization, breast feeding, mothers education, child spacing, and other socioeconomic and cultural determinants that are context specific to communities and countries. We address these issues in Chapter 8, Socioeconomic Determinants of Nutrition: Application of Quantile Regression, and apply quantile regression analysis.

Intra-household dynamics have nutritional implications, as the resources available for the individual members of the household matter. In several societies women still receive a lesser share of the household resources, and yet make most of the nutrition decisions for the family. Gender bias in nutrition intake has serious implications for emergency and long-term policy making processes (Babu et al., 1993; Babu and Chapasuka, 1997). Policies that aim at empowering women through education and targeted interventions that increase their decision-making power at the household level are key for improving the nutritional status of women and children. We introduce the binary outcome model and their estimation methods in Chapter 9, Intra-Household Allocation and Gender Bias in Nutrition: Application of Heckman Two-Step Procedure, to gain a better understanding of the nutritional implications of the variables affecting intra-household dynamics.

There is increasing recognition that the factors associated with hygiene, sanitation, access to clean water, and child care have an important role to play in determining the nutritional outcome in a population (Spears, 2012). Yet, these variables are not well studied in the nutrition literature. Further, the implications of policies affecting these variables fall under sectors other than agriculture and health, the traditional sectors held responsible for nutrition. In addition, nutrition education and behavioral changes have gained increased momentum in the study of nutrition behavior. In Chapter 10, Economics of Child Care, Water, Sanitation, Hygiene, and Health: The Application of the Blinder–Oaxaca Decomposition Method, we introduce these thematic issues and apply decomposition methods to study the differences in the outcome variables between two geographical groups of households.

Program evaluations can save resources if the pilot nutrition interventions are studies for their benefits and costs (Banerjee and Duflo, 2009). Nutrition interventions that combine different instruments are reviewed in Chapter 11, Methods of Program Evaluation: An Analytical Review and Implementation Strategies. This chapter also demonstrates various methods of program evaluations developed and applied over the years. For improving programs and their scaling up, emphasis also needs to be placed on the process lessons from pilot interventions in addition to the impact of the programs on nutritional outcomes.

Nutritional implications of social safety net programs have not been fully understood. Limited studies exist that directly address the nutritional objectives (Grosh et al., 2008; FAO, 2015). We review the literature on social safety nets in Chapter 12, Nutritional Implications of Social Protection: Application of Panel Data Method, and apply panel data methods to study the nutritional implications of social protection programs. School nutrition programs are a special type of nutritional intervention which aim to attract children to school, keep them in school, and improve their nutritional intake (Alderman, 2010). Food for education programs have been used to reduce the nutritional burden of households (Ahmed and Babu, 2006). The outcome of the program interventions, however, depend on the context in which these programs are designed, implemented, and evaluated. In Chapter 13, Economics of School Nutrition: An Application of Regression Discontinuity, we review the current status of the literature on school feeding programs and apply regression discontinuity models to explain the gains from school nutrition interventions.

Overweight and obesity coexist in communities along with under-nutrition challenges. Designing interventions in this context requires innovation (Babu, 2002; Pinstrup-Andersen, 2013). We introduce nonparametric estimation methods to study obesogenic factors and other food security indicators at the household level in Chapter 14, Economic Analysis of Obesity and Impact on Quality of Life: Application of Nonparametric Methods. How food systems could be changed to address the emerging food and nutrition issues is discussed in Chapter 15, Agriculture, Nutrition, Health: How to Bring Multiple Sectors to Work on Nutritional Goals, which demonstrates that innovations are needed in designing interventions through research and innovation for increasing dietary diversity, bio-fortification, food safety, and nutritional enhancement throughout the food value chains.

While addressing nutritional challenges, particularly micronutrient deficiencies which continue to be a major set of nutrition challenges, designing cropping patterns that meet the nutritional needs of households could be an important approach (Pinstrup-Andersen, 2013). Yet agricultural extension workers lack knowledge about nutritional needs, and approaches to connect the needs of the farming households to advice on the crops that could be grown in the context of the agroecology, resource constraints, and market considerations are required (Babu et al., 2016). In Chapter 16, Designing a Decentralized Food System to Meet Nutrition Needs: An Optimization Approach, we use a simple linear programming model incorporating the nutritional needs of farming families. Further, we show how the results developed from such farm level models could be useful in recommending cropping patterns that can help farming households to meet their nutritional needs. In the final chapter, we highlight the need for coordinated efforts to increase nutrition governance, accountability, capacity, funding, and sustainability to achieve global nutritional targets.

CONCLUSIONS

In this chapter the issues related to policies and programs that affect nutritional outcomes are identified through a conceptual framework, connecting them to larger poverty, hunger, and food security issues. The need for investment in nutrition policies and programs has evolved over the years, and has been highlighted in major conventions and declarations. The need for investing in nutrition and highlighting it as a central challenge of development has been pursued vigorously by the international and bilateral agencies over the last 30 years. However, the challenge of bringing together several sectors that contribute to the nutritional program, policy development, and implementation still remains (WHO, 2013; FAO/WHO, 2014; USAID, 2014). In order to connect various methodological approaches to address the challenges identified in Chapter 2, Global Nutrition Challenges and Targets: A Development and Policy Perspective, and Chapter 3, A Conceptual Framework for Investing in Nutrition: Issues, Challenges and Analytical Approaches, the rest of the chapters of this book introduce and demonstrate the issues, analytical methods, empirical strategies, and policy insights from the results.

Table 3.1 below summarizes the broad set of issues and the chapter contents that may help the reader to choose the policy theme and the related chapter for easy navigation through the contents in the rest of the pages of this book.

Table 3.1 Policy challenges addressed and methodological approaches used in various chapters of this book

S. No.	Thematic/Policy Issues	Chapter and Methods Applied
1	Economic growth is essential for reduction in poverty and malnutrition. However, the rate of reduction in malnutrition is less than the reduction in poverty.	Chapter 1, Why Study the Economics of Nutrition?, Chapter 2, Global Nutrition Challenges and Targets: A Development and Policy Perspective, Chapter 3, A Conceptual Framework for Investing in Nutrition: Issues, Challenges and Analytical Approaches, also Chapter 4, Microeconomic Nutrition Policy, Chapter 5, Macroeconomic Aspects of Nutrition Policy, Chapter 6, Consumer Theory and Estimation of Demand for Food; Chapter 4, Microeconomic Nutrition Policy, and Chapter 5, Macroeconomic Aspects of Nutrition Policy, address micro- and macroeconomic aspects of nutrition.
2	Malnutrition is a multi-sectoral challenge and needs to go beyond the health and agriculture sectors. Intervention strategies have to be coordinated with key ministries such as water, sanitation, gender, education, social protection, and food and agriculture.	Chapter 2, Global Nutrition Challenges and Targets: A Development and Policy Perspective, Chapter 3, A Conceptual Framework for Investing in Nutrition: Issues, Challenges and Analytical Approaches, Chapter 4, Microeconomic Nutrition Policy, Chapter 10, Economics of Child Care, Water, Sanitation, Hygiene, and Health: The Application of the Blinder–Oaxaca Decomposition Method, Chapter 12, Nutritional Implications of Social Protection: Application of Panel Data Method, Chapter 13, Economics of School Nutrition: An Application of Regression Discontinuity, and Chapter 17, Future Directions for Nutrition Policy Making and Implementation. An introduction to the issues are developed in Chapter 2, Global Nutrition Challenges and Targets: A Development and Policy Perspective, and Chapter 3, A Conceptual Framework for Investing in Nutrition: Issues, Challenges and Analytical Approaches; Chapter 8, Socioeconomic Determinants of Nutrition: Application of Quantile Regression, and Chapter 10, Economics of Child Care, Water, Sanitation, Hygiene, and Health: The Application of the Blinder–Oaxaca Decomposition Method, cover in more detail.

(Continued)

Table 3.1 Policy challenges addressed and methodological approaches used in various chapters of this book ***Continued***

S. No.	Thematic/Policy Issues	Chapter and Methods Applied
3	Policy environment, leadership, governance, coordination, financing, and sustainability of interventions are key for such coordination of multi-sectoral activities.	Chapter 2, Global Nutrition Challenges and Targets: A Development and Policy Perspective, Chapter 3, A Conceptual Framework for Investing in Nutrition: Issues, Challenges and Analytical Approaches, and Chapter 17, Future Directions for Nutrition Policy Making and Implementation. These chapters cover the issues of coordination, governance, accountability, funding, sustainability, and impact.
4	In addition to food and nutrition intake, issues related to primary health care, immunization, breast feeding, mothers education, child spacing, and other socioeconomic and cultural determinants that are context specific need to be fully understood.	Chapter 2, Global Nutrition Challenges and Targets: A Development and Policy Perspective, Chapter 3, A Conceptual Framework for Investing in Nutrition: Issues, Challenges and Analytical Approaches, Chapter 8, Socioeconomic Determinants of Nutrition: Application of Quantile Regression, and Chapter 17, Future Directions for Nutrition Policy Making and Implementation. Issues related to immediate causes (Fig. 3.2) are addressed in these chapters. In Chapter 8, Socioeconomic Determinants of Nutrition: Application of Quantile Regression, we introduce the quantile regression technique to study the determinants of nutrition.
5	Intervention strategies that help to improve service delivery in sanitation, child care, clean water, and nutrition education for behavioral change are needed to increase the effectiveness of nutritional investments.	Chapter 2, Global Nutrition Challenges and Targets: A Development and Policy Perspective, Chapter 3, A Conceptual Framework for Investing in Nutrition: Issues, Challenges and Analytical Approaches, Chapter 10, Economics of Child Care, Water, Sanitation, Hygiene, and Health: The Application of the Blinder–Oaxaca Decomposition Method, and Chapter 17, Future Directions for Nutrition Policy Making and Implementation. Issues related to understanding causes are discussed here. In Chapter 10, Economics of Child Care, Water, Sanitation, Hygiene, and Health: The Application of the Blinder–Oaxaca Decomposition Method, we introduce decomposition techniques to study the differences in outcome variables.
6	Empowering women through education and interventions that increase their decision-making power at the household level is key for improving woman and child nutrition. Continued understanding of the intra-household dynamics	Chapter 2, Global Nutrition Challenges and Targets: A Development and Policy Perspective, Chapter 3, A Conceptual Framework for Investing in Nutrition: Issues, Challenges and Analytical Approaches, and Chapter 9, Intra-Household Allocation and Gender Bias in

Table 3.1 Policy challenges addressed and methodological approaches used in various chapters of this book *Continued*

S. No.	Thematic/Policy Issues	Chapter and Methods Applied
	in resource allocation and utilization of nutrition and health services are needed.	Nutrition: Application of Heckman Two-Step Procedure. Gender relations, intra-household decision-making, and women empowerment issues are addressed, with Chapter 9, Intra-Household Allocation and Gender Bias in Nutrition: Application of Heckman Two-Step Procedure, introducing binary outcome models.
7	Monitoring and evaluation of programs implemented both for their process lessons and impact of the benefits are needed.	Chapter 11, Methods of Program Evaluation: An Analytical Review and Implementation Strategies. Program evaluation methods are presented and reviewed in Chapter 11, Methods of Program Evaluation: An Analytical Review and Implementation Strategies. Current approaches to program evaluation are demonstrated with empirical results.
8	Social safety net programs require context specific approaches and the nutritional benefits from them can only be realized if they are specifically addressed to nutritional goals during the design stage.	Chapter 11, Methods of Program Evaluation: An Analytical Review and Implementation Strategies, Chapter 12, Nutritional Implications of Social Protection: Application of Panel Data Method. Protecting vulnerable groups from abject poverty and securing future generations from having malnutrition related damage requires social safety net programs. Chapter 12, Nutritional Implications of Social Protection: Application of Panel Data Method, applies panel data methods to study the impact of social safety nets.
9	School nutrition programs continue to be the most popular intervention to attract children to school, to keep them in school, and to increase their learning abilities. However, the results differ depending on the context and program design.	Chapter 11, Methods of Program Evaluation: An Analytical Review and Implementation Strategies, Chapter 13, Economics of School Nutrition: An Application of Regression Discontinuity. School nutrition programs help children in school and provide nutrition needed for normal growth of the children. In Chapter 13, Economics of School Nutrition: An Application of Regression Discontinuity, we introduce the Regression Discontinuity Model to study the benefits of school feeding programs.
10	Overweight and obesity are increasing even in developing countries. Strategies to address over-nutrition and under-nutrition in the same community and households require innovations in all sectors.	Chapter 2, Global Nutrition Challenges and Targets: A Development and Policy Perspective, Chapter 3, A Conceptual Framework for Investing in Nutrition: Issues, Challenges and Analytical Approaches, Chapter 10, Economics of Child Care, Water, Sanitation, Hygiene, and Health: The Application of the Blinder–Oaxaca Decomposition Method, Chapter 17, Future

(*Continued*)

Table 3.1 Policy challenges addressed and methodological approaches used in various chapters of this book *Continued*

S. No.	Thematic/Policy Issues	Chapter and Methods Applied
		Directions for Nutrition Policy Making and Implementation. The increasing challenge of overweight and obesity calls for policies that can affect consumption patterns, as well as behavioral challenges. We introduce non-parametric techniques for studying factors affecting obesity.
11	Micronutrient deficiencies continue to be a major set of nutrition challenges. Continuous multipronged interventions are needed for iodine, iron, Vitamin A, and other micronutrients; agriculture and food systems have a major role to play in solving micronutrient malnutrition problems. Designing interventions through research and innovation for increasing dietary diversity, bio-fortification, food safety, and nutritional enhancement throughout food value chains is critical.	Chapter 2, Global Nutrition Challenges and Targets: A Development and Policy Perspective, Chapter 3, A Conceptual Framework for Investing in Nutrition: Issues, Challenges and Analytical Approaches, Chapter 15, Agriculture, Nutrition, Health: How to Bring Multiple Sectors to Work on Nutritional Goals, Chapter 17, Future Directions for Nutrition Policy Making and Implementation. Agriculture and food systems oriented interventions are studied in Chapter 15, Agriculture, Nutrition, Health: How to Bring Multiple Sectors to Work on Nutritional Goals. They include approaches to dietary diversity, bio-fortification, food safety, and value chain enhancements for nutrition.
12	Nutrition challenges are widespread, and yet the solutions require locality specific interventions. Designing and implementing decentralized context specific nutrition interventions that bring several sectors in a coordinated manner require local capacity at all levels.	Chapter 16, Designing a Decentralized Food System to Meet Nutrition Needs: An Optimization Approach, Chapter 17, Future Directions for Nutrition Policy Making and Implementation. Decentralized interventions are needed to tackle locality specific problems. We introduce an optimization approach to designing cropping systems.

EXERCISES

1. Review the conceptual framework in Fig. 3.2. Develop a conceptual framework for your country of choice taking into consideration the unique agroecology, natural resource base, and labor and technology constraints.
2. List the various nutrition interventions in your country of choice and demonstrate how they are interconnected to various sectors. How could their coordination at the national, subnational, and community levels improve. Suggest approaches to the stakeholders to hold the program implementers accountable.

PART

ECONOMIC ANALYSIS OF NUTRITION

B

CHAPTER

MICROECONOMIC NUTRITION POLICY

4

While psychological and behavioral theories help to account for some areas of economic behavior, in the case of the Giffen phenomenon, and of the consumption behavior of the extremely poor more generally, the standard model appears to be the right one.

Robert T. Jensen, Nolan H. Miller, 2008. *Am. Econ. Rev.*, 98 (4), 1553–1577.

INTRODUCTION

How do individuals and households make decisions related to nutritional choices? Do they? What incentives are needed to enable individuals and households to make better choices toward optimal nutrition? And how could nutrition program and policy interventions be better designed to improved nutritional and health outcomes? These are some of the basic questions that confront the development community as they try to address malnutrition in all forms.

The nutritional status of individuals and households is a result of the choices they make. From the conceptual framework we presented in Chapter 3, A Conceptual Framework for Investing in Nutrition: Issues, Challenges, and Analytical Approaches, we know that to attain optimal nutrition these choices relate to food, water, sanitation, health, and care, among other things. How households allocate their income, time, and other resources to obtain the right combinations of these key inputs will decide their nutritional status. How they allocate these key inputs among those living within a household will determine the nutritional status of the individual members of the household. One can say, then, studying the decisions households make about how to allocate their income and resources to *demand* these inputs will help in understanding the behavior of the households toward their nutritional choices. Such understanding can further help in designing programs and policies to achieve optimal nutrition for the society. Further, studying how households use these key nutrition inputs in the process of *producing* optimal levels of nutrition will help in designing programs and policy interventions that can increase the effective and efficient utilization of these inputs. In this chapter, we introduce the basic economic principles to study both the demand side issues and the production of nutrition. We will apply these principles to study the development issues related to nutrition policies and program that are presented in rest of the chapters of this book.

Let us look at the demand side of nutrition first. As mentioned in Chapter 3, A Conceptual Framework for Investing in Nutrition: Issues, Challenges, and Analytical Approaches, in addition to food, households need water, sanitation, care, health, and other inputs to produce good nutrition for all the members of the family. Thus, we know food is necessary but not sufficient to produce optimal nutrition. Yet without food, one cannot talk about nutrition, as both macro (say calories and protein) and micronutrients are derived from the food that one consumes. While we will address the demand for other nonfood inputs in the later sections of this chapter, we will begin our

Nutrition Economics. DOI: http://dx.doi.org/10.1016/B978-0-12-800878-2.00004-9

study of nutrition demand with demand for food. Specifically, we will look at how choices made related to the quantity and quality of food people consume determine their nutritional status.

The quantity of food consumed relates to food availability at the national, market, community, household and individual levels. The availability of food at all these levels depends on how much food is produced, stored, transported, exported, imported, and how much food is available in the market to purchase. Food availability in the market determines the price households pay to access food. For a household that completely depends on food produced in their backyard garden or field, market availability of food and its price may not matter much. However, for households that depend on the market to access food, their level of income and the food prices that prevail matter, along with, of course, other factors such as taste, preference for certain foods, sociocultural norms, and religious preferences for certain types of foods. The share of income that the households can afford to spend to buy food in the market, then, becomes a major determinant of household food consumption.

The food accessed using income or from their own production has to be processed, cooked, and consumed, which requires not only food but also the time and fuel to prepare the food. A household's access to clean water, sanitation, health care, and other complementary factors, along with food, determines how the household access to food is translated into the nutritional status of the individual members and the household as a whole. Studying how households decide to spend on various food and nonfood items, and how they allocate time to prepare food, for child care, and for providing clean water can help us to guide the policy makers to identify the right type of interventions to achieve optimal nutritional outcomes. In this chapter we lay the microeconomic foundations for understanding the choices households make that have implications for their nutritional outcomes.

We begin with a graphical analysis of the consumption choices of a household as the decision-making unit.[1] Then we present the demand analysis of food and nutrients which involves studying how households and individuals change their consumption of food, and hence nutrients, due to changes in their income, prices, and other noneconomic factors. In order to prepare the readers for subsequent sections that involve econometric analysis, we present some generic reduced form models of nutritional outcomes that bring nonfood factors into the analysis of nutrition policies. Finally, we present the dynamics of nutritional choices to motivate readers to think about how nutritional choices are made at different stages of the life cycle. This has implications for policy makers, as different age groups are affected by nutritional policies in different ways.

THE HOUSEHOLD AS THE DECISION-MAKING UNIT

The following sections of this chapter are based on and closely follow the exposition of Bryant (1990) and Chernichovsky and Zangwill (1990). We adopt the same notation following Bryant (1990).

In explaining the individuals' responses toward policy and market changes, it is useful to understand how households choose among various options they face by allocating or reallocating their

[1]We present only the concepts that are useful to follow the rest of the contents in this book and that are useful in the context of nutrition policy analysis. Any basic microeconomic text can help the reader to catch up on the concepts of microeconomics theory (Nicholson, W., 2000. Microeconomics Theory and Its Application, eighth ed. The Dryden Press, Fort Worth).

minimal resources. The new household economic theory assumes that the household is a harmonious unit which aims to increase collective welfare. A household can be one person or a small group of individuals who share resources for pursuing mutual well-being (Bryant, 1990). The study of economic organization of the household relates to the size, structure, and composition of the household, as well as the pattern of resource use and the activities pursued by the household (Bryant, 1990).

Household resources include human and physical resources. Human resources include the time, skills, and energy of each member. Physical resources consist of financial resources. Household activities involve the use of resources and have attributes from which the household derives satisfaction (either directly or indirectly). Consumption and leisure are activities from which satisfaction is derived directly, whereas work is an activity from which satisfaction is derived indirectly. Work activities are divided into market and nonmarket work; the distinction is based on whether work activities take place in the context of an employer – employee relationship, or in the worker's household. Activities can also be differentiated based on whether satisfaction (utility) is derived immediately or in the future. Activities that give immediate utility or satisfaction are called consumption activities, while activities pursued toward future goals are considered saving or investment activities. Activities that transfer resources from one household to other households or groups in society are transfer activities (Bryant, 1990).

Households maximize satisfaction, subject to certain constraints. Households face four types of constraints: economic, technical, legal, and sociocultural. Household resources are limited; each day has 24 hours, households have limited income and assets and a given amount of credit. Technical constraints refer to processes obeying the laws of biology, chemistry, and physics; different types of food preparation alter the color, taste, consistency, and nutrients contained in raw foods in specific ways. Household behavior is also determined and directed by the laws and regulations of the political entities in which the households participate. Laws forbid some activities and the use of some resources, and alter the prices of the resources at which they are bought or sold. In addition, social and cultural norms affect the resources, activities, and satisfaction received by the household.

HOUSEHOLD EQUILIBRIUM AND DEMAND FOR NUTRIENTS

How does a household attain optimal nutrition choices? Household demand for nutrients depends on two critical attributes. The first attribute is a set of preferences for food commodities and nonfood items that the household can afford and is willing to buy, given its income and market prices; the second attribute relates to the goals of the household expressed in terms of the preferences it has for goods.

First, let us look at the budget constraint that the household faces as it tries to meet its nutritional needs. Assume, for the sake of simplicity, we deal with one nutritional attribute of one food item—the fiber content of whole wheat bread. In this discussion we work with two commodities: let the two goods be a fiber-rich diet good (such as whole wheat bread), and a nonfood composite good (such as rent and clothing).

Let p_{fr} be the market price of a unit of food, and p_{nf} be the price of the composite nonfood good. Let the quantities of food and composite good be q_{fr} and $q_{\mathrm{nf.}}$ The total income that the house has to spend on these goods per decision period is Y. The budget constraint represents the possible combinations of fiber rich diet and composite good purchasable by the household when all income is used:

$$p_{fr}q_{fr} + p_{nf}q_{nf} = Y \tag{4.1}$$

When graphed, the slope of this line is the relative price of the fiber-rich diet good in terms of the nonfood composite good (p_{fr}/p_{nf}) (Fig. 4.1). The slope represents the rate at which households can exchange the fiber-rich diet good for the composite good. The determinants of the budget constraint are the prices of the fiber-rich good and the composite good, and income.

Next let us look at the preferences of the household. Preferences reflect the likes and dislikes of the household and its goals of increasing its well-being. The structure of household preferences follow three properties (Bryant, 1990): (1) the ranking of each combination of goods and services; (2) the household is "consistent" in its preferences, i.e., if it prefers combination X to combination Y and combination Y to Z, then it cannot prefer Z to X; and (3) more goods are preferred to less by the households.

Given the above constraints, the household preferences can be represented by a utility function:

$$U = u(q_{fr}, q_{nf}) \tag{4.2}$$

Let U denote the amount of satisfaction from consuming a fiber-rich diet (q_{fr}) and the composite good(q_{nf}).

In Fig. 4.2 these combinations are represented by points farther from the origin in the northeasterly direction, which are preferred to combinations closer to the origin. The curves I_1 and I_2 are called the "Indifference Curves" (ICs). They connect points of combination of q_{fr} and q_{nf} that the household treats with equal preference. These curves are negatively sloped, and the household prefers more to less, which implies that the points farther from the origin represent greater well-being. These curves cannot intersect one another.

In Fig. 4.2 the line in the q_{fr}–q_{nf} space connects the points such as (A, B, C, and D) which are combinations of two sets of commodities that households prefer equally, given a certain level of income. By the properties of the indifference curves, as discussed above, the household prefers the bundle through E-F-G-H to any of A-B-C-D. Since more is preferred to less, an IC cannot be

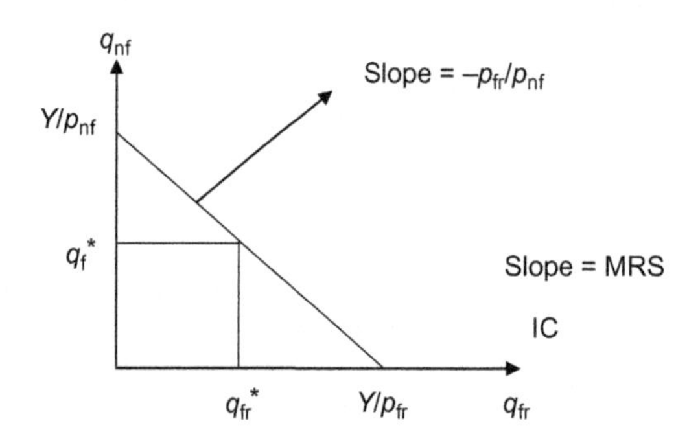

FIGURE 4.1

Graphing the household budget line.

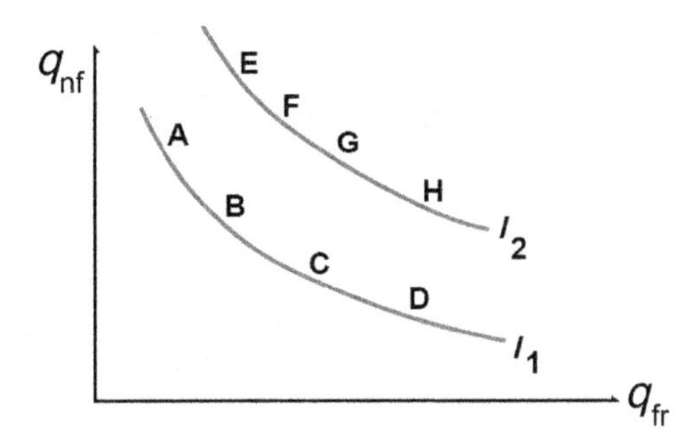

FIGURE 4.2

Indifference curves.

positively sloped. The ICs are convex to the origin, implying that the household prefers a bundle such as C over A and B, or D, that is, it prefers the average of two extremes in I_1.

We use these properties of the indifference curves to understand the behavior of consumers toward food and nutrient consumption and consumption of nonfood goods. For example, how foods containing specific nutrients are preferred to other foods which may not contain these nutrients, yet may give the same satisfaction.

One of the useful concepts in the study of consumer behavior is the Marginal Rate of Substitution (MRS) which is defined by the slope at any point on an indifference curve. It represents the rate at which a household can exchange a fiber-rich diet good for the composite good as in Fig. 4.2. The declining MRS reflects the state of relative satiation. It implies that the more of something a household possesses relative to other goods, the less it will be willing to give up the other good in order to acquire more of the relatively abundant good (Bryan, 1990). For example, a subsistence household while it produces its food and has plenty of it in harvest season, will give up less nonfood items it has in its possession in order to get more food. This behavior has implications for food policies during drought and famine conditions. During a lean season, food becomes scarce and households are willing to give up their nonfood items which are relatively more available compared to food, in order to get food which is less abundant. This results in asset erosion and puts the households in a vicious cycle of poverty and malnutrition.

Fig. 4.2 shows how the marginal rate of substitution can be studied using the properties of indifference curves. Let the movement from point A to point C denote the change in utility MUq_{nf} (Δq_{nf}), and let the movement from C to B denote change in utility MUq_{fr} (Δq_{fr}). Then the total change in utility between A and B is zero, since A and B are on the same indifference curve.

$$\text{Thus MRS} = (dq_{nf}/dq_{fr}) = (-\,MUq_{fr}/MUq_{nf}).$$

Another useful concept is Marginal Utility. It is defined as the added utility that a household obtains from consuming an extra unit of a good, holding the consumption of all other goods constant.

For a convex IC, the slope goes from high on the left to low on the right, implying that as a household has more of q_{nf}, it will be willing to give up less and less of the fiber-rich diet food in exchange for acquiring equal amounts of the composite good. This notion is also known as the "diminishing marginal rate of substitution" (Fig. 4.3).

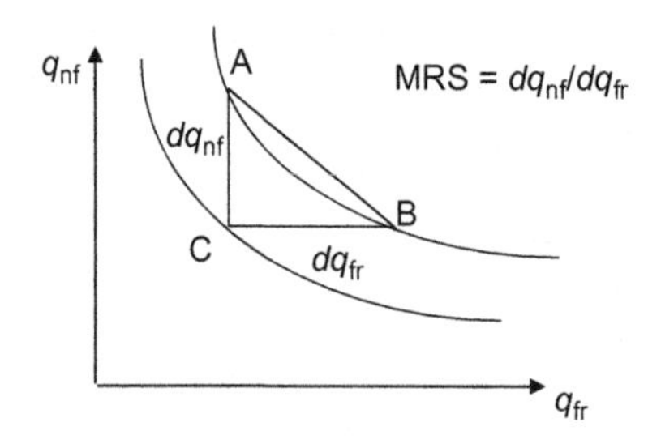

FIGURE 4.3

Marginal rate of substitution.

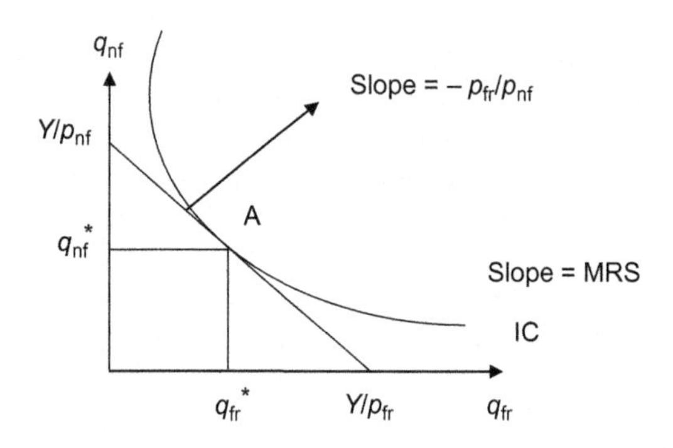

FIGURE 4.4

Household equilibrium.

A household is in "equilibrium" when it has no incentive to change its purchase pattern (Bryan, 1990), that is:

$$MU_{fr}/p_{fr} = MU_{nf}/p_{nf} \tag{4.3}$$

Where, MU_{fr} and MU_{nf} represents the marginal utility of a fiber-rich diet and the marginal utility of the composite good, respectively. Eq. (4.3) states that in equilibrium an extra dollar spent by the household on any good must yield the same added satisfaction.

We describe the household equilibrium concept using Fig. 4.4.

Given convex and smooth ICs, the household maximizes utility at point A, where the slope of the IC (MRS) is equal to the slope of the budget constraint. At this point, $p_{fr}/p_{nf} = MRS = MU_{fr}/MU_{nf}$. The optional quantities of fiber-rich food and composite good that the household will consume are q_{fr} and q_{nf}, respectively. However, the tangency can fail at the optimal point if: (1) the ICs are not smooth, as in the case of complements; and (2) if the optimal point is at a corner of the budget set.

What are the implications of the above theoretical analysis for nutrition policy making? The analysis of preferences individuals and households make given a specific level of resources has implications for their nutritional status. The tradeoff they make for one good, given the quantity of other goods they have, needs to be understood for designing holistic interventions that increase their nutritional status. Yet such basic tradeoffs are poorly understood by program designers and policy makers. As a result, the interventions for nutritional improvements have focused on single factors such as food subsidies (Pinstrup-Andersen, 1988), care (Engle, 1995), gender (Quisumbing and Meinzen-Dick, 2012), school feeding programs (Ahmed and Babu, 2006), or, recently, toilets (Spears, 2012).

All the above interventions, although well intended, have largely made a lesser dent in the programmatic perspective as other needed ingredients. The preferences households make not only depend on how much of the goods and services are available to them, but also on how much they can give up one nutrition input for the other. For example, toilets can be promoted through subsidies, and even after building toilets for every household, they could be quickly abandoned if there is no water supply to flush. In fact, this could work against nutrition since now the feces that is not flushed out is rotting next to their kitchen. Households thus go back to open defecation, even when there is a toilet attached to their home. Policies and program designs need to consider other constraints households face in translating food availability into good nutrition.

The analysis of indifference curves and the concept of declining marginal rate of substitution help us to understand such preferences in the two-dimensional world. However, nutrition is a product of a combination of several goods, including various types of foods, and the choice of the amounts and the tradeoffs they make to choose various quantities can explain much of the persistent malnutrition in societies, even where food availability is abundant.

Equipped with the basic concepts of utility function, indifference curves, marginal rate of substitution, and marginal utility we are ready to study the demand for food and nutrients.

BASIC CONCEPTS OF DEMAND FOR FOOD

In this section we begin with the basic concepts needed to study the demand for food. When a household goes to the market with a certain level of income to purchase food, the first factor they face is the price of the food commodity. How much food the household can purchase given the level of income available to buy food is determined by what price exists in the market for food. This relationship between the price and quantity demanded or purchased is depicted by the demand schedule, which is shown in graphical form in Fig. 4.5.

The demand curve in Fig. 4.5 shows the market demand curve for food which indicates various levels of food quantities that consumers will be able to buy from the market for different levels of prices. This is a depiction of a relationship between only two variables, which means we keep all other factors affecting demand for food at constant levels. Once we have this schedule, we can study, by moving along the demand curve, the change in quantity demanded of food for a particular change in price levels. Simply stated, as the price goes down (P_1 to P_2), e.g., there will be an increase in the quantity demanded (q_1 to q_2), holding other factors constant. These factors include the income of the consumers, prices of other food and nonfood items, and other factors such as tastes and preferences.

While the food price change is traced along the demand curve to locate the change in the quantity demanded, the changes in the levels of any of the nonprice factors mentioned above will shift

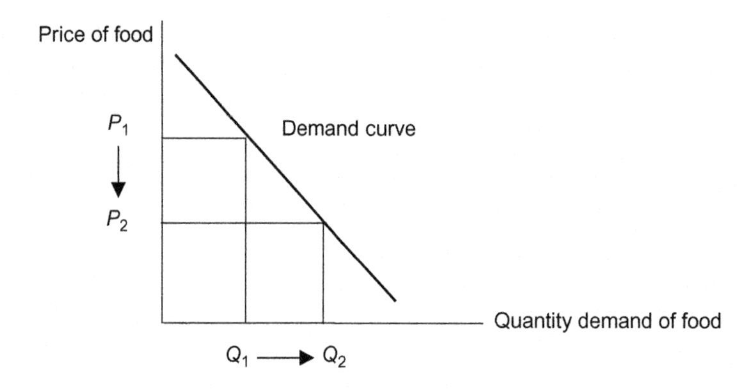

FIGURE 4.5

Market demand curve for food.

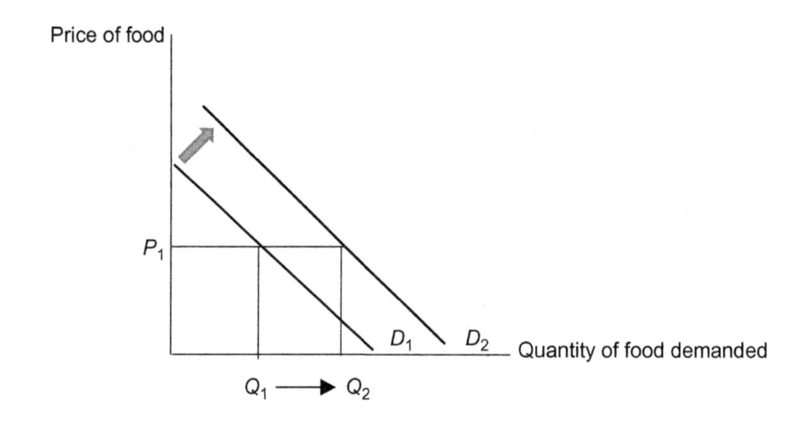

FIGURE 4.6

Shift in demand for food.

the demand curve itself. For example when the income of the consumers increases due to economic growth, the demand curve will shift to the right, and at the same price level of the food commodity, consumers will be able to buy more of the food. Thus, in Fig. 4.6 for the same price level, a shift in the demand curve brings about an increase in the quantity of food demand from Q_1 to Q_2. Similar shifts in demand can happen due to changes in other factors, such as a change in the food consumption patterns of the consumers. For example, increased education about the impact of foods rich in fat and cholesterol on chronic diseases can reduce the demand for such foods (see Exercise 1 at the end of this chapter).

Consumers often substitute one food for another when the price of other foods change. For example, in a society when both maize and rice are consumed and equally preferred, the consumption of maize can change if the price of rice increases (Fig. 4.7). Increases in the price of rice will

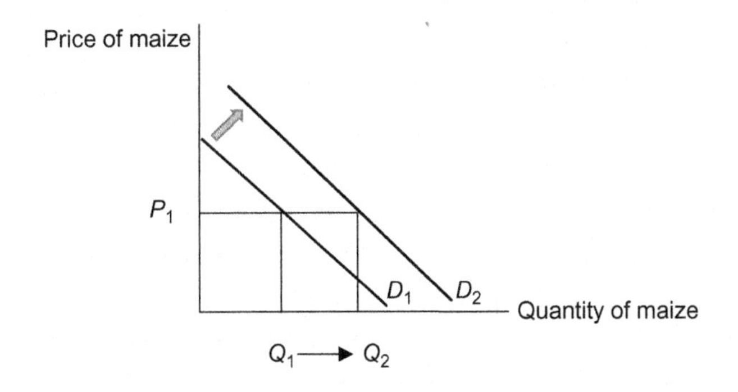

FIGURE 4.7

Shift in maize demand: price of rice.

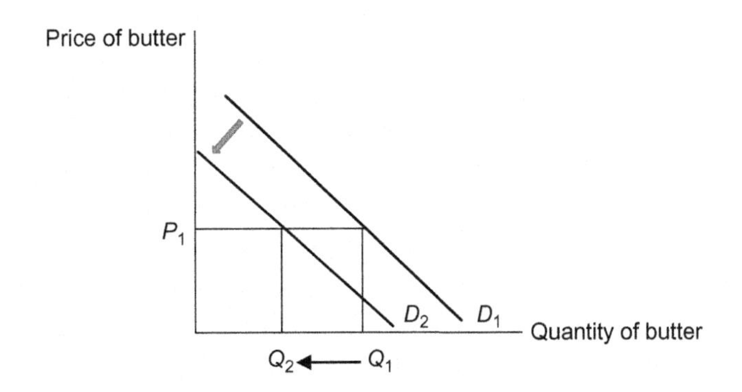

FIGURE 4.8

Shift in butter demand: price of bread.

make it more expensive compared to maize, and thus can increase the demand for maize as it is considered a good substitute for rice in certain societies, particularly in developing African countries. Similarly, the price of a food that is a complement can shift the demand curve as well. For example, a price increase in a food such as bread will shift the demand curve for butter to the left, as they are considered complements (Fig. 4.8). At price P_1 of butter which is not affected by a bread price change, the quantity of demand for butter goes down from Q_1 to Q_2.

As income increases, households tend to buy more expensive forms of nutrients and less of certain foods they bought when their income levels were low, although both the foods may give similar nutritional benefits. The foods which consumers demand less as their income goes up are called inferior foods. The demand curve for an inferior food will shift to the left as the income of the consumer goes up. For example, when household income goes up, the demand for cassava goes down, even at the same price level of cassava (Fig. 4.9).

Analysis of changes in demand for food due to changes in the food prices, prices of other commodities, income changes, and changes in other sociocultural factors can help in formulating

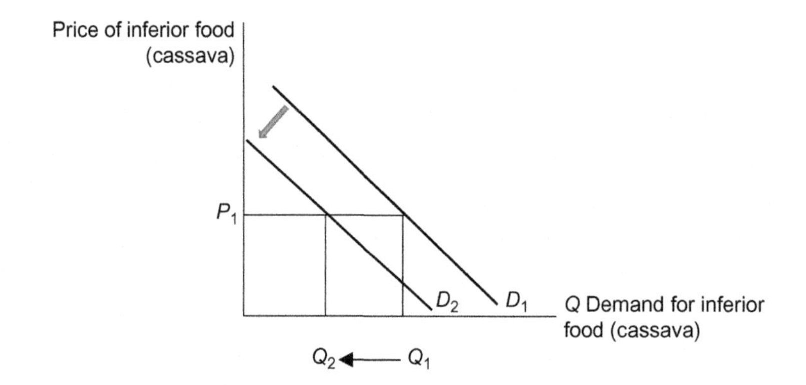

FIGURE 4.9

Demand for inferior food.

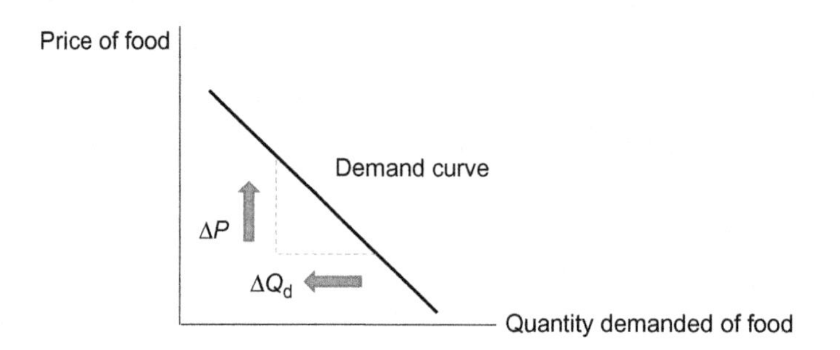

FIGURE 4.10

Slope of a demand curve.

policies that improve nutritional outcomes. This then requires information about the magnitude of change in food demand due to changes in the values of factors affecting it. In the context of the demand curve, such information can be derived from studying the slope of the demand curve (Fig. 4.10). The slope of the demand curve is the change in the value price (rise) divided by the change in quantity demanded (run). The slope of the demand curve is given by $\Delta P/\Delta qd$. The inverse of the slope of the demand curve ($\Delta qd/\Delta P$) is called the responsiveness of demand, which shows how the quantity demanded of food changes due to changes in its price.

Because the quantities of food are measured in various units, and price is measured in various currencies, we need a measure of responsiveness that is independent of the units of measurements. In economic analysis this measure is denoted as "elasticity." The elasticity is expressed in terms of percentage changes. Thus, it does not matter if the food price is measured in South African rand, Zambian kwacha, Pakistani rupees, European euros, or Canadian dollars.

The elasticities of food demand could be calculated with respect to the own price of food in demand, other food prices, nonfood prices, income, and other factors. These are useful measures for nutrition policy and program interventions, and we use them extensively in the next two chapters.

Let us begin with the own price elasticity of demand of food, which is the percentage change in quantity demanded of food for a 1% change in its own price ($\%\Delta Qd/\%\Delta P$). It is denoted as ("ε" Qd, P). For example, if the price of food changes by 10% and the quantity demanded changes by 5%, then the own price elasticity of food would be $[(\Delta Qd/Qd)/(\Delta P/P)] = |0.5|$ holding, of course, all other prices and values of other factors constant (see Exercise 3).

The own price elasticities are negative, since the demand for food decreases with its own price increase. We say $\varepsilon = -0.6$ when there is a 1% increase in food price that results in a 0.6% decrease in the quantity of food demanded. The absolute values of price elasticities are used for comparing the elasticities of two food commodities. For example, if the own price elasticities of rice and maize are, respectively, $|0.6|$ and $|0.8|$, then maize is considered more own price elastic, as the absolute value is higher than that of rice.

If the own price elasticity value for a good or a food commodity is $|1.0|$, then the elasticity is called "Unitary" meaning a 1% increase in its price will decrease the quantity demanded by 1%. If the absolute value of own price elasticities are more than unitary ($\varepsilon > 1.0$), then the food commodity is considered own price elastic. On the other hand, if the own price elasticity is less than "unitary," then the demand for the food commodity is considered own price inelastic with respect to its own price.

INCOME ELASTICITY OF DEMAND

In nutrition policy making, a frequent question policy makers ask relates to how much the income of poor households needs to be raised for them to reach optimal nutritional well-being. A good starting point to study the income – nutrition relationship is to study the income – food consumption relationship. We ask the fundamental question: as income increases what happens to the consumption of food? The measure of change in food demanded due to change in income is given by the income elasticity of demand at any given set of prices. Thus, the income elasticity demand is defined as the percentage change in quantity demanded due to a 1% change in income, holding preferences and relative prices constant:

$$\eta x = (\Delta\, qX/\Delta\, Y)\,(Y/qX) \tag{4.4}$$

where ΔqX is the change in quantity demanded of any food due to change in income (ΔY), and q and Y are the pre-change values of quantity demanded and income, respectively.

Expressed differently, income elasticity of demand is the ratio of the marginal propensity to consume ($\Delta q/\Delta y$) and the average propensity to consume (q/y). When $\eta x > 1$, demand for food increases more than proportionally to income and the food demand is income elastic, and when $\eta x < 1$, the demand for food goes down when income increases. Food demand is generally income inelastic; however, this assertion is being challenged.

Analysis of food consumption behavior of households resulting from increases in income has been of interest to policy makers. For example, Logan (2005) studied food – income relationships in the late 19th century and found that foods are in general very bad proxies for nutrients. He found that the income elasticity of dairy products, which are usually high in fat, was greater than the income elasticity of fat. Similarly, his study showed, the income elasticity of fruit and vegetables, which are high in fiber, was greater than the income elasticity of fiber. Carbohydrates and cereals were not good proxies for one another, and the income elasticity of cereals were lower than the income elasticity of carbohydrates. However, meat was a good proxy for protein, and the income elasticity of meat was very close to the income elasticity of protein.

We can now define different types of goods useful for economic and policy analysis. A "Normal Good" is any good for which an increase in income is associated with an increase in demand, with prices and preferences held constant. People demand more protein-rich food in their diet as their income increases, and thus it is a normal good. An "Inferior Good" is any good for which demand decreases as income increases and *vice versa*, with prices and preferences held constant, e.g., carbohydrates. "Giffen Goods" are those goods for which the demand curves are upward sloping. For example, a recent study by Jenson and Miller (2008) found that extremely poor households in China consumed more rice as the price of rice increased.

DERIVATION OF ENGEL CURVES FROM INCOME CONSUMPTION LINES

Income consumption lines are the locus of points representing the equilibrium purchase patterns as income changes, holding preferences and relative prices constant. Engel Curves are the locus of all points representing the quantities demanded of the goods at various levels of income, when prices and preferences are held constant (Fig. 4.11 for normal goods and Fig. 4.12 for inferior goods).

The first part of Fig. 4.13 shows the equilibrium purchase pattern of a household at two levels of income, Y_1 and Y_2, holding preferences and relative prices constant. The second part shows how the quantity of fiber-rich diet changes with income; the incomes Y_1 and Y_2 are represented by the respective budget lines. AB is the income consumption line, and CD is the analogous Engel curve. Engel curves begin at origin and intersect the income axis: with zero income, the quantity demanded for any good must be zero, as the household is unable to afford any purchases. At low incomes, the demand for particular diets (such as a fiber-rich diet) may also be zero (Fig. 4.13).

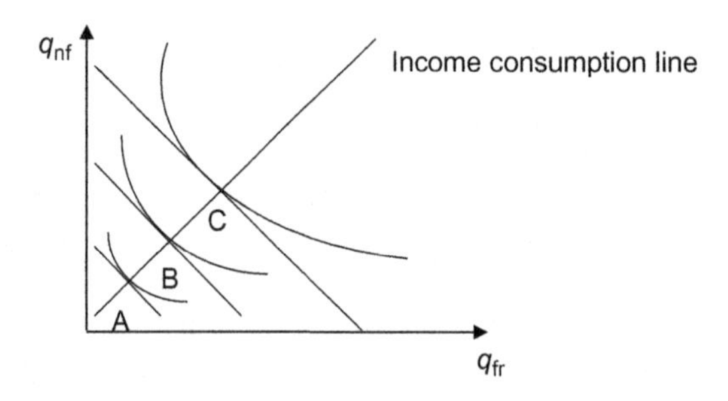

FIGURE 4.11

Both goods are normal.

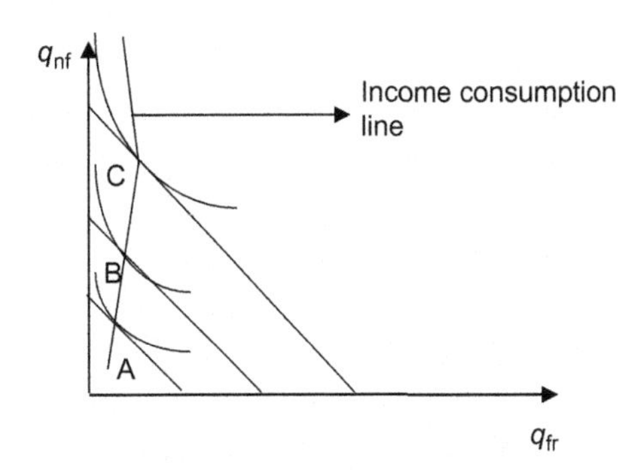

FIGURE 4.12

q_{fr} is an inferior good.

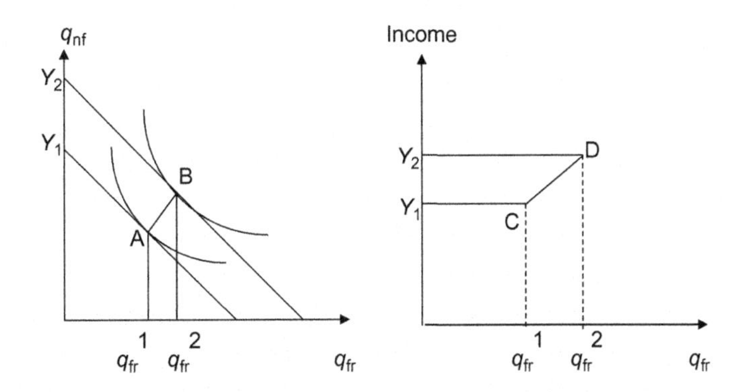

FIGURE 4.13

Deriving an Engel curve from the income consumption line.

DERIVING THE HOUSEHOLD DEMAND CURVE FOR A FIBER-RICH DIET

In the top panel of Fig. 4.14, three equilibrium purchase patterns of a fiber-rich good are depicted with income, and other prices, with preferences held constant. The bottom panel shows the price of the fiber-rich good plotted on the vertical axis, and the quantity of the fiber-rich good on the horizontal axis. As the price of the fiber-rich good declines from ${p_{fr}}^1$ to ${p_{fr}}^2$ to ${p_{fr}}^3$, the demand for the fiber-rich diet commodity increases from ${q_{fr}}^1$ to ${q_{fr}}^2$ to ${q_{fr}}^3$.Thus, the demand curve depicts the schedules of quantities that the household is willing and able to consume at different prices, holding income, other prices, and consumer preferences constant. As seen before, the slope of the demand curve is given by $(\Delta q_{fr}/\Delta p_{fr})$.

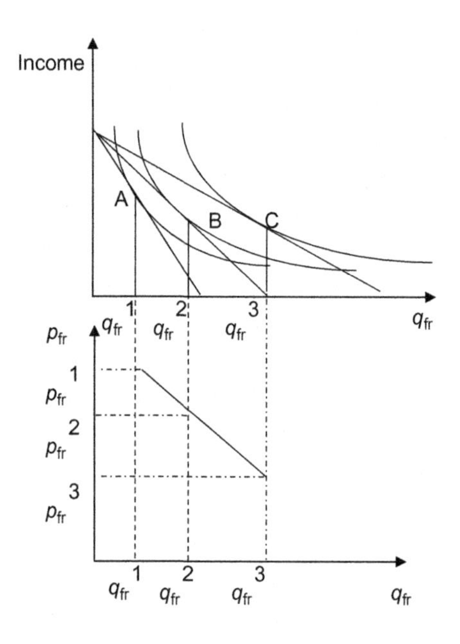

FIGURE 4.14

Derivation of the household demand curve.

INCOME AND SUBSTITUTION EFFECTS

In applied policy analysis, the implications of a nutrition policy could be understood in the context of price changes. In demand analysis the effect of relative price change can be decomposed into two effects: substitution and income effects. The substitution effect is the effect of a price change on the demand for that good, holding preferences constant. The income effect is the effect on the demand for a good of the change in income brought about by the change in the price of the good.

As the price of a fiber-rich diet good decreases, there is a tendency to consume more of it, since it becomes cheaper. Due to the price decrease, the household finds itself wealthier, and thus can buy the old bundle and still have some money left over. The increase in the consumption of the fiber-rich diet good resulting from spending the left over money is the "income effect."

In Fig. 4.15, consider the price decrease that moves the household from point A to B. The movement from A to C is the substitution effect, and the movement from C to B is the income effect. The "substitution effect" can be isolated by taking away from the household enough money to put them back at the same level of satisfaction as before the price change. The line through C has the same slope as the one through B, and is at a tangent to the IC through A. The direction of the substitution effect is always opposite the price change. In the case of the fiber-rich good (which is a normal good), the income effect is in the same direction. Thus, substitution and income effects reinforce each other, and the demand curve is downward sloping.

The above analysis helps in better understanding the implications of a nutrition policy and program intervention. If the good is normal, the income and substitution effects for a given price

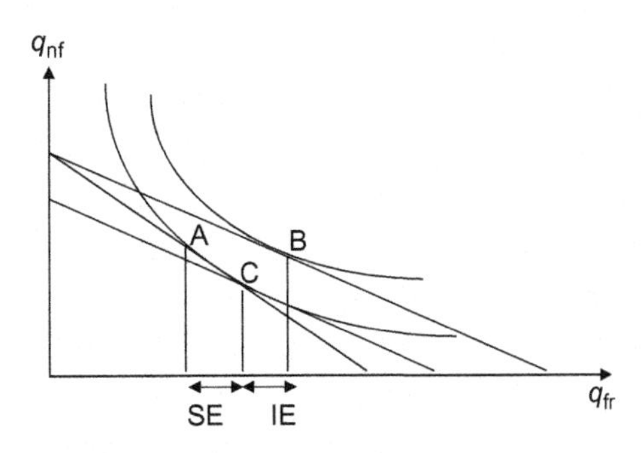

FIGURE 4.15

Decomposing the own price effect for a fiber-rich good.

change work in the same direction, making the demand curve for the good downward sloping. The more and better substitutes a commodity has, the larger is the substitution effect of any own price effect, and *ceteris paribus*, the larger is the own price effect. For example, compare price elasticities of cholesterol-rich diets and fiber-rich diets; cholesterol-rich diets (e.g., mutton) have many good substitutes (e.g., high fat cheese, shrimp, beef); however, fiber-rich diets (such as whole wheat bread) do not; thus mutton will be more price elastic than the fiber-rich diets. If the demand for a nutrient is very responsive to income, the larger is the income effect of a price change, and thus the larger is the total price effect.

Households that are extremely poor are also the ones that are not able to come out of the malnutrition trap. Even though the consumers as a whole may have the preferences exhibited by convex indifference curves, for extremely poor households, the preferences may in fact be represented by concave indifference curves, at least on a section of the curve. Studying household preferences of vulnerable groups and special groups helps policy makers understand their consumption patterns better, and this further helps to consider the incentives that they can put in place in the form of programs and policies to induce them toward optimal nutrition.

DEMAND FOR NONFOOD FACTORS FOR OPTIMAL NUTRITION

In addition to income and prices of food and nonfood goods that were used in the above sections to analyze the demand for food, several other factors, particularly nonfood factors, determine the demand for food, and hence the optimal nutrition households can obtain. Following the conceptual framework used in Chapter 3, A Conceptual Framework for Investing in Nutrition: Issues, Challenges, and Analytical Approaches, understanding the pathways through which optimal nutrition is obtained can help identify opportunities for designing program interventions that enhance the availability of such factors. For example, as already alluded to earlier, the determinants of food and nutrition consumption go beyond income and prices to the tastes and preferences of households toward specific foods (Silberberg, 1986; Babu and Rajasekaran, 1991).

In addition, family size and the composition of the household also determines the allocation of nutrients among them. Given that certain members of the household could be discriminated against during the allocation of food and other resources, the access to nutrition will differ among the household members. For example, the inherent gender bias in some societies, where women eat last after providing adequate food for the men and children, affects the nutritional status of women (Babu et al., 1993). In addition to the above factors, food and nutrition consumption, and particularly the utilization of food and its conversion into nutritional well-being, is the environment in which households and members of the households live. Environmental factors such as the availability of potable water, sanitation within the household, sanitation in the community, and the behavior of household members in terms of hygiene can affect nutritional status (Smith and Haddad, 2015). The demand for these nonfood factors and how the cost of providing them to individuals and households affect food consumption and nutrition is explored in several chapters in this book.

PRODUCTION OF NUTRITION WITH FOOD AND NONFOOD INPUTS

The nutritional status of households has often been depicted as the outcome of a production process (Chernichovsky and Zangwill, 1991; Akin et al., 1992; Baten and Blum, 2014). Measuring the effects of various inputs that help in the production of nutritional outcomes can help in the design of appropriate program interventions to improve nutrition. By identifying the most compelling factor, program managers could invest their limited resources on those factors and make their nutrition intervention more cost-effective. Thus, the knowledge of the determinants of malnutrition at household and individual levels can help in understanding the household's potential and ability to respond to specific interventions.

Reduced form models of nutrition production have been used for quite some time (Chernichovsky and Zangwill, 1990). Nutritional status indicator variables, such as body mass index, stunting, and underweight are frequently used as endogenous variables, and the causal factors shown in the conceptual framework in Chapter 3, A Conceptual Framework for Investing in Nutrition: Issues, Challenges, and Analytical Approaches, are used as exogenous variables. However, estimating these relationships has been constrained by the choice of the variables, measurement errors associated with the variables chosen for estimation, and other econometric variables. Researchers continue to debate the selection of correct models and continue to correct for these errors. Using instrumental variables has been a popular approach to improve estimates of the contributions of various factors to nutritional outcomes (Bhargava, 2016).

Recently, Peuntes et al. (2016) corrected for several econometric challenges in estimating such production functions. They found that increasing energy intake increases both height and weight in both the countries—Guatemala and the Philippines—that they studied. They also found that if that energy comes from protein sources it is even better, and that child nutrition status can increase with increases in protein intake. Recognizing this as an important nutrition puzzle, they recommend that programs designed to increase child nutrition should increase the quality of food provided to them. They also call for additional research to test if combining food-related interventions with nutritional education and behavioral changes of the households, and nutrition sensitive agriculture can improve the nutritional status of children. Researchers have spent their time looking for improved estimates of how various causal factors influence nutritional status, so that they can better advise policy makers and program managers on refining and improving nutrition intervention programs to increase cost-effectiveness. But how much refinement is necessary in these estimates, and how robust do they have to be before we invest in nutrition interventions (see Exercise 6 at the end of this chapter)? We address some of these issues and econometric challenges in the rest of the chapters of this book.

DYNAMICS OF NUTRITION CHOICES

Nutritional status depends on the choices households and individuals make at any point in time. Yet, nutritional intake today affects the health of individuals not only in the short run in terms of reducing hunger and food insecurity on a daily basis, it also affects their health in the long run. A well-nourished child who is not stunted can grow up to be a healthy and productive individual earning more money than a malnourished child. The economic cost of poor nutrition was discussed in Chapter 2, Global Nutrition Challenges and Targets: A Development and Policy Perspective. In this context, the choices individuals and households make could be considered as decisions related to investment in human capital formation (Chavas, 2013). If a household is currently starving, it will not give much attention to micronutrient intake as their current need with limited income is to meet the immediate food need, which can come from cheap calories. Further, as the income and time available is limited in poor households, they may have resources only to meet basic food needs and not to use on other inputs such as clean water, sanitation, or health care. Thus, even the food consumed may not be fully converted into adequate nutrition. In addition, even as the income of these individuals increases, further consumption of calories, fat, and sugar-rich food items contribute to obesity. Even though poor nutrition can have adverse health consequences in the future, households and individuals seem to not worry about the future, as their current need to satisfy their hunger is more crucial. The kind of choices households and individuals make have implications for economic policy making. For example, poor households spend less than optimal resources on food and nutrition, but they may send their children to school and spend their resources on educating the children in the hope that they may get out of poverty. They give more preference to educational investment now than investment in nutrition. As a result, their preference is different from a household which invests more in nutrition now in terms of adequate food, but ignores other future-oriented investments such as children's education.

Taking these issues into consideration, Chavas (2013) recently challenged the standard economic models which treat the time preferences of the household as exogenous and unchanging. Chavas (2013) developed a dynamic model that treats the rate of time preference as endogenous, and showed how spending on food now can affect future investment and capital accumulation. He argued that with decision making under malnourished conditions, current hunger income can positively influence food consumption, but can reduce investment incentives. Chavas (2013) found that under conditions of over-nutrition such as obesity, the consumer demand can be inelastic to prices. Under these circumstances, economic policies such as fat-taxes will not have much effect on reducing obesity. These results have implications on how income transfer and food transfer policies and programs affect investment decisions of households when they are under-nourished, as well as over-nourished. Chavas (2013) argues that further studies are needed to understand the role of nutritional education on the nutrition investment behavior of the household facing hunger and malnutrition. Thus, economic policy making and designing program interventions need to take into consideration the dynamic investment behavior of the households, and how they allocate resources over time jointly as households and as individuals.

CULTURAL FACTORS AFFECTING FOOD CONSUMPTION AND NUTRITION

In the previous sections the economics of food and nutrition were addressed in terms of economic choices. In addition to the economic variables, such as income and prices, food consumption and the quality of nutrition intake are determined by cultural and other social variables that are contextually specific to a particular society. The society specific issues have become policy debates, as some researchers believe nutritional requirements may vary depending on cultural norms and food consumption preferences (see the Panagaria (2014) and Gillespie (2014) debate for example, discussed in Chapter 10, Economics of Child Care, Water, Sanitation, Hygiene, and Health: The Application of the Blinder–Oaxaca Decomposition Method), and that food preferences are also determined by the religious and spiritual norms individuals follow. For example, beef and pork are not eaten in certain religious sects. While horse meat and dog meat are delicacies in some societies, they are not in others. Some spiritual groups advocate vegetarian food. Even within vegetarians, some allow eggs and milk as part of the vegetarian diet, while others do not allow any animal-based products. All these dictums people adhere to depending on their religious, spiritual, and ideological affiliations bring various forms of preferences to the food eaten by the members in a society. Thus, understanding cultural factors affecting food choices and food consumption patterns needs to go beyond economic factors. The choice of foods can be understood in the context of cultural, social, and natural circumstances, and irrational and rational consumer behavior. Nutrition is only one argument for buying certain kinds of foods (Ilmonen, 1990).

CONCLUSIONS

Food supply and its consumption contribute to the nutritional status of the population. Increased food supply in several developing countries and developed countries has helped to avoid hunger and famine. However, even in countries where food supply has increased to meet the national food needs, malnutrition continues to prevail. The nutritional status of individuals also depends on the choices they make with regard to nonfood commodities that help in the conversion of food into nutrition. In addition to income levels to purchase food, the size and composition of the households, sociocultural factors, consumption of nonfood factors such as clean water, sanitation, health, and care can increase the chances of malnutrition (Chernichovsky and Zangwill, 1990; Smith and Haddad, 2015).

Designing intervention programs and policies for improving nutritional status requires a good understanding of how households and individuals will respond to the new set of opportunities in terms of income and food prices that they face. In addition, the policies and programs have implications for the provision of nonfood inputs that are required for nutrition production. Using economic theory and modeling, the choices individuals and households make in response to policies and programs can help in designing cost-effective nutrition interventions. By understanding the behavior of households as decision-making units in terms of how they respond to various programs and policies, nutrition interventions could be targeted better to increase their efficiency. In this chapter we introduced basic economic concepts for studying the variables that influence household choices in the context of improving nutrition. In the rest of the chapters we will use these concepts to explore the implications of various policies and programs on nutritional outcomes.

In the context of nutrition policies and program interventions, food consumption patterns and nutrition intake are affected by major determinants such as incomes and prices, other resources households have, household composition and size, and environmental factors such as access to water, sanitation, and health. The access to food, its intake among household members, and its use to achieve optimal nutrition all depend on these factors. The nature and magnitude of these factors and their distribution in the society determines nutritional status (Chernichovsky and Zangwill, 1990). Policy and program interventions attempt to change these factors, and in response consumers change their demand for food and nonfood factors contributing to nutrition depending on the nature of these goods—normal, inferior, or giffen goods (Bryant, 1990, p. 76).

In this chapter, we also looked at the role of nonfood factors as they enter into economic decision-making toward nutrition. Nutritional choices people make go beyond today's benefit or harm to their health status. Understanding the dynamic implications of choices people make toward their nutritional investment is a growing area and needs further study. Considering nutritional status as a final outcome of a production process in which food quantity and quality, along with other inputs such as care, health, water, sanitation, nutrition education, and others can help in designing interventions that have direct implications for nutritional outcomes. Yet these relationships are hard to quantify, precisely due to data and methodological challenges. Nevertheless, understanding household preferences, and their changing consumption patterns in response to changes in the factors affecting their preferences constitute the fulcrum of nutrition policy making. We expand on these issues in the rest of the chapters of this book.

EXERCISES

1. We often hear that as the population of the earth increases we need to feed more people. Suppose the population of a country increases from 10 to 12 million people, what will happen to food demand in the country? What will happen to the demand curve in Fig. 4.1 and why?
2. Normally, the preferences of the decision-making units are depicted by indifference curves that are convex and they imply a declining marginal rate of substitution. Under what conditions could the indifference curves exhibit concavity at least partly, and what are the implications of such behavior for optimal diets and hence optimal nutrition?
3. We defined own price elasticity of food as a percentage change in its quantity demanded for a 1% change in its own price, holding other prices and factors constant. Calculate own-price elasticity when there is a 4% change in food price resulting in a 24% change in the quantity of food demanded.

4. Consider the definition of income elasticity of demand. Calculate income elasticity for milk when there is a 10% increase in income resulting in a 5% increase in milk demand.
5. In the context of the country you are considering to study nutrition policy, identify major food staples that are equally good in nutritive value but have differences in preference by households due to a change in income induced consumption patterns. What is the current trend in the consumption of high value nutrient-rich commodities among the groups of households whose income has been increasing? How does this affect the nutrient intake of the most vulnerable groups in the country?
6. Estimates of the effect of food and nonfood factors on the nutritional status of the population varies, depending on the data used, methods used for the estimations, and choice of the variables in the regression models, as well as the choice of instrumental variables used for correcting the violations of assumptions made in estimation of the relationship between the indicators of nutrition and their causal factors. Trace a line of research—following a paper such as Puentes et al. (2016) and check how much improvement has been made in the estimates of contribution of protein to nutritional status of children over the years. What implications do you derive for increasing program efficiency? Is it worth investing in further refining these estimates and increasing their robustness? What is the cost saving in such research compared to the time wasted in not doing something about malnutrition when nutritionist have told us "not just quantity but quality of food matters" a long time ago?

CHAPTER

MACROECONOMIC ASPECTS OF NUTRITION POLICY

5

These kids have fewer – literally fewer – neuronal connections than their non-stunted classmates... For every inch that you're below the average height, you lose 2 percent of your income... This is fundamentally an economic issue... We need to invest in gray-matter infrastructure. Neuronal infrastructure is quite possibly going to be the most important infrastructure.

Jim Yong Kim, President of the World Bank, in an interview to *Foreign Policy Magazine* (2016)

INTRODUCTION

Although food consumption and nutrition at the household level are considered as a microeconomic issue, policies designed to address the macroeconomic challenges of a country often have a profound impact on nutritional outcomes. Macroeconomic changes, such as structural adjustment and stabilization policies, undertaken as part of reviving economies in the past have had significant influence on the food and nutrition security of the population (Cornea et al., 1987; Sahn and Dorosh, 1997; Diaz-Bonilla, 2015). Considerable efforts have been made to reduce the adverse effects of such macroeconomic policies and shocks on the nutritional well-being of the poor and vulnerable segments of the population. Food price, financial, and fuel crises in the past have also been of interest to the policy makers as they affect food consumption, service provision, and nutritional status (Pinstrup-Andersen, 2015). In this chapter we introduce the basic concepts of macroeconomic policy issues and analysis as they relate to nutritional challenges, and present a broad conceptual framework to understand the pathways through which such policies that influence macroeconomic variables could affect the nutritional status of the population.

CONCEPTUAL FRAMEWORK

Per Pinstrup-Andersen (1987) made one of the earliest attempts to synthesize the available evidence and identify the research needs on the impact of macroeconomic policies on nutritional outcomes. Scobie (1989) also analyzed the nutritional implications of macroeconomic policies. We use the conceptual framework that Scobie (1989) developed as the basis for introducing macroeconomic concepts and their pathways to poverty and nutritional outcomes. Diaz-Bonilla (2015) further expanded on a similar set of pathways to agriculture and food security. The discussion below follows this literature closely.

Fig. 5.1 presents a simple framework connecting macroeconomic factors at various levels to household and individual nutrition and poverty outcomes. It begins with the global factors that govern the international economic environment, which in turn affect the terms of trade exchange rates,

Nutrition Economics. DOI: http://dx.doi.org/10.1016/B978-0-12-800878-2.00005-0

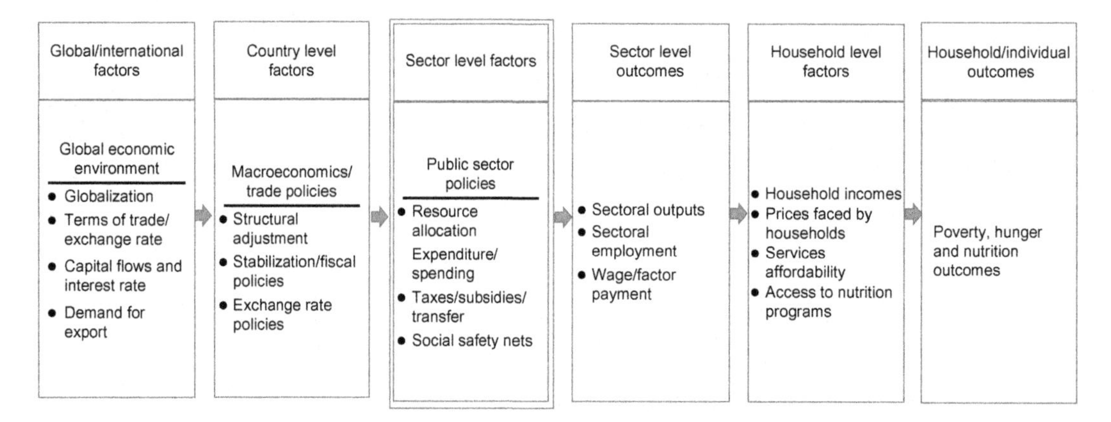

FIGURE 5.1

Macroeconomic pathways to nutrition.

Adapted from Scobie (1989); Diaz-Bonilla (2015).

capital flows including foreign direct investments, interest rates, and the demand for exports that a country faces at any point in time. These global factors induce reactions and responses at the country level. In general, such responses take the form of macroeconomic and trade policies. These policies include those influences on the structure of the economies, e.g., with an emphasis on trade and market liberalization, and moving from agricultural-based economies to industrial-based economies. They also take the form of fiscal, monetary, and exchange rate policies.

Stabilization of prices to overcome inflationary levels is also part of the macroeconomic adjustments that a country makes. Such macro policies affect public and private sectors. They also influence the sector level policies, such as agricultural policies including input and output pricing, as well as sectoral investment macro policies which also affect the budget allocation to individual sectors, the pattern of expenditure/spending within the sector, the level of taxes, and subsidies and income transfers that the sector puts in place. The nature and levels of these policy-induced factors have profound implications for the sector level outputs, employment, wage levels, and payment for factors of production such as land and capital. Further, these policies also affect the commodity prices through their influence on the cost of production of commodities in that sector. In the case of the agricultural sector, this in turn will determine the competitiveness of commodities produced in the world market. These sector level outcomes have a direct influence on household income levels, the real prices consumers face, and the accessibility and affordability of food and nutrition, which further determines the welfare outcomes of households and individuals in terms of poverty, hunger, and nutritional status (Scobie, 1989; Diaz-Bonilla, 2015).

NUTRITIONAL IMPACT OF MACROECONOMIC SHOCKS AND POLICIES

The importance of distributional issues in policy making creates the need for empirical tools to assess the impact of economic shocks and policies on the living standards of households and individuals. Macroeconomic shocks and policies and their impact on nutrition are no exception. The financial crisis in Asia in the late 1990s and the food, financial, and fuel crisis of 2007 – 08 are some examples of sudden crises that confront policy makers who must quickly adapt their policy and program interventions to protect the poor (Pinstrup-Andersen, 2015).

Developing countries face a host of macroeconomic challenges related to the design and implementation of nutritional strategies and policies. As these nutrition programs and policies depend on the macroeconomic environment that prevails in the country, understanding the pathways is key. Such a macroeconomic environment in turn depends on the fiscal policy, monetary and exchange rate policy reforms, trade liberalization policies, and terms of trade shocks. We look at these policies in detail below.

The distributional considerations from macroeconomic policies and shocks, and related outcomes, depend on the overall economic development goals and the population segments affected by them. For example, the first two Sustainable Development Goals are related to eradicating extreme poverty, malnutrition, and hunger. Distributional issues underpin the political dimension of policy making because of the heterogeneity of interests which arise from differences in income, tastes, resource endowment, or technology use. Modeling the nutritional outcomes and distributional impacts of macroeconomic shocks and policies requires a clear understanding of the transmission channels that are outlined in Fig. 5.1.

Different initiatives have attempted to study macroeconomic and nutrition – poverty linkages. Investigations into the relationship between macroeconomic policy and shocks gained momentum in the early 1980s, with UNICEFs "Adjustment with a Human Face" set of studies (Cornea et al., 1987). These studies addressed issues related to the adverse effects of structural adjustment policies and the negative impacts of such policies on children's nutritional outcomes. The scope in terms of geographical coverage was extensive, covering countries from Botswana and Zimbabwe to the Philippines and South Korea. The studies mainly focused on monitoring and collecting statistics on poverty to examine the impact of structural adjustment on poverty.

Later, during the late 1980s, research undertaken at the World Bank and the United Nations Development Programme (UNDP) under the "Living Standards Measurement Survey" (LSMS) and "Social Dimensions of Adjustment" (SDA) focused on the collection and dissemination of disaggregated statistics on poverty, micro – macro linkages, and long-term institutional development (Grootaert et al., 1991). The OECDs "Adjustment and Equity" series of studies used a "general equilibrium" framework to understand micro – macro linkages. The main focus of these studies was to determine whether "managed structural adjustment" was more or less favorable to the poor than a "no adjustment followed by a crisis" scenario (Morrison, 1992). This was followed by the International Development Research Centre (IDRC)-launched program of activities called MIMAP (Micro Impacts of Macroeconomic Adjustment Policies Project) that attempted to be more holistic. It employed both top-down quantitative economy wide analytical frameworks, such as general equilibrium (GE) models, and bottom-up household models. The approach used a range of different quantitative and qualitative methods for assessing the impact of macro policies on micro level outcomes (Lamberte et al., 1991). We discuss a few more macroeconomic frameworks that have been applied to study macroeconomic policies and shocks at household and individual levels, later in this chapter.

ELEMENTS OF MACROECONOMIC POLICY AND THEIR IMPLICATIONS[1]

We briefly review the basic elements of macroeconomic policies in the context of nutrition investments and interventions. We follow Diaz-Bonilla (2015), who presents an excellent overview of these issues in the context of agriculture and food security.

Several broad policy objectives designed and implemented at the national level constitute macroeconomic policies. Fiscal policy is the set of decisions that determine how much revenue the government collects and the expenditures that it incurs among sectors of the economy. Revenues come from government income taxes, profits of public sector industries, and borrowings to finance productive investment. They fund government expenditures which constitute both current and capital expenditures among different sectors of the economy. Various categories of expenditure include military and defense expenditures, welfare programs for disadvantaged consumers, education and health investments (including social safety nets), nutrition intervention programs, public and private sector industries, and employment. If revenues exceed current expenditures, the difference is a budget surplus.

Governments face a budget deficit when the current expenditures are more than the revenues collected. Such calculations are important, since they reveal the accommodating policy actions that governments must take to resolve budget imbalances. For example, welfare program spending affects such imbalances. With regard to the composition of public expenditure toward nutritional improvement of the population, policy makers need to assess not only the appropriateness of proposed nutrition improvement spending programs, but also of planned nondiscretionary and discretionary nonpriority spending. In addition, particular consideration needs to be given to the distribution and growth effects of such spending programs in health and nutrition improvement.

Policy makers must address whether the provision of public goods and services, such as knowledge of better child care practices to caregivers, can be delivered efficiently by targeting the intended beneficiaries, such as working mothers, and if not, whether appropriate incentives can be put in place to ensure such efficient delivery. In the context of medium-term planning, the quality of public expenditure in health and nutrition improvement could be assessed with the help of donors. Policy makers could attempt to rank health and nutrition programs in order of relative importance within the country's economic and social priorities, and the country's absorptive capacity in light of existing institutional and administrative constraints.

Monetary policies are concerned with expansion of the money supply and domestic credit availability, and measures to influence interest rates, while "exchange rate policies" are used to maintain the exchange rate at a desired level that balances the national supply of foreign exchange to the demand for it. These policies affect the nutritionally insecure through three main channels: food prices, agricultural outputs, and the real exchange rate. First, increases in food prices hurt the poor by decreasing their purchasing power, and thus can increase the incidence of malnutrition through a lower quality diet or an increase in susceptibility to disease, or by a combination of the two. Second, fluctuations in output affect the nutritionally insecure through income changes. Finally, the output monetary and exchange rate policies can affect these fluctuations in two ways: a short-run

[1]The concepts of macroeconomic policies discussed in this section are based on and can be found in several basic macroeconomic texts, see for example Mankiw (2011), and Agenor and Montiel (2008).

effect on the real interest rate, which in turn affects output; and a chosen exchange rate regime that can eliminate or increase exogenous shocks.

The real exchange rate changes can affect the nutritionally insecure through two main pathways, the country's external competitiveness and thus growth rate; and change in the real exchange rate which has a direct impact on the nutritionally insecure through changing the relative price of tradable to nontradable commodities. It is well known that targeting variables such as a low and stable rate of inflation, and especially stable food prices, are critical for monetary and exchange rate policies to be effective instruments.

The impact of trade liberalization policies can be transmitted to the nutritionally insecure through price changes and through the domain of trade. The degree of price change is transmitted to the economy through the domain of trade, depending on whether the good is traded internationally, nationally, or locally. For example, if the Stolper – Samuelson Theorem were to hold, and trade liberalization increased demand for labor-intensive goods, unskilled wage income would increase, and this could be transmitted to the household through an improvement in their diet. Further, if the impact of trade liberalization on growth is positive, it will increase demand and generate higher government revenue, and increased government revenue could then be spent on various health and nutrition programs.

In designing trade liberalization policies the length of the planning horizon is important, as they may have an adverse effect on nutrition in the short run. Terms of trade shocks depend on the choice of exchange rate regime (fixed or flexible), which in turn depends on the "nature of economic shocks" affecting an economy, as well as "structural features" of the economy. The terms of trade can be defined as the ratio of the price of tradables to that of nontradables in an economy. For example, the income of the nutritionally insecure is more associated with tradable goods, while consumption is associated with nontradable goods. A depreciation of the domestic currency will make a country's exports more attractive and increase the demand for tradables, which will increase their income. This will have an improved distributive impact on the nutritionally insecure. In an economy, if the predominant source of influences on the poor is terms of trade, a flexible exchange rate regime can be best, since the nominal exchange rate can respond to shocks and bring the real exchange rate to a new equilibrium. However, if there are shocks to the demand for money, a fixed exchange rate regime can be best, since the shocks can be absorbed by fluctuations in international reserves.

Structural features refer to how shocks in an economy affect the system, as well as the insulating properties of an exchange rate regime. If an economy is characterized by a significant degree of nominal wage rigidity, wages will not adjust in the short run in response to real shocks, and thus the effect of those shocks will be amplified. In these situations, a flexible exchange rate system is preferable. Further, the greater the degree of openness of an economy, the greater the exposure to external shocks. If an open economy is sufficiently diversified, and if its prices are sufficiently flexible, then a fixed exchange rate system may be preferable since the volatility of a flexible exchange rate may impede trade and thus lower external demand. Exchange rate policies are key for determining how households will be able to afford optimal diets and other nutrition-related inputs. This is particularly important for the poor and vulnerable segments of the society.

These concepts are useful in the discussion related to macroeconomic policies and their nutritional implications. We apply them in the rest of the chapter.

HOW DO MACROECONOMIC ADJUSTMENT POLICIES AFFECT NUTRITIONAL OUTCOMES?

Understanding the pathways through which macro-adjustment policies influence nutritional outcomes requires study of the linkages between policy variables and their intermediary effects on various outcomes before they affect household welfare indicators such as poverty, hunger, and malnutrition (Pinstrup-Andersen, 1987; Scobie, 1989; Orbeta, 1994). We describe one such pathway for illustrative purposes.

Let us consider trade regulations which impose export and import taxes or subsidies on specific commodities. Such regulations affect trade volumes, and hence the balance of payments. They further affect and are influenced by the exchange rates that prevail, and could be motivated by the balance of payments situation. The exchange rates in turn affect the relative prices of tradable and nontradable goods, which determine the volume of trade and balance of payments. The domestic money supply can influence the budget deficit of the government through deficit financing. It can also influence taxes, transfers, and subsidies which affect the real income of a poor and nutritionally vulnerable population. In a developing country context, the government budget and expenditure allocation also affects public sector employment and wages, which in turn affect household income levels.

Relative prices of tradable and nontradable goods influence output, employment, and factor payments directly, and influence the real income of poor households. In the context of developing countries, the government budget is also supported by foreign aid and external borrowing. These can affect the balance of payments as well. In addition, internal sources of borrowing can help with deficit financing, credit availability, and interest rates in the domestic economy. This can result in improved investment, higher private sector borrowing, and increases in output and employment, thereby enhancing real income. Finally, domestic money supply and exchange rate policies affect product and commodity prices. Along with such policies, price policies influence output, employment, and factor returns, which further can influence the real incomes of households, and welfare outcomes such as poverty and malnutrition. Of course, the translation of real income changes at household level is influenced by expenditures on food, health, water, sanitation, and other factors, as shown in Chapter 3, A Conceptual Framework for Investing in Nutrition: Issues, Challenges, and Analytical Approaches. In what follows we review specific analytical approaches to studying macro policy changes on the welfare outcomes at household levels.

ANALYTICAL METHODS FOR STUDYING THE IMPACT OF MACROECONOMIC POLICIES ON NUTRITION

In this section we briefly review selected methods of analyzing policy impacts on nutritional outcomes.

PARTIAL EQUILIBRIUM ANALYSIS

The partial equilibrium method equates supply and demand in one or more markets so that prices stabilize at their equilibrium level. Using this approach, the prices become endogenous in contrast to the demand functions for food studied in Chapter 3, A Conceptual Framework for Investing in Nutrition: Issues, Challenges, and Analytical Approaches. This approach is distinguished from GE models (see below), since it does not capture all production and consumption accounts in an economy. It also does not capture all markets and prices in an economy, and will not capture the impact of changes in one market on other key markets in the economy. Partial equilibrium analysis is more suitable for analyzing sectoral reforms. Specific tools for partial equilibrium include "multi-market models" and "reduced form techniques." The multi-market models estimate systems of demand and supply relationships, and study how the impact of policies in one sector can be translated into other related sectors. Multi-market models are used in a number of contexts to examine the welfare impact of technical changes in agriculture, e.g., input subsidies in India (Binswagner and Quizon, 1984), and gainers and losers in trade reform in Morocco (Ravallion and Lokshin, 2004).

Reduced form techniques are used to simulate the impact of different policy variables on social outcomes, such as poverty and nutritional status. We review an example of a reduced form technique for Tanzania. Tanzania achieved rapid per capita GDP growth during the period 1995 – 2001. However, household surveys showed that the decline in poverty was relatively small. Studying this situation, Demombynes and Hoogeveen (2004) argued that a possible explanation for this outcome was that poverty increased during the early 1990s, while economic growth could only offset part of the early rise in poverty. They showed that, under a variety of scenarios, poverty incidence first increased to more than 40% during the early 1990s, and then declined to below 36% by 2000 – 01. Their sectoral simulations suggested that the poverty reduction impact of economic growth in Tanzania was more significant in urban areas than in rural areas. Their sectoral decomposition of the poverty outcomes indicated that a small part (11.6%) of the decline in headcount poverty at the national level could be explained by a shift in the population from the poorer rural areas to the wealthier urban areas. They concluded that achieving the Millennium Development Goals would thus require changing patterns of growth in the rural areas.

More recently, Holmes and Dharmasena (2016) used the monthly national US data for the period 1997–2012 to study the linkages between macroeconomic shocks and participation in food assistance programs. Their modeling involved polynomial distributed lags, vector autoregression approaches, and directed acyclic graphs. Such approaches can be used to develop better predictions of participation rates in food assistance programs at the time of shocks induced by macroeconomic variables, can help in improved assessment of the costs involved in food assistance programs, and can save public resources through making the intervention cost-effective.

GENERAL EQUILIBRIUM MODELING

A GE model is a representation of a socioeconomic system where the behavior of all agents is compatible (Dervis et al., 1982). In GE analysis, the key modeling issues involve: (1) identification of the participants; (2) specification of individual behavior; (3) interactions among socioeconomic agents; and (4) characterization of compatibility. In general, the Walrasian framework serves as template for GE models. The GE models involve two basic categories of economic agents: consumers and producers, also referred to as households and firms. Further, the behavior of economic agents should conform to the optimization principle, in which each agent implements the best feasible action. For example, households and firms interact through a network of perfectly competitive markets. The market participants are buyers and sellers whose supply and demand behavior is a consequence of optimization; thus, behavioral compatibility is achieved through market equilibrium in GE models. The GE is achieved through relative prices so that for each market, the amount demanded is equal to the amount supplied of that commodity. The comparative statistics involve a change in equilibrium states associated with changes in shocks or policy reforms, discussed earlier in this chapter.

Empirical implementation of the GE models is briefly described here. Applied GE models are represented by a system of equations. Three sets of equations form this system. The demand equations from the optimizing behavior of consumers are the first set. The supply equations from the optimizing behavior of firms forms the second set. Finally, the equilibrium conditions for all markets are described by a third set of equations.

All supply and demand equations are homogenous to a zero degree. This means if all commodity and factor prices are multiplied by a constant **k**, the equilibrium supply and demand will not change. Thus money is neutral, and only relative prices can be determined. The GE model also satisfies Walras' Law, i.e., if all economic agents satisfy their budget constraints and all but one markets are in equilibrium, then the last market must also be in equilibrium. Applications of GE models are facilitated by computational methods that involve a social accounting matrix (SAM).

A social accounting matrix (SAM) as an accounting system: a conceptual framework

A SAM is a comprehensive and disaggregated snapshot of the economy for any given year that reflects various aspects of the economy, such as production, consumption, trade, accumulation, and income distribution. A SAM is a square matrix in which each transaction or account has its own row and column. The payments are listed in columns, and receipts are recorded in rows. The sum of all expenditures by a given account must equal the total sum of receipts or income for the corresponding account. This also implies that in a n-dimensional matrix, if the $(n-1)$ accounts are in balance, then so is the last one (this section is based on Thorbecke, 2000; Thorbecke and Jung, 1996) (see Table 5.1).

There are six kinds of accounts in SAM. The production activities produce sectoral goods and services by buying raw materials, intermediate goods, and services, and also pay indirect taxes to the government. The remainder is the value added distributed to the factors of production. Its receipts come from sales to households, exports, and the government. The factors of production account includes the labor and capital subaccounts. They receive income from the sale of their services to production activities in the form of wages, rent, and net factor received from abroad or from other regions. These revenues are distributed to households as labor incomes, and to companies as profits. Institutions include households, firms, and the government. Households receive factor income (wages and other labor income, rent, interest, and profits) as well as transfers from the government and remittances. Households' expenditure, on the other hand, consists of consumption of goods and services, and income taxes, with residual savings transferred to capital accounts. Companies receive profits and transfers and spend on taxes and transfers.

The government account allocates its current expenditure on buying the services provided by the production activities; it buys intermediate goods, pays wages, and delivers public and administrative services. Its income includes tax revenues from a variety of sources, and current transfers from abroad. The combined capital account on the income side collects savings from households, companies, and government, as well as foreign savings and channels these aggregate savings into investment. The rest of the world account includes households' consumption expenditure on

Table 5.1 Simplified social accounting matrix

		Expenditures				
		Endogenous accounts			Exogenous	Total
		Factor	Institutions (households and companies)	Production activities	Sum of other accounts	
		1	2	3	4	5
Receipts		Endogenous accounts				
Factors	1	0	0	T_{13}	x_1	y_1
Institutions, i.e., households and companies	2	T_{21}	T_{22}	0	x_2	y_2
Production activities	3	0	T_{32}	T_{33}	x_3	y_3
		Endogenous accounts				
Sum of other accounts	4	$\int_1'$	$\int_2'$	$\int_3'$	t	y_x
Total	5	y_1'	y_2'	y_3'	y_x'	

T_{13}, *Matrix that allocates the value added from various production activities into income accruing to the factors of production;* T_{33}, *Matrix showing intermediate input requirements;* T_{32}, *Expenditure of the various institutions of the different household groups on the commodities which they consume;* T_{21}, *Maps the factorial income distribution into household income distribution;* T_{22}, *Inter-institutional transfers among different types of households or between companies and households;* x_1, *Total exogenous demand for production activities resulting from government expenditure, investment, and export demand;* x_2, x_3, *Total exogenous demand for factors and total exogenous demand accruing to different household groups and companies from government subsidies and remittances from abroad; and* $\int_i'$ *are leakages from savings, imports, and taxation.*
From Thorbecke (2000).

imported final goods, as well as imports of capital goods and raw materials, while its income comes from the nation and the world from exports, and factor and nonfactor income earned.

The major assumptions of the SAM application to the multiplier analysis include: prices and expenditure propensities of endogenous accounts are constant; the government, rest of the world, and capital accounts are exogenous, while the factors, institutions and sectoral production activities are endogenous; and production technology and resource endowments are given for a time period.

Thorbecke (2000) and Thorbecke and Jung (1996) also developed procedures to analyze the poverty and nutritional implications of macroeconomic policies. In their approach, e.g., the total protein deficiency effects of an increase in the output can be decomposed into: (1) mean income change of the nutritionally deficient across all household groups, which are further decomposed into interdependency and three types of distributional effects; and (2) sensitivity of the protein deficiency measure to the mean income change. Parra and Woden (2008) provide a comparative analysis of the multiplier impact of both oil price and food price shocks using a recent Social Accounting Matrix for Ghana. The potentially large direct food price impact and indirect oil price impact suggested that these two factors play a role in designing social protection policies to help the consumer recover from price shocks.

In a recent study Springmann et al. (2016) analyze the implications of climate change on the nutritional status of the population. Specifically, they look at excess mortality attributable to agriculturally mediated changes from climate changes in dietary and weight-related risk factors. They assess changes in the consumption of various food commodities and associated deaths resulting from poor diets for different emission and socioeconomic pathways under various climatic inputs. They find substantial health effects of climate change, and suggest strengthening public health systems to prevent possible deaths resulting from risk factors arising from poor diets.

POVERTY ANALYSIS MACROECONOMICS SIMULATOR

The Poverty Analysis Macroeconomic Simulator (PAMS) developed by Pereira et al. (2002), links household surveys to macroeconomic variables and can infer changes in disposable income for specific categories of workers from expected changes in aggregate variables such as GDP by sector. The linkages allow one to conduct poverty and distribution analysis in a way that makes income growth, transfers, employment, and nutritional changes consistent with the macroeconomic framework (Pereira et al., 2002).

Linking the macroeconomic framework to the household sector involves multiplying the incomes/expenditures of each household of a specific group by the relevant growth rate. Growth rate is derived based on changes in disposable income of that specific group induced by changes in aggregate variables. PAMS has three recursive "layers," as shown in Fig. 5.2.

Layer 1 features the aggregate macro-framework; objective of this layer is to project GDP, national accounts, the balance of payments, and the aggregate price level. Layer 2 features the meso framework, and is a simplified earnings and labor market model that calculates disposable income across representative groups of households (RHs). Each of the RHs match one, and only one, economic productive sector. Within each RH group, a representative household will have its disposable income decomposed into labor income, nonlabor income, average taxes, and average transfers. Thus, overall, a RH-specific growth rate of disposable income can be computed. Layer 3 (micro) is the poverty and distribution simulator (nutritional improvement can also be incorporated).

The underlying principle is that changes in taxes or transfers can be transmitted to each household's income or expenditure according to pre-defined characteristics of the household. Further, PAMS calculates the average growth rate of disposable income for each RH (as determined in Layer 2), to simulate the income growth for each individual or family unit inside its own RH. A key assumption is that change in the disposable income for each individual or family unit in any RH group will be the same as for the RH itself. Thus, income distribution within each RH group remains constant.

The development and application of PAMS involves the following:

- Constructing appropriate "linkage aggregate variables" (LAVs). The LAVs used are GDP, GDP and employment by economic sectors, the general price level, overall tax revenue, and budgetary transfers to households. The number of RH groups corresponds to a breakdown by economic sectors that reflects the structure of the economy.
- PAMS works in a mechanical top-down fashion. The decomposition of GDP into its sectoral components will accurately depict the functioning of the economy. This is key to determining the accuracy of distribution analysis.

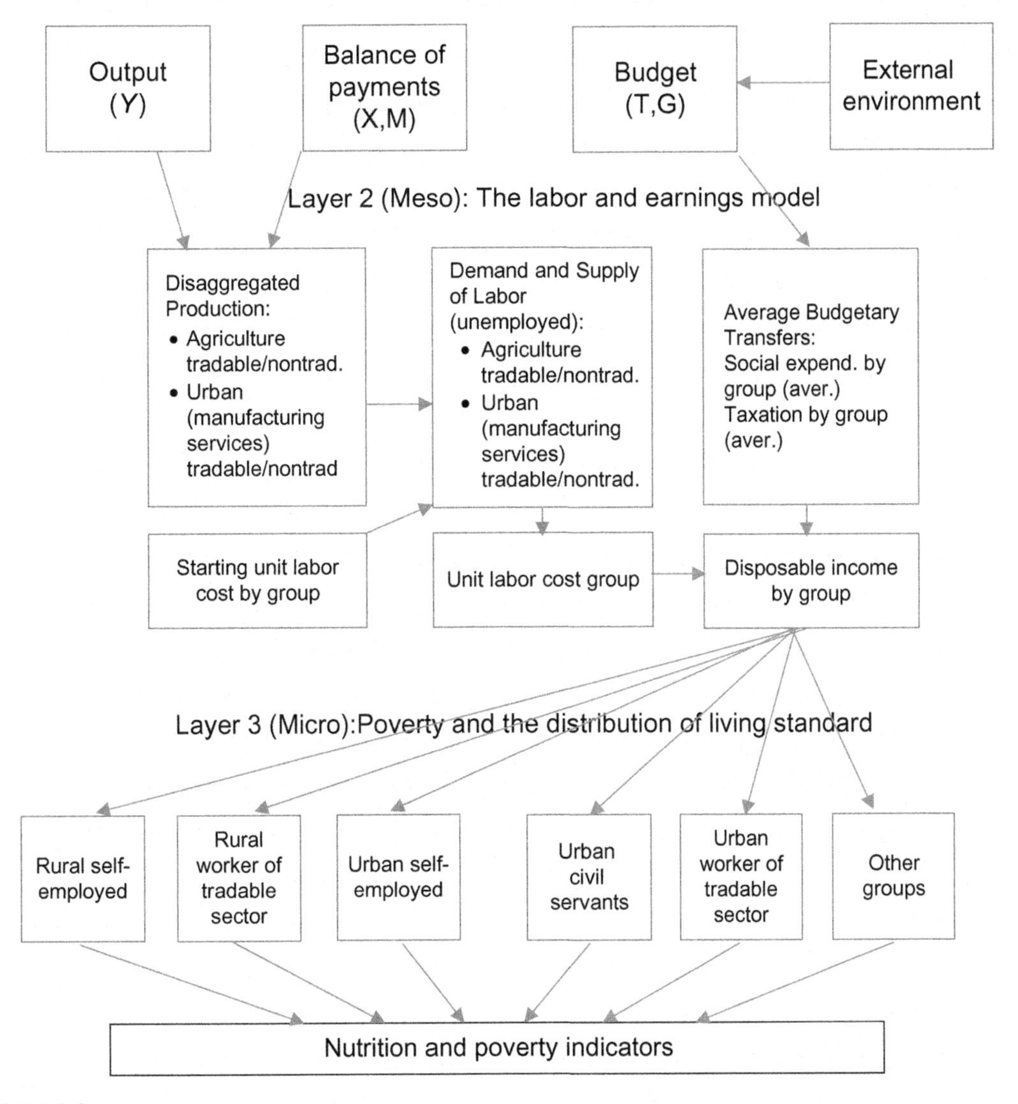

FIGURE 5.2

Structure of PAMS framework (Pereira et al., 2002).

- PAMS does not take into account some degree of substitution between the factors of production due to relative price changes.
- The labor market that is the key to transforming macroeconomic variables into factor prices is kept very simple. For example, each sector only uses one category of workers and no mobility of workers between sectors is allowed (Box 5.1).

BOX 5.1 KEY EQUATIONS OF THE LAYERS OF THE PAMS FRAMEWORK (PEREIRA ET AL., 2002)

Layer 1 (Macro): macroeconomic consistency framework

$$Y_t = C_t + I_t + G_t + (X_t\ IM_t) \tag{5.1}$$

$$X_t\ IM_t = \Delta\ R_t \tag{5.2}$$

$$T_t - G_t = \Delta\ M_t \tag{5.3}$$

$$p_t = f(\Delta\ Y_t, \Delta\ M_t) \tag{5.4}$$

Layer 2 (Meso): labor and earnings model

$$Y_t = p_t \sum_k Y_{k,t} \tag{5.5}$$

$$G_t = \sum G_{k,t} \tag{5.6}$$

$$T_t = \sum T_{k,t} \tag{5.7}$$

$$\Delta D\ Y_{k,t} = (p_t\ w_{k,t})\ L^d_{k,t} T_{k,t} + G_{k,t} + NL_{k,t} \tag{5.8}$$

$$\sum D\ Y_{k,t} = Y_t\ T_t \tag{5.9}$$

$$L^d_{k,t} = \gamma_k Y^{\alpha}_{k,,t}\ w^{\beta}_{k,t} \tag{5.10}$$

$$L^s_{k,t} = L^s_{k,t\text{-}1}(1 + n) - MIGR_{k,t} \tag{5.11}$$

$$\mathrm{L}^s_t = \theta_t\ \mathrm{POP}_t = \sum \mathrm{L}^s_{k,t} \tag{5.12}$$

$$w_{k,t} = \eta_k w_k^{-\varepsilon} (L^d_{k,t}/L^s_{k,t})^{-\delta} \tag{5.13}$$

Layer 3 (Micro): nutrition and poverty equations

$$NS_t = f(POP_t,\ L^d_{k,t},\ \Delta D\ Y_{k,t},\ Z_t) \tag{5.14}$$

$$Z_t = Z_{t-1}(1 + p_t) \tag{5.15}$$

Linking the macroeconomic framework to nutrition and poverty

The macro framework of PAMS consists of Eqs. (5.1–5.4). They provide national accounts consistency and predict changes in the key macroeconomic variables: GDP, public expenditures (G), overall taxes (T), private consumption (C), savings and investment (I), balance of payments (exports, X, and imports, IM), and aggregated price level (p), which is used to predict the protein deficiency line (z). The fiscal deficit (T-G) is financed by an increase in domestic financial liabilities (ΔM), and the current account deficit (X-M) has to be financed by an increase in external financial liabilities (change in reserves ΔR).

The meso layer is summarized in Eqs. (5.5–5.13). This layer models the functioning of the labor market. The framework disaggregates GDP into k sectors used in modeling labor demand, k components of average nominal spending accruing to each of the k groups, and k components of average nominal taxes paid by each of the households in k representative groups. The disposable income of the household consists of taxes, transfers, and social expenditures. Labor demand (Eq. 5.10) is broken down by k sectoral components. PAMS assumes that each sector hires only one kind of labor. There is no substitution between types of labor in the production process, and the labor demand depends on the level of activity in the sector and the sector's real unit labor cost. Labor supply for each group is determined by demographic considerations, and the real labor income (Eq. 5.13) in each of the sectors is determined by a sector-specific trend and by the unemployment rate in the sector. The excess of total disposable income (*Y-T*) over total labor income represents profits that are included in other incomes.

The micro layer is summarized by Eqs. (5.14) and (5.15). The main assumption made is that income or expenditure of each individual/household shifts in the same proportion to the group they belong to. Prices (in the case of nutrition, it is food prices) are the linkage variables from the macro to the micro layers of PAMS (Pereira et al., 2002).

Recently Laborde et al. (2016) developed a new analytical framework to identify and analyze the various roles of key long-term drivers of food and nutrition security at the aggregate level. They trace the key variables contributing to food and nutrition security at the micro level (household), and at the macro level (country level). Further, they identify a set of factors both endogenous and exogenous that affect the main set of variables driving the food and nutrition security outcomes. They fall under two broad categories, namely the aggregate supply of and aggregate demand for food. In their framework, the aggregate food demand is driven by demographic factors, income growth, changes in dietary patterns, aggregated domestic distortions, and overall quality of the food system; and the aggregate food supply is driven by land available for food production, share of food wastes and losses in the food system, and the productivity of crops and livestock captured by the normalized yields (Laborde et al., 2016).

Physical yields of crops are combined with the quality of the production to generate the amount of nutrients that could be produced from a land unit. Yields change with production patterns and are connected to the dietary patterns and changes in cropping patterns that are driven by exogenous phenomenon such as climate change. Further, the supply and demand for food and nutrients at the macro level is affected by macro and trade policies. For example, trade policies that are more liberal can improve food and nutrition security by working through the variables such as terms of trade, returns to the factors of production, foreign direct investment, and output prices. The framework also suggests that trade liberalization can leave some of the households less food and nutrition secure (Laborde et al., 2016). Such a modeling framework allows the policy makers to design interventions that affect selected drivers to achieve specific food and nutrition security goals. For example, we use a similar approach in demonstrating how cropping patterns could be designed to achieve nutritional security using optimization techniques in Chapter 16, Designing a Decentralized Food System to Meet Nutrition Needs: An Optimization Approach.

MACRO–MESO–MICRO POLICY LINKAGES IN THE CONTEXT OF NUTRITION

Macro and trade policies can have a profound influence on the nutritional outcomes at household and individual levels. For example, food self-sufficiency policies mostly focusing on procuring cereal crops provide little incentive for diversification of farming into nutrient rich crops. Minimum support prices announced for rice and wheat (which are sometimes more than the market prices) and assured procurement of such crops from farmers for storage and price stabilization purposes do not allow market forces to play their role in ensuring best allocation of land and water resources to crops that are more profitable and nutritionally enriching. Further, limited opportunities to trade in food commodities can restrict the local production of nutrient-rich foods. Restrictions on the movement of food commodities and infrastructural challenges result in high prices in one region of the country due to high demand, while in another part of the country the produce can be rotting without buyers. Such imbalances in market policies can be detrimental to diversification away from cereal crops to perishable high value commodities, affecting dietary diversity and nutritional intake at the household level. Low incomes and high inflation in salient commodities, such as pulses in south Asia, and among vegetarians can affect nutritional status as well.

Macro level issues need to be studied in the context of the micro level nutritional impacts. These linkages and challenges are brought out in a recent paper by Heady et al. (2015) in the context of Bangladesh. They find farm level constraints in diversification, weak market and trade infrastructure, and low level demand for nutrient-rich foods are basic problems of macro level policies having a micro level nutritional impact. They identify several causes that could be addressed through policy interventions.

Heady et al. (2015) argue that competing uses for land and water resources in a highly populated country where per capita availability of land is shrinking over time, limits the expansion of land to nutrient-rich crops. Further, emphasis on rice production for food security goals has focused the research and extension investments to major food crops such as rice and wheat. In addition, poor development of the value chains, including the marketing infrastructure and cold storage facilities for the perishable commodities, have limited crop diversification. This is compounded by the low level of demand for micronutrients. These factors resulted in limited crop diversification and continue to constraint households. As a result, Bangladesh has one of the least diversified diets in the world. Given that income is increasing among the middle income groups and demand for high value commodities is increasing due to changes in the dietary patterns, the limited crop diversification away from cereals crops has resulted in high prices of high value commodities pushing Bangladesh to rely on imports of these commodities.

CONCLUSION

Macroeconomic policies and the changes in global economic trends have a profound impact on human welfare as indicated by poverty, hunger, and malnutrition levels. Understanding the pathways through which such changes affect nutritional outcomes is important for designing policies that will improve nutritional outcomes, and protect vulnerable sections of the population from sliding into poverty and malnutrition. Yet such analysis continues to be the domain of very few macroeconomists. The purpose of this chapter, which complements Chapter 3, A Conceptual Framework for Investing in Nutrition: Issues, Challenges, and Analytical Approaches on microeconomic issues and their application to food consumption and nutrition, is to expose the readers to the approaches to studying the nutritional implications of macroeconomic policies.

The conceptual frameworks reviewed above help to connect macro policy levels to the sector level outcomes, which in turn influence household and individual welfare indicators. The applications of the analytical methods to the study of nutritional impacts were also presented to provide an introduction and motivation to readers to explore the methods further. For example, the recently announced Social Development Goals and their achievement can be tested under various globalization, macroeconomic, and trade policy scenarios. Finally, the concepts introduced in this chapter help understand the methods and results presented in the rest of the chapters of this book in a macroeconomic context of the country in which micro-level analytical studies are undertaken.

EXERCISES

1. Select a country of interest to you. Identify a set of recent macroeconomic policy announcements in this country. For a specific macroeconomic policy change, trace the impact pathways through various sectors of the economy, to the welfare outcomes such as poverty, food insecurity, and malnutrition (under-nutrition or obesity) by developing a conceptual framework, similar to the one presented by Pinstrup-Andersen (1987).
2. Prepare a review of the studies that have looked at the recent food and financial crisis from the context of health and nutrition implications. What variables stand out in terms of policy interventions to protect the nutritionally vulnerable population when you apply the results to the country of your choice?

PART

ECONOMICS OF NUTRIENT DEMAND

CHAPTER

CONSUMER THEORY AND ESTIMATION OF DEMAND FOR FOOD

6

If people behave in their own interests, we can infer something from their behavior about how well they are doing.

Angus Deaton, Nobel Lecture on Measuring and understanding behavior, welfare and poverty, Stockholm, December 8, 2015

This chapter applies consumer theory to develop an understanding of the demand for food. Understanding what factors can affect changes in the demand for food and nutrients is important for designing policies that can enhance or reduce the intake of specific nutrients, and for influencing healthy living through such choices. We introduce the theory of consumer demand in this chapter to lay the foundation for understanding the methods of estimating demand functions. We demonstrate specific methods of estimation of a system of demand equations using STATA, and apply the parameter estimates to study how changes in the various factors influence the demand for food and nonfood commodities.

Food policy analysis is based on economic theory. Studying the theory of demand helps in the formulation of models of consumer behavior toward food. Using the theory of demand for various commodities (food and nonfood), a set of demand equations can be specified when the prices of commodities, disposable income of the consumers, and their preferences are known.

As we saw in Chapter 4, Microeconomic Nutrition Policy, the utility maximization framework (Varian, 1978; Deaton and Muellbauer, 1980b; Philips, 1983; Powell, 1974; and Theil, 1975, 1976; Beattie and LaFrance, 2006) helps in the formation of how to solve the problem of how consumers allocate their income to various food and nonfood commodities. In what follows, we present the theory of utility maximization to derive a set of demand functions that depict consumers' choice toward food and nonfood commodities. The discussions below follow a standard text book approach to demand theory (see e.g., Deaton and Muellbauer, 1980b). The theoretical framework presented in the rest of this chapter closely follows Deaton and Muellbauer (1980b) and Johnson et al. (1984), and is updated from Babu (1989). To begin with, we review specific axioms based on which the theory of consumer demand is developed. The first of these axioms are known as completeness assumptions: when there are two sets of commodity bundles $q'_1 = (q'_1, \ldots q'_n)$ and another $q''_1 = (q''_1, \ldots q''_n)$, the consumer will be able to rank them and select one of these bundles over the other ($q' \gtrsim q''$ *or* $q'' \gtrsim q'$; $\gtrsim$ meaning "is preferred to").

The second axiom, known as the reflexivity assumption, states that each bundle is as good as itself $q' \gtrsim q'$. The third axiom, called the transitivity axiom, states that if $q' \gtrsim q''$, and $q'' \gtrsim q'''$, then $q' \gtrsim q'''$, indicating that the consumers preference for the commodity bundles is consistent. These

Nutrition Economics. DOI: http://dx.doi.org/10.1016/B978-0-12-800878-2.00006-2

three axioms ensure that consumers will choose a set of commodity bundles which he or she would be indifferent to in preferring one over the other (Deaton and Muellbauer, 1980a; Sproule, 2013). Finally, for the commodity bundles to be transformed into utility, the continuity assumption states that for any bundle q', define $A(q')$ a set of bundles preferred to $q'(q{:}q \gtrsim q'$ and $B(q')$ a set of bundles not preferred to $q'(q{:}q' \gtrsim q)$; then for any q in the choice set, $A(q')$ and $B(q')$ are closed sets that are contained in their own boundaries. This axiom guarantees that the commodity bundles chosen are transformed into utility through the existence of a utility function.

To maximize utility, we make use of two more axioms. That consumers prefer more of any good to less of it is stated as the nonsatiation axiom: if q' is greater than q'', then $q' > q''$. The convexity axiom states that a linear combination of q' and q'' is preferred to q'' if $q' \gtrsim q''$, and then for any $0 \leq \lambda \leq 1, \lambda q' + (1-\lambda)q'' > q''$. In maximizing the utility function and deriving demand functions, specific functional forms of utility functions are used. When such functional forms are strictly quasiconcave, the convexity assumption is satisfied. The functional forms of utility functions are also assumed to be twice differentiable differentiated in economic analysis of consumer demand (Deaton, 1986; Shirai, 2013).

DERIVATION OF CONSUMER DEMAND FUNCTIONS

Given a vector of commodities q, we specify a utility function $U(q)$. The consumer now chooses q by allocation of his or her disposable income x. To solve this utility maximizing problem we make several assumptions about $U(q)$. First, it is an increasing function, meaning $(U'\ (q) > 0)$. Second, $U(q)$ is a continuous function, strictly quasiconcave and twice differentiable. The budget constraint the consumer faces is $P'q = x$, where P is a column vector of commodity prices. We use the Lagrangian method to solve this problem:

$$\max_{q,\lambda} L = U(q) + \lambda[x - P'q] \tag{6.1}$$

where λ is the Lagrangian multiplier. Differentiating Eq. (6.1) with respect to q_j and λ yields $n+1$ first order conditions:

$$\frac{\partial U}{\partial q_j} - \lambda P_J = j = 1, \ldots, n \tag{6.2}$$

$$x - P'q = 0. \tag{6.3}$$

This is a set of $n+1$ simultaneous equations. We solve them by applying implicit function theorem (Apostol, 1957; Sproule, 2013), to get a set of Marshallian demand functions (they are also called ordinary demand functions):

$$q_j^* = q_j(P_1, P_2, \ldots P_n, X);\ j = 1, 2, \ldots, n \tag{6.4}$$

$$\lambda^* = \lambda(P_1, P_2, \ldots P_n,\ X) \tag{6.5}$$

That these functions are unique in prices and income is ensured by the regularity conditions of the utility functions. Now we define the following terms. The term $\lambda = \frac{1}{P_i}\frac{\partial U}{\partial q_i}$ is the marginal utility of income; $\frac{1}{P_i}$ is the number of units of q_i which can be purchased with one unit of income; and, $\frac{\partial U}{\partial q_i}$ is the marginal utility of q_i.

We now invoke the strict quasiconcavity assumption and derive the second order conditions of Eqs. (6.2) and (6.3), express as $q'Uq \geq 0$ for all q such that $P'q = X$, and verify the equations in (6.2) and (6.3), the first order conditions of the utility maximization problem yields a maximum.

We now study the properties of the Marshallian demand functions in Eqs. (6.4) and (6.5). Understanding these properties helps in applied policy analysis and in studying the behaviors of the consumers observed through collecting data on their choices and preferences. In general, these properties are derived by Eqs. (6.2) and (6.3), and placing certain mathematical restrictions on them. The partial derivatives of Eqs. (6.2) and (6.3) can explain the resulting preferences from changes in prices and income. In order to develop testable hypotheses for testing the consumers preferences, four properties of the demand functions are analyzed.

The *Adding up* property states that the values of the Marshallian demand functions should add up to the total income or expenditure on the commodities demanded:

$$(P_1q_1 + P_2q_2 + \cdots P_nQ_n = x).$$

This is the linear budget constraint given in Eq. (6.1). Substituting q_i^* (P, x) for q_i, we get:

$$\sum_i P_iq_i\,(P,x) = x \tag{6.6}$$

We can further derive two other subproperties from Eq. (6.6).

First, we differentiate Eq. (6.6) with respect to income x, to get:

$$P_1\frac{\partial q_1}{\partial x} + P_2\frac{\partial q_2}{\partial x} + \cdots Pn\frac{\partial n}{\partial x} = 1$$

We then multiply each term by $\left(\frac{q1}{x}\frac{x_1}{q1}\right)$, to get:

$$\left(\frac{q_1P_1}{x}\right)\left(\frac{\partial q_1}{\partial x}\frac{x}{q_1}\right) + \cdots + \frac{q_nP_n}{x}\left(\frac{\partial q_n}{\partial x}\frac{x}{q_n}\right) = 1.$$

Using the concept of elasticity this equation could be written as:

$$w_1e_{1x} + w_ne_{nx} = 1 \tag{6.7}$$

This is obtained where w_i is the budget share of good, and e_{ix} is the income elasticity of that good. Eq. (6.7), also called the *Engel aggregation*, shows that the sum of weighted shares of income elasticities is equal to one, the weights being the budget shares of the commodities. Second, we derive a property called Cournot Aggregation by differentiating Eq. (6.6) with respect to any price Pj:

$$P_1\frac{\partial q_1}{\partial P_j} + P_2\frac{\partial q_2}{\partial P_j} + \cdots q_j + P_j\frac{\partial q_j}{\partial Pj} + \cdots P_n\frac{\partial q_n}{\partial q_j} = 0.$$

We then multiply each term by $\frac{qi}{qi}$ and by $\frac{Pj}{x}$ to get:

$$\frac{q_1P_1}{x}\frac{\partial q_1}{\partial P_j}\frac{P_j}{q_1} + \cdots \frac{q_jP_j}{x} + \cdots \frac{q_jP_j}{x}\frac{\partial q_j}{\partial Pj}\frac{P_j}{q_j} + \frac{P_nq_n}{x}\frac{\partial q_n}{\partial P_j}\frac{P_j}{q_n} = 0;$$

By writing in elasticity form, we obtain:

$$w_1e_{1j} + w_2e_{2j} + \cdots w_je_{jj} + w_j + \cdots w_ne_{nj} = 0 \tag{6.8}$$

$$\sum_{i} w_i e_{ij} = -w_j$$

where w_i is the budget share of good i, e_{ij} is the cross-price elasticity of the ith and jth commodity, and e_{jj} is the own price elasticity. It shows the Cournot Aggregation: the sum of the own price and cross-price elasticities weighted by their budget shares due to changes in the price of the jth commodity is equal to the negative of the budget share of the jth commodity.

The next property of the Marshallian demand functions is known as the "Homogeneity" property: if we multiply all the prices and income by a constant k, the optimal quantity of commodities demanded is unchanged. Thus, the demand functions are homogenous to degree zero in income and prices. This property is based on Euler's theorem which states that, if a function $f(y)$ is homogeneous to degree θ, then derivatives of this function satisfy the following properties:

$$\frac{\partial f}{\partial y_1} y_1 + \frac{\partial f}{\partial y_2} y_2 + \cdots \frac{\partial f}{\partial y_k} y_k = \theta f(y)$$

We apply this property to Eq. (6.4), $q_i\,(P_1 \ldots P_n,\, x)$ to get:

$$\frac{\partial q_i}{\partial P_1} P_1 + \frac{\partial q_i}{\partial P_2} P_2 \ldots \frac{\partial q_i}{\partial P_n} + \cdots \frac{\partial q_i}{\partial x} x = 0.q_i(P,x) = 0. \tag{6.9}$$

We then divide Eq. (6.6) by q_i to get:

$$\frac{\partial q_i}{\partial P_1} \frac{P_1}{q_i} + \frac{\partial q_i}{\partial P_2} \frac{P_2}{q_i} + \cdots \frac{\partial q_i}{\partial P_n} \frac{Pn}{q_i} + \frac{\partial q_i}{\partial x} \frac{x}{q_i} = 0,$$

Writing in the elasticity form:

$$e_{i1} + e_{i2} + \cdots e_{in} + e_{ix} = 0 \tag{6.10}$$

We now define these terms which are useful in policy analysis:

- e_{ij} is the elasticity of demand i with respect to a change in price of good j;
- e_{ii} is referred to as own price elasticity;

e_{ij} is also referred to as cross-price elasticity; and e_{ix} is the elasticity of demand with respect to a change in income known as income elasticity.

The homogeneity property also gives the "Row constraint" by rewriting Eq. (6.7): $\sum_j e_{ij} = -e_{ix}$which states that the sum of all own and cross-price elasticities is equal to the negative of the income elasticity. Thus, if there are "n" demand equations then there will be "n" restrictions on the demand systems based on row constraints.

The "*Negativity*" property states that the $(n \times n)$ matrix formed by $\frac{\partial q_i}{\partial P_j}$, is negative semi-definite. The elements $\frac{\partial q_i}{\partial P_j}$ are denoted by S_{ij}, and the matrix $S\,(S_{ij})$ is called the Slutsky matrix or substitution matrix of compensated price responses. The negativity property places the following inequality restriction on s_{ij}; the diagonal elements of this matrix must be nonpositive for all i, $s_{ii} \cdot \left(\frac{\partial q_i}{\partial P_j} |\overline{u}\right) > 0$. The assumption of quasiconcavity of the utility function by which the second derivative with respect to any price is negative enables this property to hold.

The "Symmetry" property of demand functions follows from the Slutsky equation, for any price j:$\frac{\partial q_i}{\partial P_j}\Big|\overline{u} = \frac{\partial q_i}{\partial P_j}\Big|\overline{x} + q_j \frac{\partial q_i}{\partial x}$.

Writing in elasticity form:

$$\frac{\partial q_i}{\partial P_j}\Bigg|\overline{u} = \frac{\partial q_i}{\partial P_j} \frac{P_1}{q_i} \frac{q_i}{P_j} + \frac{q_i q_j}{x} \frac{\partial q_i}{\partial x} \frac{x}{q_i} = e_{ij} \frac{q_i}{P_j} + \frac{q_i q_j}{x} e_{ix} \tag{6.11}$$

Similarly, for the own price P_i:

$$\frac{\partial q_i}{\partial P_j}\bigg|_{\overline{u}} = e_{ij}\frac{q_j}{P_i} + \frac{q_i q_j}{x} e_{jx} \tag{6.12}$$

Equating (6.11) and (6.12) and using Young's theorem yields:

$$e_{ij}\frac{q_i}{P_j} + \frac{q_i q_j}{X} e_{ix} = e_{ij}\frac{q_j}{P_j} + \frac{q_i q_j}{X} e_{jx}$$

$$e_{ij}\frac{q_i}{P_j} = e_{ji}\frac{q_j}{i} + \frac{q_i q_j}{x}(e_{jx} - e_{ix})$$

$$e_{ij} = e_{ji}\frac{q_j P_j}{q_i P_i} + \frac{P_j q_j}{x}(e_{jx} - e_{ix})$$

$$e_{ij} = e_{ji}\frac{w_j}{w_i} + w_j(e_{jx} - e_{ix})$$

The above symmetry condition can be used to calculate a set of cross-price elasticities when the budget shares and another set of cross-price elasticities, and the income and own price elasticities are known.

In testing a hypothesis related to food consumption and in estimating the parameters of the demand functions, the properties discussed above can be useful by placing restrictions on parameters, and on the expected sign of the elasticity. For example, by the adding up the property $\sum_j w_i + e_{ix} = 1$, also called the Engel aggregation, only $n - 1$ of the income elasticities are independent. And by the homogeneity property $\sum_j w_i + e_{ix} = 0$, for each demand function there is one redundant elasticity and, therefore, n redundant elasticities for n equations.

By the symmetry property, knowing the budget shares and one set of off-diagonal elements, the other set of off-diagonal elements can be calculated, which reduces the number of independent elasticities by $1/2(n^2 - n)$. In practice, to derive all price and income elasticities we need to estimate $n^2 + n$ parameters (n^2 price and n income elasticities). However, by using the properties of demand functions, namely homogeneity, Engel aggregation, and symmetry, the number of independent parameters to be estimated can be reduced to:

$$\left[(n^2 - n) - n - 1 - \left(\frac{n^2 - n}{2}\right)\right] = \frac{n^2 + n}{2} - 1.$$

We use these properties and restrictions in estimating demand functions later in this chapter.

In specifying the forms of utility functions, analysts make two more assumptions: separability and additivity. These also help in placing specific restrictions on the estimation of the demand parameter. As consumers use a large number of goods, and some are similar in nature and use, it is possible to aggregate some of these goods for the case of analysis. When goods are aggregated, their prices are also aggregated in the form of an index. This is done based on the assumption that the utility function is separable, and still gives the same level of utility as if all the individual goods are identified in the utility function. The implication of separability in decision-making is that the level of consumption of a good in one group does not affect the marginal rate of substitution between any two goods in another group (Philips, 1983). The consumer is able to make optional decisions in stages, and this allows the analysts to use aggregated data on a specific group of commodities.

Under the condition of weak separability, the utility function could be written as:

$$U(q) = f[U_1(q_1), U_2(q_2)\ldots U_G(q_G)\ldots U_m(q_m)] \tag{6.13}$$

where $U_1(q_1), U_2(q_2)\ldots$ are subutility functions, and the functions q_1, q_2 and f are increasing in the subutility levels. The n commodities are partitioned into $m<n$ groups, and each group has n_r commodities.

The necessary and sufficient conditions for $U(q)$ to be weakly separable in m groups is as follows:

$$\frac{\partial[U_i(q_G)/U_j(q_G)}{\partial q_k} = 0 \tag{6.14}$$

where i and j belong to group G, k belongs to group K, and ($G \neq K$). The compensated demand for a good in a subgroup is given by:

$$q_{iG} = q_{iG}(P_G, X_G) \text{ and } X_G = X_G(P, X)$$

where P_G is the price index of group G, X_G is expenditure on group G, P is the aggregate price index, and X is total expenditure.

Eq. (6.14) has important implications for the demand functions. To analyze the consequences of this assumption on substitution effects, consider a good q_i in subgroup G, and a good q_j in subgroup H, where $H \neq G$. The pure substitution terms are:

$$\left.\frac{\partial q_i}{\partial P_J}\right|_{\overline{u}} = \frac{\partial q_i}{\partial x_G} \cdot \left.\frac{\partial x_G}{\partial P_J}\right|_{\overline{u}} \overset{=}{=} S_{ij} \tag{6.15}$$

and similarly:

$$\Big|_{\overline{u}} = \frac{\partial q_j}{\partial x_H} \cdot \left.\frac{\partial x_H}{\partial P_i}\right|_{\overline{u}} \overset{=}{=} S_{ji} \tag{6.16}$$

By the symmetry condition discussed earlier, $S.. = S..$; so that rewriting the above equations gives:

$$\left.\frac{\partial q_i}{\partial x_G}\frac{\partial x_G}{\partial P_J}\right|_{\overline{u}} = \frac{\partial q_j}{\partial x_H} \cdot \left.\frac{\partial x_H}{\partial P_i}\right|_{\overline{u}}; \left[\left(\frac{\partial x_G}{\partial P_J}\right) \Big/ \left(\frac{\partial q_j}{\partial x_H}\right)\right] = \left[\left(\frac{\partial x_H}{\partial P_i}\right) \Big/ \left(\frac{\partial q_i}{\partial x_G}\right)\right] = \lambda_{GH}$$

note that λ_{GH} is a constant for any two commodities from subgroups G and H. Thus:

$$\left(\left.\frac{\partial x G}{\partial P j}\right|_{\overline{u}}\right) \Big/ \left(\frac{\partial q_i}{\partial x_H}\right) = \lambda\text{GH (or) } \left.\frac{\partial x_G}{\partial P_j}\right|_{\overline{u}} = \lambda_{GH} \cdot \frac{\partial q_j}{\partial x_H} \tag{6.17}$$

Now, substituting in Eq. (6.15) $S_{ij} = \frac{\partial q_i}{\partial x_G} \cdot \frac{\partial q_j}{\partial x_H} \cdot \lambda_{GH}$ for i from G and j from H.

For any good q_r and any group K:

$$\frac{\partial q_r}{\partial x} = \frac{\partial q_r}{\partial x_k} \cdot \frac{\partial x_k}{\partial x};$$

$$\frac{\partial q_r}{\partial x} = \frac{\partial q_r}{\partial x_k} \Big| \frac{\partial x_k}{\partial x};$$

Substituting in Eq. (6.15):

$$S_{ij} = \lambda_{GH} \cdot \frac{\frac{\partial q_i}{\partial x}}{\frac{\partial x_G}{\partial x}} \cdot \frac{\frac{\partial q_j}{\partial x}}{\frac{\partial x_H}{\partial x}}; \quad S_{ij} = \mu_{GH} \cdot \frac{\partial q_i}{\partial x}\frac{\partial q_j}{\partial x}$$

where $\mu_{GH} = \lambda_{GH} / \frac{\partial x_G}{\partial x} \cdot \frac{\partial x_H}{\partial x}$; is the parameter summarizing the pattern of substitution between branches G and H. Rewriting in elasticity terms using the Slutsky equation:

$$\frac{\partial q_i}{\partial P_J} = \frac{\partial q_i}{\partial P_J} \mid \bar{u} - q_j \frac{\partial q_j}{\partial x} = S_{ij} - q_j \frac{\partial qj}{\partial x};$$

$$\mu_{GH} \cdot \frac{\partial q_i}{\partial x} \cdot \frac{j}{\partial x} - q_j \frac{\partial q_i}{\partial x}; \quad \frac{\partial q_i}{j} \frac{P_j}{q_i} = \frac{\mu_{GH}}{x} \cdot \frac{\partial q_i}{\partial x} \frac{x}{q_i} \frac{x}{q_j} \frac{\partial q_i}{\partial x} \cdot \frac{P_J q_j}{x} \cdot \frac{q_j P_j}{x} \frac{\partial q_i}{\partial x} \frac{x}{q_i} \tag{6.18}$$

Writing in elasticity notation:

$$e_{ij=\theta_{GH}} \cdot e_{ix} e_{jx} w_j - w_j e_{ix} \quad \text{for } i \text{ from } G;\ j \text{ from } H;\ \text{and } G \neq H \tag{6.19}$$

where θ_{GH} is called the coefficient of proportionality. Eq. (6.19) implies that all elements of the matrix of demand elasticities $[l/2(m^2 - m)]$ could be identified if the income elasticities (n) and the group substitution terms θ_{GH}'s are known. By employing the assumption of weak separability, the analyst can reduce the number of parameter estimates to $n + \frac{1}{2}(m^2 - m)$.

Strong separability is a special case of weak separability where the overall utility function is of the following form:

$$U(q) = f[U_1(q^1) + U_2(q^2) + \ldots U_m(q^m)] \tag{6.20}$$

where each U_i is a function of the q^1 in that group. The utility function $U(q)$ is strongly separable with respect to partition $(U_1, U_2, \ldots U_m)$, if the marginal rate of substitution between two commodities i and j from different groups I and J, respectively, does not depend on the quantities of commodities outside of I and J. That is:

$$\frac{\partial(U'_i/U'_j)}{\partial q_k} = 0 \tag{6.21}$$

where U'_i and U'_j indicate a partial derivative of U with respect to q_i and q_2, respectively. For three different goods from three groups, i, j, and k belonging to I, J, and K, respectively, this implies:

$$S_{ij} - \mu_{IK} \frac{\partial q_i}{\partial x} \frac{\partial q_j}{\partial x} = \mu_{IL} \frac{\partial q_i}{\partial x} \frac{\partial q_j}{\partial x}$$

where L is the union of subsets J and K. Also:

$$S_{ik} - \mu_{IK} \frac{\partial q_i}{\partial x} \frac{\partial q_k}{\partial x} = IL \frac{\partial q_i}{\partial x} \frac{\partial q_k}{\partial x}$$

since k belongs to the new group L. Rewriting:

$$\frac{\mu_{IJ}}{\mu_{IK}} \left(\frac{\partial q_j}{\partial x} / \frac{\partial q_k}{\partial x} = \frac{\partial q_j}{\partial x} / \frac{\partial q_k}{\partial x} \right)$$

which means $\mu_{IJ} = \mu_{IK} = \mu_I$. It follows that $S_{ij} = \mu_I \frac{\partial q_i}{\partial x} \cdot \frac{\partial q_j}{\partial x}$, and by anology

$$S_{ij} = \mu_J \frac{\partial q_i}{\partial x} \cdot \frac{\partial q_j}{\partial x}.$$

Also, since $S_{ji} = S_{ji}$, $\mu_I = \mu_J = \mu$. Thus, for strong separability:

$$S_{ij} = \mu \frac{\partial q_i}{\partial x} \cdot \frac{\partial q_j}{\partial x} \text{ for } I \text{ belongs to } I, j \text{ belongs to } J \text{ and } I \neq J. \tag{6.22}$$

where μ is independent of the commodity groups to which *i* and *j* belong. Because of additivity, the commodity groups are not interconnected, i.e., there are no unique utility branches.

With strong separability, the Slutsky equation for cross-price effects in elasticity terms is:

$$e_{ij} = \theta\, e_{ix} e_{jx} w_j - w_j e_{ix} \tag{6.23}$$

for i belongs to I, j belongs to J, and $I \neq J$. Thus, for the complete demand system, only $n + 1$ parameters need to be estimated: n income elasticities, and a value for θ.

BLOCK ADDITIVITY

This is a special case of strong separability where the function f in Eq. (6.20) is set to unity. Under this assumption, the marginal utility of the ith commodity depends on the quantity of the jth commodity only if i and j belong to the same block. The block utility function $U(q)$ can be written as:

$$U(q) = U_1(q_1) + U_2(q_2) + \ldots U_m(q_m) \tag{6.24}$$

where q^i is the vector of commodities in the ith block. The number of parameters to be estimated in a complete system of demand equations under this assumption depends on the number of groups and the number of commodities in each group. After imposing the usual restrictions given by demand function properties, the number of parameters to be estimated reduces to:

$$\left(n + \sum^{m} n_m^2\right)/2$$

where m is the number of blocks, and n_m is the number of commodities in the mth block.

DIRECT ADDITIVITY

Also known as point-wise separability, this is also a special case of strong separability in which the function f is defined as an identity function and each component of the utility function, $U_i(q^i)$ contains only one element. The condition on the cross-partial derivatives of the utility function becomes: $\frac{\partial U_u}{\partial U_j} = 0$ for all i and j.

The Slutsky term S_{ij}, reduces to a general substitution relation:

$$S_{ij} = \tilde{\mu} \frac{\partial q_i}{\partial x} \frac{\partial q_j}{\partial x}$$

where $\tilde{\mu} = -\lambda\left(\frac{\partial \lambda}{\partial x}\right)$. The expression for the uncompensated cross-price elasticities is, in this case:

$$e_{ij} = w_j \tilde{\theta}\, e_{ix} e_{jx} - w_j e_{ix} \tag{6.25}$$

where $\tilde{\theta} = -\frac{\tilde{\mu}}{x}$ and $\frac{1}{\theta}$ is the money flexibility parameter (Frisch, 1959), which describes how the level of marginal utility of money changes in relation to its rate of change with the level of money or income x. This simple preference structure requires only estimates of n income elasticities and the income flexibility parameter ($\tilde{\mu}$) for a complete characterization of the demand system. Also, the substitution matrix S_{ij} is positive if $\mu > 0$, and the income elasticities are all positive.

Thus, under the direct additivity assumption, inferior and complementary goods are excluded. These behavioral implications suggest that direct additivity should be used only for broad groups of commodities (Johnson et al., 1984). For estimating food demand systems, the food commodities are assumed to be separable from nonfoods. Among food commodities, the food grains are assumed to be separable from nonfood grains. A procedure for testing weak separability is presented below.

TESTING FOR SEPARABILITY

An initial analysis of the budget share for rural consumers showed that they spend almost as much on food grains as on all other foods combined. Also, the data used in this study have detailed figures on the consumption and prices of different food grains. The analysis of how consumers allocate their income between food grains, other food, and nonfood items using separability assumptions could have major implications for policy formulations. This section demonstrates how the separability hypothesis could be tested using parametric restrictions.

Barten (1964, 1967), Brown and Heien (1972), and Jorgenson and Lau (1975) have developed separable demand models. In most of these studies, where tested, the separability restrictions have been rejected. Eales and Unnevehr (1988) derived a restriction to test for weak separability in a first difference almost ideal demand system (AIDS) model. In this study, a restriction that is easy to impose on the data and allows the use of the convenient likelihood ratio techniques is used to test separability (Hayes et al., 1988; Winters, 1984).

$$C = G[g_1(P_1, U), \ldots, g_m(P_m, U)] \tag{6.26}$$

where $G(.)$ and the function $g_m(.)$ also have the general properties of cost functions, goods are grouped into r groups with $P_1 \ldots P_m$ price subvectors, and the functions $g_m(P_m, U)$ are increasing in U and G.

The group budget shares $\left(w_i = \frac{X_i}{X}\right)$ and the intragroup budget shares $\left(w_{ij} = \frac{q_j P_j}{X_i}\right)$ may be derived from Eq. (6.26) (Deaton and Muellbauer, 1980b):

$$w_i = \frac{\partial \ln G}{\partial \ln g_i} \tag{6.27}$$

$$w_{ij} = \frac{\partial \ln g_i(P_i, U)}{\partial \ln P_{ij}} \tag{6.28}$$

where $g_i(P_i, U)$ could be considered as group price indices that depend on the level of utility U. The share of subgroup J within the total expenditure X given by $W_{GJ} = \frac{q_j P_j^2}{X}$ can be derived using Shepard's lemma:

$$W_{GJ} \frac{\partial \ln G}{\partial \ln g_i} \frac{\partial \ln g_i}{\partial \ln P_{ij}} = W_i W_{ij} \tag{6.29}$$

The Slutsky terms between the goods can be derived by differentiating the above Eq. (6.29) with respect to the price of good r in group $S(P_{sr})$ holding U constrained:

$$\gamma_{ijsr} = \frac{\partial W_{GJ}}{\partial P_{sr}} = \frac{\partial^2 \ln G}{\partial \ln g_i \ln g_i} \cdot \frac{\partial \ln g_i}{\partial P_{ij}} \cdot \frac{\partial \ln g_s}{\partial \ln P_{sr}} \tag{6.30}$$

where

$$\frac{\partial \ln G}{\partial \ln g_i} = W_i; \quad \frac{\partial \ln W_i}{\partial \ln g_s} = \gamma_{is}; \quad \frac{\partial \ln g_i}{\partial \ln P_{ij}} = W_{ij}; \quad \frac{\partial \ln g_s}{\partial \ln P_{sr}} = W_{st}$$

and γ_{is} is the estimated cross-price parameter between groups i and s, estimated from an aggregate AIDS model that has shares W_i and W_s as dependent variables (say food grains and other foods in our analysis). This is derived by differentiating the W (food grains) equation with respect to the group price index for s (nonfood grains), holding the level of utility constant. The restriction that is implied by the quasi-separability of the cost function could be written in terms of known shares and estimated parameters as:

$$\gamma_{ijsr} = W_{ij} W_{sr} \gamma_{is} \tag{6.31}$$

Two groups of food grains i, and other foods s, may be considered separable if the compensated cross-price effects between the share of good j (rice) in group i (food grains), and the price of good r (milk) in group s (nonfood grains) satisfy the above condition (Hayes et al., 1988).

For the present analysis, W_{ij}, W_{sr} and γ_{is} are, respectively, the expenditure share of each food grain (rice, wheat, millets, pulses) in the food grain group, the share of expenditure for a particular food item (milk) in a nonfood grain group, and the cross-price parameter between the food grains and nonfood grain groups from an aggregated AIDS model. If $W_{sr} = 1$, the above restriction becomes:

$$\gamma_{irs} = W_{ij} \gamma_{is} \tag{6.32}$$

To implement the separability test using this restriction, the following procedure is used.

First, an estimate of γ_{is} was obtained using the AIDS model for food grains and nonfood grains in a two equation system. Second, another AIDS model was estimated in which individual food grain shares and the nonfood grain group share were dependent variables. The mean share of the individual food grains within the food grain group was multiplied by *is* to obtain a set of parametric restrictions. These restrictions were then placed on the cross-price terms between each food grain price and nonfood grain group price.

To determine whether these separability restrictions were accepted by the data, a likelihood ratio test was performed. The separability between the food grains and nonfood grains, as well as the separability between nonfood grains and nonfood groups were also tested using the same procedure.

AGGREGATING CONSUMER DEMAND FOR POLICY ANALYSIS

Understanding the changes in food demand due to changes in income and prices which can be influenced by policies requires parameter estimates of market demand functions. The market demand functions are defined as the horizontal summation of the individual demand functions of all the consumers under competitive market conditions. If the individual household's demand function for the *i*th good is given by:

$$q_i^h = q_i^h(P, x^h); h = 1, \ldots H \tag{6.33}$$

where h is the household, then the market demand would be given by:

$$\begin{aligned}\overline{q}_i &= \sum_{h=1}^{H} q_i^h(P,x^h)/H \\ &= f_i(P,x^1,x^2,\ldots x^H)\end{aligned} \tag{6.34}$$

The aggregation of individual demand functions involves transition from individual household behavior in food consumption to the analysis of the market demand for food. To understand this transition, and the aggregation analysis, we explore the common properties of demand functions $f_i\ (P, x^1, x^2, \ldots x^h) = \overline{q}_i(p, \overline{x})$, and $\overline{q}_i(P, \overline{x})$ in relation to the household demand functions, $q_i^h(P, x^h)$.

One factor that may change the aggregate demand is when the marginal propensity to consume (MPC) changes due to income redistribution. This change may affect the aggregation. For example:

$$q_i^h = a_i^h(P) + b_i(P)\, x^h$$

when all consumers in a market face the same marginal propensity to consume $[b_i(P)]$. This will satisfy the condition: $f_i\ (P, x^1, x^2, \ldots x^H) = \overline{q}_i(P,\overline{x})$. To check this:

$$\begin{aligned}\overline{q}_i \frac{\sum_h q_i^h}{H} &= \frac{1}{H}(\Sigma_h a_i^h + \Sigma_h\, b_i\, x^h) \\ &= \frac{\sum_h a_i^h}{H} + \frac{b_i \sum_h x^h}{H} \\ \overline{q}_i &= a_i(P) + b_i\, \overline{x}\end{aligned} \tag{6.35}$$

This shows that under linear demand curves, the individual is equivalent to aggregate demand, if the MPC is same for all consumers and the allocation of income is mean-preserving.

The income elasticity of demand e_{ix} becomes 1 as the income level reaches α. Thus, quasi-homothetic demand curves are required to have linear Engel curves. If the individual household demand function for good i given by Eq. (6.33) is linear in income, and the individual expenditure functions are quasi-homothetic given by $C^h(P, U^h) = a^h(P) + Ub(P)$, then the resulting market cost function is $\overline{C}(P, U) = \overline{a}(P) + U\overline{b}(P)$. In general, there is no need for the linearity of Engel curves when representative budget level (x_o) is used for consumers at market level. Then, for the demand function specified in Eq. (6.33), the exact nonlinear aggregation will hold when:

$$\frac{\sum_{n=1}^{H} q_i^h(P,x^h)}{H} = \overline{q}_i(P,x_o) \tag{6.36}$$

DUALITY, INDIRECT UTILITY, AND COST MINIMIZATION

The solutions of utility maximization $U(q)$ subject to a given budget constraint $Pq \leq X$ are given by a set of Marshallian demand functions denoted as $q^*(P,X)$. Using these optimal quantities in the original utility function, the maximized utility function is given by $U^* = U[q^*(P,X)] = V(P,X)$.

This is called the indirect utility function and has the following properties: (1) $V(P,X)$ is continuous in P and X; (2) it is nonincreasing in vector P, and nondecreasing in X (monotonicity property); (3) $V(P,X)$ is quasiconvex; (4) $V(P,X)$ is homogeneous to degree zero in (P,X); and (5) by the derivative property, the Marshallian demand function could be retrieved from indirect utility functions using Roy's identity (Roy, 1947):

$$q_i(P,X) = -\left[\frac{\partial V(P,X)}{\partial P_i} / \frac{\partial V(P.X)}{\partial X}\right]$$

Roy's identity is derived as follows:
The derivative of $V(P,X)$ with respect to any price P_i:

$$\frac{\partial V}{\partial P_i} = \sum_{j=1}^{n} \frac{\partial U}{\partial q_j}\frac{\partial q_j}{\partial P_i}$$

from the first order conditions of Eq. (6.1):

$$\frac{\partial U}{\partial q_i} = \lambda P_j$$

Thus:

$$\frac{\partial V}{\partial P_i} = \lambda \sum_j P_j \frac{\partial q_j}{\partial P_i}$$

using the adding up (Cournot aggregation) property of the Marshallian demand function:

$$\sum_j P_j \frac{\partial qj}{\partial P_i} = -q_i$$

$$\frac{\partial V}{\partial P_i} = -\lambda q_i \tag{6.37}$$

The derivative of $V(P,x)$ with respect to x:

$$\frac{\partial V}{\partial X} = \sum_{i=1}^{n} \frac{\partial U}{\partial q_j}\frac{\partial q_i}{\partial x}$$

$$= \lambda \cdot \sum_{i=1}^{n} P_i \frac{\partial q_i}{\partial x};$$

by the Engel aggregation property:

$$P_i \frac{\partial q_i}{\partial X} = 1;$$

thus:

$\frac{V}{X} = \lambda$ is the marginal utility of money.

Also:

$$\frac{\partial V}{\partial P_i} = -\frac{\partial V}{\partial X} qi; q_i = -\left[\frac{\partial V}{\partial P_i} / \frac{\partial V}{\partial X}\right]. \tag{6.38}$$

This result has important implications in applied demand analysis. If a functional form is assumed for $V(P,X)$, then the estimable form of Marshallian demand equations could be derived using Roy's identity and will have the same structure as the ones derived from the direct utility function (Barten and Bohm, 1982; Fuchs-Seliger, 1999). The approach to derive demand functions using the indirect utility function is also amenable for applications in welfare economics and index number analysis, since it represents the allocations to achieve the maximum utility levels under different prices and income (Jorgenson et al., 1988; Fuchs-Seliger, 1999).

COST MINIMIZATION AND HICKSIAN DEMAND FUNCTIONS

In the utility maximization approach to deriving demand functions for commodities, the consumer's problem was to maximize utility for a given level of income. The optimizing solution for this problem was used to attain some utility level of U. If reformulated to choose the commodities to minimize the total expenditure to reach the same level of U, then this problem is described as a "dual" problem of the former approach. The expenditure minimization problem is given by:

$$\min_q P'q;\ \text{subject to } U(q) = \overline{U} \tag{6.39}$$

and the solutions to this constrained optimization problem are a set of quantity demand functions which are functions of P and U, $q_i^* = h^i\ (P,U)$, called Hicksian or compensated demand functions. The minimized expenditure function of this problem is given by substituting the optimal values of q_i^* into $P'q$. Thus, the minimized expenditure or cost to achieve a certain level of utility U given the price vector P is $[p'q^* = Ph(P,U) = C^*(P,U)]$. This is known as the expenditure function or cost function. The properties of $C^*(P,U)$ are useful in understanding the restrictions on the demand functions. They are summarized as: (1) $C(P,U)$ is continuous in P and U; (2) it is nondecreasing in P and U (monotonicity); (3) homogeneous to degree one in prices; (4) concave in prices; and (5) by the derivative property the Hicksian demand functions could be retrieved from cost functions using Shephard's lemma (Shephard, 1953): $h_i\ (P,U) = \frac{\partial C(P,U)}{\partial P_i}$.

The indirect utility function could be derived by inverting the cost function, and vice versa, using the Shephard − Uzawa duality theorem (McFadden, 1978; Diewert, 1974, 1980). The Marshallian demand functions could be derived by substituting the inverse of the expenditure function into the Hicksian demand function (Deaton and Muellbauer, 1980b). Using appropriate functional forms in the cost function, the demand functions could be derived for applied empirical work (Deaton and Muellbauer, 1980a).

These relations from duality results can be combined together as in Fig. 6.1.

INTRODUCTION TO EMPIRICAL DEMAND SYSTEMS

In this section, we briefly review empirical demand systems as the basis for understanding, specifying, and estimating demand parameters. One way to specify the demand equations is to derive them using specific forms of utility function, and maximize them subject to a budget constraint that consumers face. The linear expenditure system and indirect addilog model discussed later are examples of this approach. Another approach is to specify functional form arbitrarily and approximate them. The Rotterdam model, transcedental logarithmic system, and AIDS are all examples of this approach. Yet another approach is to impose theoretical restrictions on ad hoc specifications

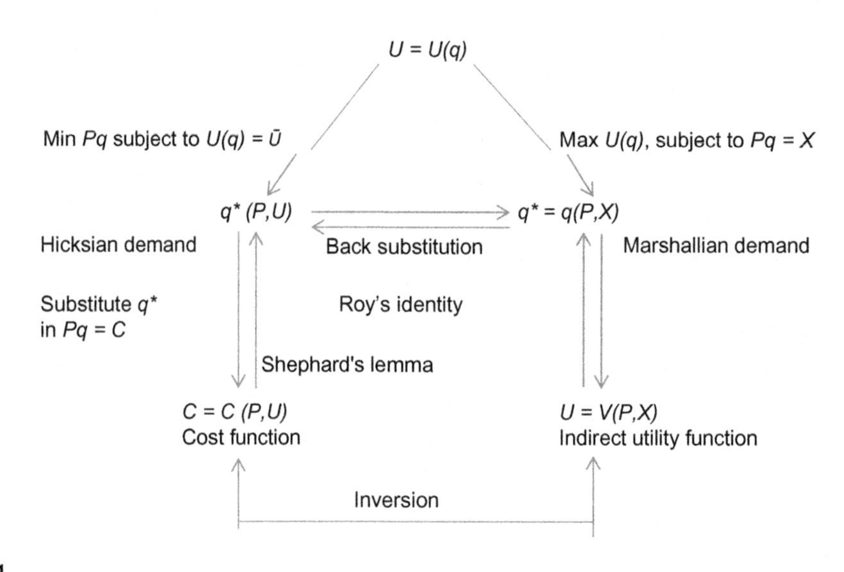

FIGURE 6.1

Relations From Duality Results (based on Deaton and Muellbauer, 1980b).

of the models. The generalized addilog model and Theil's multinomial extension of the linear logit model are good examples of this approach.

Several reviews of the demand system specifications are available in the literature (Barten, 1977; Johnson et al., 1984; Deaton, 1986; Bewley, 1986). Further, these specifications have been compared and evaluated for their appropriateness by a number of studies (Parks, 1969; Yoshihara, 1969; Goldberger and Gamaletsos, 1970, 1973; Deaton, 1974; Theil, 1975, 1976; Lybeck, 1976; Pollack and Wales, 1978; Klevmarken, 1979; Hansen and Sienknecht, 1989; Mazzocchi, 2003; LaFrance and Pope, 2010). We briefly review the most commonly used demand system models below.

LINEAR EXPENDITURE SYSTEM (LES)

Based on Stone (1954), the direct utility function for the LES is given by:

$$U(q) = \prod (q_i - C_i)^{bi}; \ \sum_j b_i = 1 \tag{6.40}$$

where C is a subsistent requirement of q_i, and b_i is marginal budget share.

Maximizing Eq. (6.32) subject to budget constraints results in Marshallian demand equations given by:

$$q_i = C_i + \frac{b_i}{P_i}(X - \sum_j P_j C_j) \tag{6.41}$$

where $(X - \sum P_j C_j)$ is the available income to allocate among goods in fixed proportions b_i, and is termed as supernumerary expenditure.

Substituting Eq. (6.41) in $U(q)$, the indirect utility function is given by:

$$V = \frac{(X - \sum P_j C_j)}{bo \prod P_j^{bj}}; \; bo = \prod b_i^{bi} \tag{6.42}$$

The cost function derived from the inversion of the above indirect utility function can be written as (Deaton and Muellbauer, 1980b; Mazzocchi, 2003; LaFrance and Pope, 2010):

$$C(U,P) = \sum_k P_k \, C_k + U \prod_{k=1}^{n} P_k^{bk} \tag{6.43}$$

where $\sum P_k C_k$ is the fixed cost of a subsistence requirement with no substitution, and $\prod P_k^{bk}$ is the term that allows for the utility to be attained at a constant price per unit.

The linear expenditure system can be derived by differentiating Eq. (6.43) with respect to prices using Sheppard's lemma, and substituting in the indirect utility function:

$$P_i q_i = C_i P_i + b_i (X - \sum\nolimits_j P_j C_j) \tag{6.44}$$

$i = 1, \ldots, n$ goods. The number of parameters to be estimated are $2n$ (n of C and n of b). Given the constraint imposed by $\Sigma b_i = 1$, the LES needs only $(2n - 1)$ parameters to be chosen independently. When the following restrictions on the parameters of the utility function (6.32), $\Sigma b_i = 1$, $0 < b_i < 1$, and $q_i > C_i'$, respectively, are imposed the "adding up" and "symmetry" conditions are met. The income elasticity for commodity i is given by:

$$e_{ix} = \frac{b_i}{w_i}$$

where $w_i = \frac{P_i q_i}{x}$ is the average budget share. The marginal budget shares are given by $b_i = e_{ix} \, w_i$. The own price elasticity for commodity i is given by:

$$e_{ii} = -1 + (1 - bi)\left(\frac{C_i}{q_i}\right) \tag{6.45}$$

and the cross-price elasticity between good i and the price of good j is given by:

$$e_{ij} = -bi \frac{P_j C_j}{P_i q_i} \tag{6.46}$$

so that for LES, all goods are gross complements (Johnson et al., 1984). Deaton (1974), Lluch et al. (1977), and Philips (1983) elaborate on the properties of LES and its application in demand system estimation.

Wales (1971) incorporated the concept of elasticity of substitution between the uncommitted expenditures and developed a generalized version of LES termed GLES.

$$w_i = C_i \frac{P_i}{X} + \left[\frac{q_i P_i^{-\sigma}}{\sum_{k-1}^{n} q_k \, P_k^{1-\sigma}}\right] \frac{P_i (X - \sum_{k-1}^{n} P_i \, c_i)}{X} \tag{6.47}$$

where σ is the elasticity of substitution. When $\sigma = 1$, LES becomes a special form of GLES.

Lluch (1973) incorporates inter-temporal effects in LES. This model is specified as:

$$P_{it} q_{it} = P_{it} C_{it} + b_{it} \left(W - \sum_t \sum_k P_{tk}^{*} C_{tk} + V_{it} \right) \tag{6.48}$$

where C_{it} and b_{it} are parameters specific to periods t, which vary over the life cycle, W is the current discounted value of present and future income and current financial assets, P^*_{tk} is the current discounted price of good k in the future period t, and V_{it} is the error term. Blundell and Ray (1982), Ray (1985), Green et al. (1980), and Blanciforti et al. (1986) have extended LES to include non-additive and dynamic properties.

INDIRECT ADDILOG MODEL (IAD)

Houthakker (1960) used an additive indirect utility function to derive a demand system called the indirect addilog model.

$$V(P,x) = \sum_{i=1}^{n} a_i(\frac{x}{p_i})^{bi} \tag{6.49}$$

where a and b are parameters with $\sum_{ai} = -1$, $a_i\ b_i > 0$ and $0 < b_i < -1$.

Roy's identity could be used to derive addilog demand functions:

$$q_i(P,x) = \frac{a_i b_i \left(\frac{x}{P_j}\right)^{1+bi}}{\sum_{k=1}^{n} a_k b_k \left(\frac{x}{P_k}\right)^{bk}} \tag{6.50}$$

in the log form the addilog model is written as (Somermeyer, 1974; Theil, 1975):

$$\text{ln}\ q_i = \text{ln}\ a_i b_i + (1+b_i)\ \text{ln}\left(\frac{x}{P_i}\right) - \text{ln} \sum a_i b_i \left(\frac{x}{P_i}\right)^{bi} \tag{6.51}$$

which satisfies Engel aggregation and Cournot aggregation, and the substitution matrix is negative semi-definite given $b_i > -1$. However, only quasiconcavity of the utility function is required for negative definiteness, which implies that at most one b can be equal to -1 (Murty, 1982; Conniffe, 2006). The income and price elasticities of addilog demand systems are given by Johnson et al. (1984).

The income elasticity is:

$$e_{ix} = (1+b_i) - b_i\ w_i \text{ for all } i \tag{6.52}$$

where $e_{ix} \lesseqgtr b_j w_j$

where the own-price elasticity is:

$$e_{ij} = (1+b_i) - b_i\ w_i \text{ for all } i \tag{6.53}$$

where $-1 < e_{ii} < 0$, with $-1 < b_i < 0$ and $w_i > 0$;

and the cross-price elasticity is:

$$e_{ij} = b_j w_j \text{ for all } i \neq j \tag{6.54}$$

which depends only on the commodity whose price is changing, and not on the good whose quantity is responding. The complete set of demand parameters in an addilog demand system can be estimated with $2n - 1$ independent coefficients (n for b_i's, and $n - 1$ for a_i's).

ROTTERDAM MODEL

The demand system is started with a specific algebraic demand system, and then the general demand restrictions are imposed to make it consistent with the theory of consumer demand (Theil, 1965; Barten, 1964; Barnett et al., 2013). The relative price version of this system begins with Stone's (1954) logarithmic demand function:

$$\ln q_i = a_i + e_{ix} \ln X + \sum e_{ij} \ln P_j \tag{6.55}$$

Writing the above Eq. (6.55) in differential form yields:

$$d \ln q_i = e_{ix}\, d \ln X + \sum e_{ij}\, d \ln P_j \tag{6.56}$$

multiplying by the budget share, Eq. (6.56) can be expressed as:

$$w_i d[\ln q_i] = b_i\, d \ln \overline{X} - \sum_j C_{ij} d \ln P_j \tag{6.57}$$

where $d \ln \overline{X} = d \ln X - \sum w_k\, d \ln \mathrm{P}_k$

$$b_i = w_i\, e_{ix} = P_i \frac{\partial q_i}{\partial X}$$

and $C_{ij} = w_i\, e_{ij} = \frac{P_i P_j S_{ij}}{X}$, S_{ij} is the (i,j) term in the Slutsky substitution matrix. The total differential of the budget constraint is:

$$\sum_{i=1}^{n} P_i\, dq_i + \sum_{i=1}^{n} q_i\, dp_i = dX \tag{6.58}$$

in logarithmic terms:

$$\sum_{i=1}^{n} w_i d[\ln q_i] + \sum_{i=1}^{n} w_i d[\ln P_i] = d \ln X \tag{6.59}$$

A typical equation of the absolute price version of the Rotterdam model can be written from Eq. (6.59):

$$\overline{w}_{it}\, \Delta \ln q_{it} = a_i + b_i \sum_k \overline{w}_{kt}\, \Delta \ln q_{kt} + \sum_k C_{ik}\, \Delta \ln P_{kt} \tag{6.60}$$

where Δ stands for the first difference operator over time, q_{it} and P_{it} are, respectively, the quantity consumed and the price paid for the ith commodity in period t, and $\overline{w}_{it}$ is the average budget share of the ith commodity in period t and $t-1$. The parameters a_i, b_i, and C_{ik} $(i, k = 1, 2 \ldots n)$ are interpreted as the intercepts, the income, and the price coefficients, respectively (Barten, 1969; Barnett et al., 2013).

The adding up the restrictions in the Rotterdam model implies:

$$\sum_{i=1}^{n} b_i = 1; \sum_{i=1}^{n} C_{ik} = 0; k = 1, \ldots, n.$$

The homogeneity can be enforced by imposing the restriction $\sum_{i=1}^{n} C_i = 0$, and the Slutsky symmetry restriction is given by $S_{ik} = S_{ki}$. The income elasticity is given by $e_{ix} = \frac{b_i}{w_i}$. The income elasticities are positive if b_i's are positive. Also, it could be noted that $e_{ix} \gtreqless 1$ when $b_i \gtreqless 1$. Thus, if $b_i > w_j$ the commodity is a luxury item. The own price and cross-price elasticities are given by:

$$e_{ii} = \frac{(C_{ii} - C_{ii}b_i - b_i w_i)}{w_i} \tag{6.61}$$

$$e_{ij} = \frac{(C_{ij}\, b_i - b_i w_i)}{w_i} \tag{6.62}$$

The parameters of the demand system can be significantly reduced if additivity restrictions are further imposed (Johnson et al., 1984). Then, the required number of parameters are only $(n+1)$ to form a complete set of demand elasticities.

TRANSCENDENTAL LOGARITHMIC DEMAND SYSTEM

Christensen et al. (1975) approximated the true indirect utility function with a second order Taylor series expansion. The indirect utility function of the translog model is given by:

$$\ln V = \alpha_o + \sum_{i-1}^{n} a_{ji} \ln\left(\frac{P_j}{x}\right) + \frac{1}{2} b_{ji} \ln\left(\frac{P_j}{x}\right) \ln\left(\frac{P_i}{x}\right) \tag{6.63}$$

Using Roy's identity, the translog demand system can be written as:

$$w_j = \frac{a_j + \sum_{i=j}^{n} b_{ji} \ln\left(\frac{P_i}{x}\right)}{a_m + \sum_{i=j}^{n} b_{mi} \ln\left(\frac{P_i}{x}\right)} \tag{6.64}$$

where $a_m = \sum_{j=1}^{n} a_j; b_m = \sum_{j}^{n} b_{ji}$. Thus, the demand system uses normalized prices with respect to income. A normalization $a_m = -1$ is imposed to identify the parameters of consumer demand or expenditure share equations in Eq. (6.64).

The income elasticity for the indirect translog demand system is given by:

$$e_{jx} = 1 + \frac{-\sum_i b_{ij}/w_j + \sum_j \sum_i b_{ij}}{-1 + \sum_j \sum_i b_{ji} \ln(P_i/x)} \tag{6.65}$$

The own price elasticity is given by:

$$e_{jj} = -1 + \frac{b_{jj}/w_j + \sum_j b_{ij}}{-1 + \sum_j \sum_i b_{ji} \ln(P_i/x)} \tag{6.66}$$

and the cross-price elasticity is given by:

$$e_{ji} = \frac{b_{ji}/w_j + \sum_j b_{ij}}{-1 + \sum_j \sum_i b_{ji} \ln(P_i/x)} \tag{6.67}$$

The translog demand system has been widely utilized in applied demand analysis (Christensen et al., 1975; Jorgenson and Lau, 1975).

GENERALIZED ADDILOG DEMAND SYSTEM

The generalized addilog demand system is a representative from a group of arbitrarily specified demand systems, and is derived from a consumer utility maximization. It is specified in an ad hoc manner imposing the theoretical restrictions directly.

The generalized addilog model (GAD) is given by n:

$$w_i = \frac{\exp[a_i + \sum_{j=1}^{n} b_{ij} \ln(P_j) + b_{ix} \ln(x) + u_i]}{\sum_{k=1}^{n} \exp[a_k + \sum_{j=1}^{n} b_{kj} \ln(P_j) + b_{kx} \ln(x) + u_k]} \tag{6.68}$$

where w_i is the average budget share of the ith commodity, P_j are the prices, x is the total expenditure, and u_i are the disturbance terms. This model reduces to the addilog model discussed earlier when $b_{ij} = 0$, $i = j$, $j = 1, \ldots, n$, and $b_{ix} = -b_{ii}$, $i = 1, \ldots, n$. Bewley (1986) has shown that the addilog model derived from GAD satisfies the Slutsky symmetry restrictions, and could be derived from a utility function. Theil (1965), Tyrrel and Mount (1982), and Considine and Mount (1984) used the linearized version of GAD with the logarithm of the ratio of pairs of budget shares:

$$\frac{\ln w_i}{w_n} = (a_i - a_n) + \sum_{k=1}^{n} (b_{ik} - b_{nk}) \ln(P_k) + (b_{ix} - b_{nx}) \ln x + (u_i - u_n), \quad i = n, \; i = 1, \ldots, n \tag{6.69}$$

Teklu and Johnson (1988) applied GAD to Indonesian data. The GAD models are also referred to as multinominal logit models, and represent two important propositions. In GAD models, the average budget shares sum to unity (adding-up), and the average budget shares predicted by the model are nonnegative. The own price elasticity is given by:

$$e_{ii} = b_{ii} - \delta_{ii} - \sum_{k=1}^{n} w_k b_{ik}; \; i = 1, \ldots n \tag{6.70}$$

$$e_{ij} = b_{ij} - \delta_{ij} - \sum_{k=1}^{n} w_k b_{ik} \tag{6.71}$$

where δ_{ij} is the Kronecker delta. The expenditure elasticity is given by:

$$e_{ix} = 1 + b_{ix} - \sum_{k=1}^{n} w_k b_{kx} \tag{6.72}$$

Finally, we now look at a most commonly used demand system in empirical estimation.

ALMOST IDEAL DEMAND SYSTEM

The AIDS model proposed by Deaton and Muellbaur (1980a) gives an arbitrary first-order approximation for any demand system. It satisfies the axioms of choice, aggregates perfectly over consumers, has a functional form which is consistent with household budget data, and is simple to estimate and test the true restrictions of demand theory. It also combines the best of the theoretical features of both the Rotterdam and translog models (Barnett et al., 2013). The formulation of AIDS uses the duality theory and expenditure function instead of utility or the indirect utility function. The expenditure function is specified as:

$$\ln C\,(U, P) = a_o + \sum_i a_i \ln P_i + \frac{1}{2} \sum_i \sum_j \gamma_{ij}^{*} \ln P_j \ln P_i + \overline{U}\, b_o \sum_i P_i^{bi} \tag{6.73}$$

where a_i, b_i, and γ_{ij}^{*} are parameters, $\overline{U}$ is utility level, and P_j are prices. This expenditure function is linearly homogeneous in P, provided $\sum a_i = 1, \sum_{jij}^{*} = \sum_j b_j = 0$. It is also consistent with aggregation over consumers. Differentiating the expenditure function using Shephard's lemma, yields:

$$w_j = a_j + \sum_{\gamma_{ij}} \ln P_i + \overline{U} b_o b_{j\exp}[b_{jx} \ln(P_{ij})] \tag{6.74}$$

Substituting for $\overline{U}$, which is the indirect utility derived from the expenditure function:

$$w_j = a_j + \sum a_{ij} \ln P_i + b_j \ln\left(\frac{X}{P*}\right) \tag{6.75}$$

where

$$\ln P* = a_o + \sum a_i \ln P_i + \frac{1}{2} \sum_i \sum_j \gamma_{ij}^* \ln P_i \ln P_j \tag{6.76}$$

is an overall price index, which could be replaced by Stone's (1954) index in empirical applications since Eq. (6.70) is highly nonlinear. Stone's index is given by:

$$\ln P* = \sum_i w_i \ln P_i \tag{6.77}$$

When Stone's index is used in Eq. (6.71), the model is termed as a linear approximation of an almost ideal demand system (LA/AIDS). The sets of restrictions on the AIDS model are given by:

$$\sum_{i=1}^{n} a_i = 1; \quad \sum_{i=1ij}^{n} = 0; \quad \sum_{i=1}^{n} b_i = 0; \quad \sum_j \gamma_{ij} = 0; \quad \text{and} \quad \gamma_{ij} = \gamma_{ji}$$

which should hold for the AIDS model to represent a system of demand equations (which add up to total expenditure ($\Sigma wi = 1$)) which are homogeneous to degree zero in prices and total expenditure, and satisfy Slutsky symmetry. The Slutsky coefficients are given by:

$$S_{ij} = \gamma_{ij} + w_i w_j - w_i \delta_{ij} \tag{6.78}$$

where δ_{ij} is the Kronecker delta (i.e., $\delta_{ij} = 1$when $i = j$, and equals 0 otherwise). The Marshallian and Hicksian measures of elasticities can be computed from estimated parameters of the LA/AIDS model as follows:

$$e_{ii} = -1 + \frac{\gamma_{ii}}{w_i} - b_i \tag{6.79}$$

$$e_{ij} = \frac{\gamma_{ij}}{w_i} - b_i\left(\frac{w_j}{w_i}\right) \tag{6.80}$$

$$d_{ii} = -1 + \frac{\gamma_{ii}}{w_i} - w_i \quad \text{and} \tag{6.81}$$

$$d_{ij} = \frac{\gamma_{ij}}{w_i + w_j} + w_j \tag{6.82}$$

where e denotes Marshallian elasticities, and d denotes the income compensated or Hicksian measure. Expenditure elasticities can be obtained using:

$$e_{ix} = 1 + \frac{b_i}{w_i} \tag{6.83}$$

The restrictions of demand theory can be imposed during estimation and tested easily with AIDS. There are several applied studies using AIDS including Deaton and Muellbauer (1980a) for Great Britain; Sergenson and Mount (1985), Hein and Pompelli (1985), Blanciforti and Green (1983), and Hayes et al. (1988) for the United States; Mergos and Donatos (1988) for Greece; Fulponi (1989) for France; Ray (1980, 1982) for India; Verbic et al. (2014) for Slovenia; Lasarte et al. (2014) for Spain; and Bilgic and Yen (2013) for Turkey.

In addition, in the last 15 years the Quadratic Almost Ideal Demand System (QUAIDS) model has been used to estimate food demand parameters. We use this model later in the chapter to explain the STATA procedures. The debate on the use of the above models in policy analysis continues. In general, the data availability dictates the choice of the model. In the section below we review applied food demand studies.

FOOD DEMAND STUDIES IN DEVELOPING AND DEVELOPED COUNTRIES

Food demand studies have used different functional forms, data types, and estimation procedures to generate demand parameters, and have used them in policy analysis and discussion. Demand studies have addressed the challenges of estimating income and price elasticities of segregated food commodities for the past 60 years, including those by Houthakker (1957); Weisskoff (1971); Lluch, Powell, and Williams (1977); Theil and Clements (1987); and more recently Dubois et al. (2014), and LaFrance (2008).

Houthakker (1957) was the first study to compare the expenditure elasticities for different (17 developed and 15 developing) countries. The expenditure elasticities calculated using double log form of demand functions in this study are compared with more recent studies. This would enable a comparison over the years where similar sample and functional forms are used. Weisskoff (1971) reported income and price elasticities of food and 5 other groups of expenditures for 21 countries based on double log equations using time series data. Lluch, Powell, and Williams (1977) conducted a cross-country comparison of 18 countries using the linear expenditure system, with both system least squares and maximum likelihood procedures.

Theil and Clements (1987) presented cross-country estimates of income and price elasticities based on Frish, Cournot, and Slutsky forms of elasticity equations using Working's model (Working, 1943). They used the country level aggregate data presented in Kravis et al. (1982). In general, the estimates of these studies varied depending on whether cross-section or time series data were used, whether the data was from rural or urban households, the nature of price aggregation, the expenditure classes of the consumers, and the period of data recall during the household surveys (Behrman, 1988; Dubois et al. 2014).

A recent study by Dubois et al. (2014) is an interesting cross-country comparison for developing countries. Dubois et al. (2014) note that there are substantial differences in food purchases between the United States and France. In order to explain these differences, the researchers estimate a demand system for food and nutrients. Interestingly, they perform counterfactual simulations, wherein they calculate the households' responses, given prices and nutritional characteristics from other countries. While prices and nutritional characteristics are an important part of the purchase decisions, there are also many other compelling forces at work, such as the economic environment and tastes.

Fat-taxes have also become an important consideration for policy makers to tackle the nutrition problem. For instance, in the United States, where dairy products contain harmful types of fat, policy makers are seriously interested in taxing these products. Chouinard et al. (2007) estimate a demand system for dairy products, and use simulations to capture the effect of taxes. Interestingly, their study finds that even a tax of 10% on the percentage of fat content reduces fat consumption by less than 1%. Although fat consumption is inelastic and it can generate a lot of tax revenue, the dead-weight loss of the tax will fall unfairly on the poor and elderly.

Jensen (2011) provides a good overview regarding the food supply and nutrition situation in the US, and for a few other countries. Jensen (2011) also links the issue of nutrition adequacy to health policies, and to the importance of economic analysis for policy recommendations. Bessler et al. (2010) provide an exhaustive evaluation of the literature in the area of agricultural and resource economics covering the last 100 years. They note how econometric techniques have evolved over time, and how agricultural economists are at the forefront of developing methodology that is appropriate for the unique problems confronted in this discipline.

Major contributions to the demand for food and nutrition have been from seminal papers by LaFrance and Roulon (2008, 2009, 2010), Beatty and LaFrance (2005), and Beattie and LaFrance (2006). In a series of papers, these authors have exploited the microeconomic foundations of consumer choice, duality, and econometric methodology to establish the superiority of Gorman Engle forms. The application by Beatty and LaFrance (2005) estimates a flexible Gorman form to demand for food and nutrients for US data from 1910 to 2000, excluding the World War II years. This exhaustive study estimates the price and income elasticities of food and nutrient demand for the selected sample time frame. Importantly, the studies cited above link diet and health to farm and food policy in the US. For instance, in the United States, Food Stamps, Aid to Families with Dependent Children, and the Women, Infants, and Children Program are all targeted food aid programs meant to create incentives for food purchases and consumption. However, the results may not be consistent with those originally envisioned for these programs. Consequently, it is possible that food aid recipients spend more on food, but are presented with incentives to eat unhealthy diets due to policy-induced price distortions. These studies also investigate such issues for 17 nutrients during the sample period (Beatty and LaFrance, 2005). They find that all foods except butter are either income normal or essentially independent of income. Butter is increasingly income inferior through the last half of the century. During the time frame, the authors note that most notably other red meat, fish and shellfish, all fresh fruits and vegetables, coffee, tea, and cocoa display marked increases in income elasticity, while poultry's own elasticity decreases from near unity to near zero. While there has been a change in nutrient consumption, the nutrient responses to food prices have been small. The authors attribute this to the availability of a wide range of substitute foods from which equivalent total nutrients can be obtained. However, the authors also point out that fat-taxes or sugar-taxes may not substantially affect nutrient consumption in the United States (Beatty and LaFrance, 2005).

Rask and Rask (2011) note that the rising affluence in major developing countries, mainly China and India, and the increasing diversion of agricultural resources for energy production, as in the United States and Brazil, have sharply increased agricultural resource demand. For instance, Bennett and Birol (2010) provide a sequence of studies covering food and related issues in Africa, Asia, Latin America, and the Caribbean, which are all policy-oriented. Rask and Rask (2011) examine food consumption and production changes during development using resource-based cereal-equivalent measures. They note that diet upgrades to livestock products require fivefold increases in per capita food resource use, reflecting a consistent pattern which is only marginally affected by land base.

Among overall trends, they note that food consumption increases exceed production during early development, leading to the need for imports. Consumption eventually stabilizes at high incomes, but production falls short in land-scarce countries. Pork and poultry consumption increase the most; less efficient beef and dairy production command a majority of agricultural resources (Rask and Rask, 2011).

The issue of inflation of food prices during 2007–08 has attracted a lot of researchers in agricultural economics to relate these to undernourishment. Recently, Anriquez et al. (2013) undertook a cross-country inquiry to analyze the short-term effects of a staple food price increase on nutritional attainments. Using demand systems and simulations they find that food price spikes not only reduce the mean consumption of dietary energy, but also worsen the distribution of food calories, further worsening the nutritional status of populations. Access to agricultural land helps to obtain adequate nutritional levels in both urban and rural areas. In this context, Abler (2010) examines structural changes in the demand for agricultural products arising from economic growth in emerging economies from the BRIIC group of countries (Brazil, Russia, India, Indonesia, and China). Abler (2010) also concurs with the view that commodity prices and price volatility are key drivers of food consumption and nutritional adequacy. We now explore specific findings from different economies.

In the context of South Asia, where a majority of malnourished populations live, Razzaque et al. (1997) estimate demand elasticities for eight different occupational groups in Bangladesh. Consumers are assumed to maximize utility from not just quantity, but also from variety and energy production. They test the theory that increases in incomes will increase the consumption of non-cereal food items for most occupational groups. Indeed, they find the income elasticities of many items such as fish, beef, milk, and vegetables to be very high, and suggest that, if the trend continues, then the government emphasis on moving away from noncereal crops will lead to serious supply constraints in the long run. They suggest policy measures to actively change the pattern of existing food production in Bangladesh.

Interestingly, in the case of Bangladesh, the same idea is indirectly supported through cointegration techniques, when Sanjuan and Dawson (2004) apply war-time data to their analysis. Sanjuan and Dawson (2004) found that the daily per capita intake in Bangladesh averaged almost 2100 kilocalories in the 1960s, but fell to 1840 in 1972 following the war of independence in 1971. The catch-up occurred only around 1987, when intake reached about 2000 kilocalories. Using data from 1962 to 97, Sanjuan and Dawson (2004) analyze the long-run relationship between per capita income, food prices, and per capita calorie intake. Their time-series findings show that a long-run relationship exists, and that the war reduced average calorie intake permanently by 10%. Importantly, they show that income Granger-causes affect calorie intake but not vice versa or, in other words, the time series data on income is able to predict future calorie intake. The causality tests adopt statistical testing of the dependent variable on lagged values of the independent variables. The Nobel laureate Clive Granger is credited with this econometric methodology, and Granger causality is a popular method to test causality between two data series.

Aziz et al. (2011) conducted a detailed LA AIDS estimation from household survey data in Pakistan. They also distinguished between rural and urban income and price elasticities, controlling for consumption quantiles. The main focus of their study was to estimate rural – urban income and own price elasticities across a range of consumption quantiles. The Linear Approximate Almost Ideal Demand System (LA AIDS) was used to estimate the parameters of aggregate food commodity groups. Overall, they found that households exhibit an increase in their consumption of vegetables, fruits, milk, and meats with higher income. The expenditure elasticities are larger in rural areas compared to urban areas, and expenditures on most food groups increase at a decreasing rate as income increases. Expenditure elasticities for all food groups were positive and less than one, except for fruits, meats, and milk which are considered as luxuries. Their results also show that cereals tend to have the lowest expenditure elasticity of demand. Demand for all food groups except meats is inelastic. The authors suggest that the high price elasticities of demand for many food items stress the importance of food price changes for households. Consequently, the government should take note of the importance of pricing policy in the development of comprehensive agricultural and food policies in Pakistan.

In a related study for the rural – urban differences for Pakistan, Rehman et al. (2014) estimate income and price elasticities controlling for income levels. The researchers use cross-sectional data from the Pakistan Social and Living Standard Measurement Survey (PSLM) 2010–11 by the Federal Bureau of Statistics (FBS), Government of Pakistan, Islamabad. They also found all the income and household size elasticities to be positive. Further, urban food consumption is higher in the upper income group, while households belonging to other classes in rural areas are more food responsive. Urban households depict higher elasticities. However, households belonging to upper – middle and upper income groups in rural areas are also more food responsive. Similar to the above studies, Kasteridis et al. (2011) also showed that for Pakistan, household characteristics play a role in food expenditures, and regional differences exist. They adopted an interesting econometric estimation based on a linear AIDS for food with a Bayesian Markov chain Monte Carlo procedure, for a sample of urban households using census data from Pakistan.

In countries where the length of the available time series data prevents the use of time series consumption expenditure data, household cross-section data have been used, deriving the prices by dividing the value of expenditure from the quantity purchased. This is termed as a unit price approach. Most of the studies in South Asia are based on household expenditure surveys with some exceptions in India. Pitt (1983) used unit prices to estimate a food demand system with a linear Tobit specification for rural Bangladesh households. He also calculated nutrient elasticities of prices and income for 10 nutrients. Rice is found to be closer to normal goods (0.94), with an inelastic own price response (−0.83). Similar results have also been reported by Ahmed (1981) using a linear expenditure system with the same data.

Most of the Indian food demand studies use representative data from the National Sample Survey (NSS) for their estimation. Swamy and Binswanger (1983) estimated their own consumption data on individual grains from production figures for 10 states, since the NSS only reported cereal aggregates. Ghatak (1984) used similar data from the National Accounts Statistics, and Tendulkar (1969) used data from the state level household expenditure survey. Earlier studies that were conducted by Iyengar and Jain (1974) were based on single equation models and Engel functions with a large commodity aggregation level. Though Iyengar and Jain (1974) used an indirect addilog model with system least squares, they estimated the model only for food and nonfood groups with 10 observations for aggregate consumers. The nonfood elasticities were higher than the food elasticities. Murty (1980) used 24 rounds of NSS data from 1953 to 74 for both rural and urban consumers with six commodity groups, food grains being one of them. They rejected homogeneity restrictions for rural symmetry and homogeneity for urban consumers using the Rotterdam model. Ray (1980) was the earliest study using the Almost Ideal Demand System following Deaton and Muellbauer (1980a) for any country other than Great Britain. Four aggregates, namely food, clothing, fuel, and lighting, and other nonfoods were analyzed. Homogeneity and symmetry could not be rejected for rural and urban consumers. Food, fuel and lighting were found to be necessities, while clothing and other nonfood items were luxuries.

Murty and Radhakrishna (1982), using 24 rounds of NSS data for five expenditure classes in rural and urban areas, estimated an extended linear expenditure system for food grains, milk, oil, meat, sugar, other foods, clothing, fuel and lighting, and other nonfoods. All commodities other than food grains and fuel and lighting were found to be luxuries for both rural and urban consumers. Ray (1982) extended his 1980 study to incorporate household characteristics such as household size based on population census estimates using the AIDS model, and found significant family size effects on the budget share of an item.

Swamy and Binswanger (1983) was the first study to disaggregate food grains into rice, wheat, and other cereals which included pulses and coarse grains using the data estimated based on production data for 10 states over 20 years. They used a covariance transformation to pool the cross-section and time series data. Using the transcendental logarithmic model they could not reject homogeneity and symmetry. However, not satisfied with the data set in their study, they used NSS data for the years 1961 and 1974 (rounds 16 and 27) for 15 states in their next study (Binswanger et al., 1984). Food grains were disaggregated using separability assumptions into rice, wheat, sorghum, and other cereals in a second stage, and using a transcendental logarithmic model, they reported negative income elasticity for sorghum and other cereals. However, they did not consider rural and urban differences and different groups of households. Their data from the 16th round has been criticized for being an over-estimation of cereal intake due to double counting among rural households (Srinivasan et al., 1974). Ghatak (1984), using data from the Central Statistical Organization (CSO), pooled a time series of 18 years and cross-sections across states, and concluded that the model based on pooled data performed better than the time series data alone.

Majumder (1986) compared the estimates from aggregated NSS data for India and three income classes with the AIDS model for rural and urban consumers, and argued that there is a remarkable variation in demand behavior among the income classes. Coondoo and Majumder (1987) used a price-independent generalized logarithmic (PIGLOG) functional form with the same data, and presented estimates of elasticities for both rural and urban consumers.

Gaiha et al. (2010) in a related paper examined Indian data regarding nutrient intake, dietary changes, and deprivation using a demand model. They showed that calorie deprivation and poverty are closely related, and found evidence for child under-nutrition and the existence of the double burden in the Indian context. Their study is important in the Indian context, because it stresses the importance of increasing food entitlements, rather than the simple approach of state provision of food. In a related study, Sinha (2005) used a nonparametric approach to capture the distribution of calorific intake for the Indian data. Sinha's results (2005) indicate that calorie consumption responds to various factors based on the observed levels of calorie consumption. For instance, the magnitude of calorie income elasticity is different for under-nourished and over-nourished households.

Other notable food demand studies in South Asia include: Pitt (1983) and Ahmed (1981) for Bangladesh; Ahmad et al. (1988), Shankat (1985), and Alderman (1988) for Pakistan; Sahn (1988) and Garanand Chandrasekara (1979) for Sri Lanka; and Swamy and Binswanger (1983), Ghatak (1984), Tendulkar (1969), Iyengar and Jain (1974), Murty (1980), Ray (1980), Majumder (1987) for India.

Ozer Huseyin (2003) used a LES for cross-section budget survey data for Turkey and showed that household behavior has important consequences for food intake, and suggested appropriate policy measures. Ingco (1991) used an LA/AIDS model to estimate for the Philippines time-series data from 1961 to 88 for rice, wheat, maize, meat, fish, fruits and vegetables, other foods, and nonfood commodities. Interestingly, Ingco (1991) found that the situation in the Philippines was similar to countries such as Japan, the Republic of Korea, and Taiwan, where rice is the basic staple, and where consumers have started to consume more wheat and wheat products. There is also a shift toward increased consumption of meats, dairy products, vegetable oils, and fruits and vegetables. Similar to Japan, Taiwan, Malaysia, Singapore, Thailand, and Nepal, Ingco (1991) found that rice is an inferior good in the Philippines. The result is attributed to urbanization and dynamic factors such as habit formation. The findings are very close to those of Dadgostar (1988) for Thailand.

Huang and Bouis (2001) include structural shifts such as changes in marketing systems, occupational changes, and changes in tastes and lifestyles to model the demand for food in Taiwan. They note that the effects of income on food demand may be overestimated, if structural changes are ignored. Zheng and Henneberry (2011) estimate a LinQuaid model for 10 major food groups for major income groups using household survey data from Jiangsu province. Their study also shows that the majority of the food categories exhibit high income and own price elasticity, particularly for low-income groups. Similar demand estimation for China, by Hovhannisyan and Bozic (2013), shows that without properly accounting for price endogeneity, the demand estimates will be biased upward.

In a recent World Bank study, Fukase and Martin (2014) provide detailed estimation of China's demand and supply for food. They note that the ongoing dietary shift to animal-based foods, induced by income growth, is likely to impose considerable pressure on supply. They note that China's demand growth is similar to the global trend. On the supply side, output of food depends strongly on the productivity growth associated with income growth, and on the country's agricultural land endowment, with China appearing to be an out-performer. From a policy point of view, the researchers note that agricultural productivity growth through further investment in research and development, and expansion in farm size and increased mechanization, as well as sustainable management of agricultural resources, are vital for feeding China in the 21st century.

In the context of sub-Saharan Africa, notable earlier food demand studies include Strauss (1982) for Sierra Leone, Savadogo and Brandt (1988) for Burkina Faso, and Deaton (1987) reported a lower price elasticity of meat in absolute values for rural households (−0.38) than urban households (−1.47) in Cote d'lvoire. For the same country, using a measurement error model and adjusting for cluster effects due to spatial variation in prices, Deaton (1998) shows that beef is a luxury commodity with high price elasticity (−1.91). Alderman and Von Braun (1984) found rice to be very price inelastic for Egypt.

More recently, Charles and Appleton (2005) examine food consumption in Ghana, with particular emphasis on the impact of trade and agricultural reforms on food demand. In contrast to the other studies for Egypt and Malawi, Charles and Appleton (2005) are interested in measuring the total welfare effect, which includes both static and dynamic responses. They estimate a complete demand system, with AIDS specification for household survey data. The estimated parameters are used to simulate the consequences of price changes on distribution. Their results indicate that the distributional burden of higher food prices falls mainly on the urban poor. Charles and Appleton (2005) rule out that trade liberalization policies are not the main cause for welfare loss. Their analysis indicates that tariff liberalization would offset welfare losses, particularly for poor and rural households. Strauss (1982), using a quadratic expenditure system, estimated the demand for rice, cereals, and four other food groups for rural households in Sierra Leone. The price elasticity of rice was found to be −0.68. An income elasticity of 0.99 for the same data using ordinary least squares was reported by Knight and Byerlee (1978).

Rather than using nominal prices, Hassan (1989) used real prices in an AIDS model for Sudan, and found the income elasticity for sorghum to be 0.64 while its own price elasticity was −1.97, a higher elasticity contrary to the fact that sorghum is the staple food in the Sudan. Mahran (2000) presents a detailed study about the food insecurity problem in Sudan. Mahran (2000) notes that this issue has prevailed in Sudan since the mid-1970s. Using OLS, Mahran (2000) assesses the performance of the national development strategies between 1970–71 and 1992–93, and examines whether Sudan is self-sufficient through both vertical and horizontal expansion in food production. The study examines the trends in area, production, and productivity for three major staple crops, namely sorghum, wheat, and millet using annual time series data covering the period 1970–95. The results provide clear evidence that vertical expansion alone does not pay off in terms of output. Instead, Mahran (2000) indicates that policies should focus more on improving agricultural productivity via the introduction of new varieties and the application of technological packages. Policies must also include infrastructure improvement, with an overall emphasis on health and education.

Similar concerns for Sudan's self-sufficiency have also been raised by Hassan et al. (2000). They note that a shift in Sudan's food production strategy toward more reliance on the irrigated sector for food supply was due to the severe food shortages following the early 1980s drought and reductions in food aid. The researchers use domestic resource cost analysis to examine whether the expansion in irrigated wheat production represents the most efficient option for using Sudan's irrigated land resources as compared to cotton, the country's most important cash crop competitor. Results indicate that expanding irrigated wheat production in Gezira for food self-sufficiency at the expense of cotton reduces employment opportunities, in addition to compromising economic efficiency. The researchers conclude that policy should concentrate on investment in research on improving cotton production technology, marketing, and lint quality. It is also vital to move toward improved wheat technologies to close the gap between potential and current yield levels.

Ahmed (2001) produced an excellent study about Egypt's food subsidy system. He investigates the economic, political, and technical feasibility and efficiency of a variety of strategic reforms that would reduce costs but maintain, or even improve, the welfare of the poor. More recently, Fabiosa (2008a,b) has used survey data to estimate demand functions for Egypt using different data sets. For example, using survey data for two periods, 1999–2000 and 2004–05, Fabiosa (2008) estimates Engle elasticities for Egypt. Fabiosa (2008) also distinguishes between rural and urban consumers, and finds that rural households have a higher expenditure share for food categories, but a lower share for nonfood categories compared to urban households.

In general, Fabiosa's (2008) estimation shows that households with lower incomes are more responsive to changes in income for food categories, and less responsive for nonfood categories. Further, the income elasticity of lower-income rural households is higher, compared to urban households for food categories. Moreover, elasticities in the 2004–05 survey period are higher compared to the 1999–2000 period. Per capita real income declined by 37.2% in 2004–05. From a policy point of view, this finding indicates that Egypt's consumption expenditure pattern has an alleviating effect on the impact of a food crisis, since a lower real income associated with a food crisis is accompanied by greater responsiveness of households to reduce their demand for food as their real incomes shrink. In particular, this is most obvious in the case of bread and cereals in rural areas, where the expenditure elasticity increased from 0.50 to 0.91 as per capita income declined.

In another interesting piece of research that examines Egypt's food-away-from-home (FAFH) expenditure pattern, Fabiosa (2008) estimates a Tobit model, where expenditure patterns are related to household demographic characteristics. Tobit estimation shows that the proportion of households with a positive FAFH expenditure is small, at 36–38% of the total number of households that spend 5–8% of their total expenditure on FAFH. Urban households with more family members, and whose household head is young and male, have higher levels of FAFH expenditure. The estimated conditional income elasticity is only 0.02, and the unconditional income elasticity is 0.52, suggesting that most of the growth in this sector will be driven by new households participating for the first time in FAFH expenditures. From these results, Fabiosa (2008) points out that high income elasticity is consistent with the expansion of the sector of hotels, restaurants, and other institutions in Egypt.

In a sequence of studies, Ogundari (2014) and Ogundari and Arifalo (2013) examine food intake and household behavior for Nigeria. In Ogundari and Arifalo (2013) a double-hurdle model is estimated for fruit and vegetables for the 2003–04 Nigeria Living Standard Survey (NLSS). It turns out that fruits and vegetable for the households in the sample are luxury items. However, viewing across income groups, the results show that for households in the low and high-income groups, fresh fruit is a necessity and a luxury, while all households irrespective of which income groups they belonged to, considered the demand for fresh vegetables to be a luxury good in the study. They also find that the demand for these goods is higher among households with younger members, compared to households with older members.

In Ogundari's (2014) study for Nigeria, a seemingly unrelated regression (SUR) estimation is performed for selected food groups and examines the relationship between share of nutrient and per capita income. Ogundari (2014) shows that the average calorie, protein, and fat intakes were still below the recommended daily allowance since the 1960s, as diets in Nigeria remained very much cereal-based over the years. Further, calorie, protein, and fat share of animal products respond positively but are inelastic to the per capita income growth in Nigeria over the years.

Zant (2012) investigates the effect of food aid on price and production of staple food, with a partial equilibrium model with nonseparable production and consumption. Using simulations, Zant (2012) is able to generate negative and also positive food aid elasticities of production. Conditions are identified which mitigate the negative impact and support a positive impact. Price and production equations, estimated with a panel of district data from the Malawi maize market for the period 1999–2010, show a small positive impact of food aid. Further, large negative impacts of food aid are not likely, given production and income shares, and behavioral responses.

In a recent and interesting study on choice behavior, Meenakshi et al. (2012) estimate willingness to pay for bio-fortified orange maize in rural Zambia. The study includes the impact of nutrition information, comparing the use of simulated radio versus community leaders in transmitting the nutrition message, on willingness to pay, and to account for possible novelty effects in the magnitude of premiums or discounts. Meenakshi et al. (2012) provide three important results: first, orange maize is not confused with yellow maize, and has the potential to compete with white maize in the absence of a nutrition campaign; second, there is a premium for orange maize with nutrition information; and third, different modes of nutritional message dissemination have the same impact on consumer acceptance.

In the context of Latin American countries, Robles and Torero (2010) have produced a detailed study which examines the impact of high food prices on food purchases. Particularly, the 2007–08 food crisis in Latin America affected consumption negatively. They note that while the share of households that were net consumers before the crisis was 68.2% in El Salvador, it fell in 2008 to 56%. Robles and Torero (2010) point out how Latin American governments must prepare well for the future, given that international food prices propagate to rural areas and affect the poorest sections of the population.

Coelho et al. (2010) also estimate a QUAIDS demand system for 18 food products using data from a Brazilian Household Budget Survey for the years 2002 and 2003. They show that purchase probabilities of staple foods were negatively related to family monthly income, while meat, milk, and other products showed a positive relation. They also find that regional, educational, and urbanization variables are also important.

In the next section, we use STATA to demonstrate how demand estimates are derived from food and nonfood consumption data, and show how this could be used in the analysis of food and nutrition policies. Some of the notable food consumption studies for Latin America include: Williamson-Gray (1982) for Brazil; Per Pinstrup-Andersen et al. (1976) for Columbia; Yen and Roe (1989) for the Dominican Republican; and Musgrove (1985) for Guatemala.

EMPIRICAL IMPLEMENTATION IN STATA

In this section we produce the estimable forms of the share equations for the quadratic AIDS or QUAIDS models, and reproduce Poi's (2012) estimation methods and calibration using STATA. Assume we have two goods 1 and 2, with prices p_1 and p_2, and total income m. Then the indirect utility function $V(p_1, p_2, m)$ in this context is from Banks, Blundell, and Lewbel (1997):

$$\ln V(p_1, p_2, m) = \left[\left\{ \frac{\ln m - \ln a(p)}{b(p)} \right\}^{-1} + \lambda(p) \right]^{-1}$$

where the terms inside the expression are all short-hand expressions:

$$\ln a(p) = \alpha_0 + \alpha_1 \ln p_1 + \alpha_2 \ln p_2 + \frac{1}{2}(\gamma_{11}(\ln P_1)^2 + \gamma_{22}(\ln P_2)^2 + 2\,\gamma_{12} \ln P_1 \ln P_2)$$

$$b(p) = P_1^{\beta_1} P_2^{\beta_2}$$

and

$$\lambda(p) = \lambda_1 \ln p_1 + \lambda_2 \ln p_2$$

Notice that the term ln $a(p)$ is the translog expansion, and $b(p)$ is the Cobb–Douglas price aggregator. The following restrictions of homogeneity, symmetry, and adding-up from consumer theory are also imposed:

$$\alpha_1 + \alpha_2 = 1$$

$$\beta_1 + \beta_2 = 0$$

$$\lambda_1 + \lambda_2 = 0$$

$$\gamma_{11} + \gamma_{12} = 0$$

$$\gamma_{22} + \gamma_{12} = 0$$

Using Roy's identity, the two share equations can be written as follows:

$$w_i = \alpha_i + \sum_{j=1}^{2} \gamma_{ij} + \ln P_j + \beta_i \ln \ln\left\{ \frac{m}{\alpha(p)} \right\} + \frac{\lambda_i}{b(p)} \left[\ln\left\{ \frac{m}{\alpha(p)} \right\} \right]^2$$

The purpose of this section is to reproduce Poi (2012), using the commands in STATA. Poi (2012) also has data available within the STATA code, taken from Poi (2002), which contains prices and shares for four commodities. This data is from the 1987–88 Nationwide Food

Consumption Survey, and has information about 4048 households. We first access the data and generate the descriptive statistics using the summarize command in STATA:

```
. webuse food

. summarize

    Variable |       Obs        Mean    Std. Dev.       Min        Max
-------------+---------------------------------------------------------
          p1 |      4048    1.736437    .6890672   .3333333   9.258823
          p2 |      4048    .5481831    .1937395   .1173621   2.672269
          p3 |      4048    1.494309     .676298   .1656758   7.202127
          p4 |      4048    .6133038    .3643831   .1087113     8.0625
       expfd |      4048    49.10378    29.36265         10     339.39
-------------+---------------------------------------------------------
          w1 |      4048    .4008241    .1416505          0   .9206853
          w2 |      4048    .2401759    .1065029          0     .82085
          w3 |      4048     .102189    .0580385          0   .4984845
          w4 |      4048     .256811    .1092822          0          1
        lnp1 |      4048    .4839956    .3645113  -1.098612   2.225577
-------------+---------------------------------------------------------
        lnp2 |      4048    -.656284    .3306715  -2.142492   .9829279
        lnp3 |      4048    .2939312    .4862892  -1.797723   1.974376
        lnp4 |      4048   -.5952874    .4353174   -2.21906   2.087224
       lnexp |      4048    3.732202    .5765527   2.302585    5.82715
```

As mentioned, we have four prices and shares (w_i) along with total expenditure. To estimate the QUIADS model in STATA, we use the code developed by Poi (2012):

```
. quaids w1-w4, anot(10) prices(p1-p4) expenditure(expfd) nolog
```

The command `quaids w1 − w4` is a convenient algorithm that estimates the parameters of the share equations. Further, `anot(10)` refers to the parameter $\propto_0$ in the share equation that is given a preset value equal to 10. Poi (2012) notes that $\propto_0$ could be estimated along with the other parameters, but that is a difficult task, and researchers usually set that to a value lower than the smallest ln m in the data. Finally, `nolog` is used to save space by suppressing the iterations in the log file. The results from the above command from STATA are produced below:

```
(obs = 4048)
Calculating NLS estimates...
Calculating FGNLS estimates...
FGNLS iteration 2...
FGNLS iteration 3...

Quadratic AIDS model
____________________

Number of obs            =        4048
Number of demographics   =           0
Alpha_0                  =          10
Log-likelihood           =   13098.227
```

	Coef.	Std. Err.	z	P>|z|	[95% Conf.	Interval]
alpha						
alpha_1	-.1527277	.1089	-1.40	0.161	-.3661678	.0607123
alpha_2	.0966575	.0971681	0.99	0.320	-.0937884	.2871035
alpha_3	.2412799	.0520561	4.63	0.000	.1392518	.3433079
alpha_4	.8147903	.0817843	9.96	0.000	.654496	.9750847
beta						
beta_1	-.1506037	.0305464	-4.93	0.000	-.2104735	-.0907339
beta_2	-.0394186	.0292707	-1.35	0.178	-.0967881	.017951
beta_3	.0409757	.0157213	2.61	0.009	.0101625	.071789
beta_4	.1490465	.0237392	6.28	0.000	.1025185	.1955746
gamma						
gamma_1_1	.1925378	.0280082	6.87	0.000	.1376427	.2474328
gamma_2_1	-.0353897	.0126463	-2.80	0.005	-.0601761	-.0106033
gamma_3_1	-.054241	.0093367	-5.81	0.000	-.0725407	-.0359413
gamma_4_1	-.1029071	.0203041	-5.07	0.000	-.1427023	-.0631119
gamma_2_2	.0702349	.0082637	8.50	0.000	.0540385	.0864314
gamma_3_2	-.0060445	.0048505	-1.25	0.213	-.0155512	.0034622
gamma_4_2	-.0288007	.0150965	-1.91	0.056	-.0583893	.0007878
gamma_3_3	.0476765	.0043425	10.98	0.000	.0391654	.0561876
gamma_4_3	.012609	.0076378	1.65	0.099	-.0023609	.0275789
gamma_4_4	.1190989	.0217369	5.48	0.000	.0764953	.1617024
lambda						
lambda_1	-.0125242	.002112	-5.93	0.000	-.0166636	-.0083848
lambda_2	-.0009866	.0022074	-0.45	0.655	-.005313	.0033397
lambda_3	.0029766	.0011864	2.51	0.012	.0006512	.005302
lambda_4	.0105343	.0017306	6.09	0.000	.0071424	.0139261

CALCULATING ELASTICITIES

The advantage of using Poi (2012) and STATA is that the calculation of different elasticities is direct and easy. For instance, the next three lines:

```
. estat expenditure, atmeans

. matrix elas = r(expelas)

. matrix list elas
```

The `estat expenditure` computes the expenditure elasticities for each observation. Usually, researchers report the elasticities at the mean value of prices. This can be calculated by using `estat expenditure, atmeans`. The elasticities are saved as a vector under `r(expelas)`. We can refer to this in the second line, where we create a matrix `elas` to refer to the saved values. The last line is the `matrix list` command, which prints the elasticities. The output for the procedure is given below:

```
elas[1,4]
            c1          c2          c3          c4
r1   1.0343105    .8897855     1.01873   1.0420714
```

Poi (2012) has established procedures in STATA that make the calculation of uncompensated and compensated elasticities straightforward. See Exercise 2 for more details on these procedures. Finally, Poi (2012) has incorporated Ray's (1983) ideas of incorporating demographic variables in the analysis, and we refer to these procedures in the exercises. We end this section with a calculation of uncompensated elasticities. The STATA commands are as follows:

```
. estat uncompensated, atmeans

. matrix uncomp = r(uncompelas)

. matrix list uncomp
```

The `estat uncompensated, atmeans` computes the uncompensated elasticities at the mean values which are saved as a matrix under `r(uncompelas)`, which we print in the next line using the `matrix list` command. The output for the procedure is given below:

```
uncomp[4,4]
             c1           c2           c3           c4
r1   -.69898157   -.13682488   -.09435155   -.10415247
r2   -.16989394   -.70835039    .00609046   -.01763163
r3    -.3640677   -.01638257   -.58307864    -.0552011
r4   -.16606501   -.05268609   -.02433375   -.79898654
```

The 4×4 matrix provides the percentage response of quantity of good in row i to a 1% change in the price of good in column j. A 1% increase in the price of good 1 reduces the quantity of good 3 by 0.36%, which indicates that these goods are complements. We use similar analysis in the next chapter to demonstrate how food and nutrient demand elasticities could be used in policy analysis related to food and nutrition interventions.

CONCLUSIONS

In this chapter we presented the theory of demand systems and reviewed various models specified to estimate food demand parameters. We also reviewed specific studies from various regions of the world to give an exposure to the current and earlier notable literature. We used a QUAIDS model to demonstrate how the estimation procedures could be implemented in STATA.

Food policy analysts are interested in the factors that influence the demand for food and nutrients. The major focus of this chapter is to use the foundations of the microeconomic theory of consumer behavior to develop testable implications, using market data. The functional forms and estimable equations are useful for policy makers, because important information via elasticities can be computed from these specifications. These elasticity expressions from the Indirect Addilog Model, The Rotterdam Model, the translog, and AIDS are all a few examples of these applications.

The application of STATA is useful in this context, because it provides useful information such as compensated and uncompensated price and income elasticities for all these functional forms. Our STATA implementation of the QUAIDS model shows that income effects are very important, when we consider nonfood items along with nutrients in the demand function. We explore this angle further in the next chapter. Imagine a situation where junk food items are consumed alongside healthy foods. How can the empirical foundations from this chapter be applied to evaluate a tax on junk foods? These are important policy questions that have received a lot of attention, and we focus on these aspects in our exercises and in some of the later chapters. The theoretical and STATA background from this chapter helps to provide a good framework to address these issues.

We encourage readers to explore using STATA or similar software to see how model specifications, the nature of the data, and estimation methods result in parameters that may differ in their magnitudes and study their implications for policy making.

EXERCISES

1. An applied researcher is faced with the problem of a small number of observations. For example, if the number of goods to be analyzed for their demand is 10, then the total number of elasticities to be estimated is 110. However, show how many can be saved in the estimation process by imposing the properties of demand functions. Use the `webuse food` command and derive the summary statistics for the data from Poi (2012). Use the `estat compensated, atmeans` command to derive the compensated elasticities. These values are saved under `r (compelas)`. Use the matrix command to print these elasticities, and note if there are major differences between the compensated and uncompensated elasticities.
2. Poi's (2012) STATA code is very useful because it also allows us to incorporate demographic variables into the analysis. Ray (1983) criticizes the standard approach because of its failure to capture the "costs of children," thereby creating difficulties in measuring welfare costs. Poi (2012) incorporates Ray's (1983) study, and illustrates the usefulness of STATA in measuring the impact of demographic variables. After the `webuse food` command, type the following three lines, to generate a random integer for the number of children in each household, and another random binary variable to distinguish between rural and urban households:

```
. set seed 1
. generate nkids = int(runiform()*4)
. generate rural = (runiform() > 0.7)
```

   After this, type the following command to estimate the quaids model:

```
quaids w1-w4, anot(10) prices (p1-p4) expenditure(expfd) demographics(nkids
rural) nolog
```

   What additional information does this estimation produce?
3. After the above procedure, type the following commands to produce the uncompensated elasticities for the rural and urban households.

```
.estat uncompensated if rural, atmeans
.matrix uprural = r(uncompelas)
.estat uncompensated if !rural, atmeans
.matrix upurban = r(uncompelas)
```

   Use the **matrix list** command and check if the elasticities differ across these groups.
4. We have observations for 27 households for their consumption of fruits (Q_1), diary (Q_2), and junk food (Q_3), with their prices in the table below. We also have a binary variable indicating whether the household is in the rural ($R = 1$), or in the urban area ($R = 0$). We also have information on the number of children (kids) and household income (m).
   a. Estimate the QUAIDS model for this data and check if the uncompensated elasticities vary across the rural–urban households.
   b. Check if all the expenditure elasticities are positive. Interpret these values.

#	Q_1	Q_2	Q_3	P_1	P_2	P_3	R	kids	m
1	129.6	0.875	0.975	0.736	0.76	0.374	1	4	6036
2	131.2	0.864	0.975	0.769	0.796	0.378	1	4	6113
3	137	0.869	0.971	0.848	0.824	0.387	1	3	6271
4	141.5	0.868	0.965	0.86	0.835	0.393	1	3	6378
5	148.7	0.864	0.962	0.901	0.851	0.4	1	2	6727
6	155.8	0.899	0.939	0.894	0.869	0.401	1	2	7027
7	164.8	0.92	0.921	0.87	0.902	0.405	1	2	7280
8	170.9	0.95	0.93	0.9	0.95	0.413	1	3	7513
9	183.3	0.964	0.958	0.928	0.996	0.431	1	3	7728
10	195.7	0.997	0.974	0.931	1.077	0.444	1	4	7891
11	207.3	1.006	1.006	0.943	1.235	0.457	1	4	8134
12	218.2	1.013	1.05	1.002	1.327	0.477	1	5	8322
13	226.7	1.026	1.04	1.005	1.384	0.485	1	3	8562
14	237.8	1.131	1.041	1.076	1.398	0.496	1	3	9042
15	225.7	1.549	1.105	1.126	1.43	0.534	0	2	8867
16	232.3	1.658	1.206	1.364	1.536	0.589	0	4	8944
17	241.6	1.729	1.287	1.579	1.692	0.625	0	3	9175
18	249.1	1.832	1.359	1.728	1.774	0.657	0	3	9381
19	261.2	1.913	1.468	1.765	1.828	0.699	0	2	9735
20	248.8	2.606	1.59	1.91	1.953	0.751	0	2	9829
21	226.7	3.641	1.723	1.981	2.466	0.822	0	4	9722
22	225.5	4.059	1.832	2.469	3.07	0.887	0	4	9769
23	228.7	3.844	1.906	2.864	3.41	0.93	0	3	9725
24	239.5	3.714	1.956	3.197	3.576	0.971	0	3	9930
25	244.6	3.657	2.015	3.657	3.802	0.968	0	4	10421
26	245.7	3.688	2.082	3.697	3.978	0.975	0	4	10563
27	269.3	2.871	2.17	3.532	4.214	0.983	0	3	10780

CHAPTER

7 DEMAND FOR NUTRIENTS AND POLICY IMPLICATIONS

How do governments respond to abrupt food price changes and why do they respond as they do? These two questions are important to help us understand policy-making, to predict how policy makers are likely to respond to future food price volatility and to support policy makers as they confront such volatility.

Per Pinstrup-Andersen (2015)

INTRODUCTION

In the previous chapter we used consumer demand theory to study the demand behavior of households toward food commodities and how understanding such behavior can help in devising food and nutrition policies that can help reduce hunger and malnutrition. While such an analysis is a starting point for improving the nutritional status of the population, understanding the role of policies, particularly policies that enhance the income of households, and understanding the effects of a change in the prices of the commodities on the nutritional status of the population is fundamental for improving the nutritional well-being of the population. In this chapter we extend the analysis of the last chapter to the analysis of demand for nutrients, and apply this analysis to the study of the implications of policies that aim to reduce malnutrition.

STUDIES ON NUTRIENT DEMAND

The nutritional implications of public policy have received a high level of attention recently in the context of continued under-nutrition and increasing over-nutrition challenges in both the developing and developed worlds. Further, the third type of malnutrition called "hidden hunger," resulting from micronutrient malnutrition, has also been recognized. For the past 30 years, researchers have given emphasis to estimating the changes in nutritional intake due to changes in household income and the price levels of food commodities (Behrman and Deolalikar, 1987; Jensen and Miller, 2011). One of the objectives of such studies in general is to analyze changes in nutritional status due to changes in income and prices of goods for different groups of households. The studies that have estimated nutritional elasticities with respect to change in income and prices are briefly reviewed below.

In general, studies on nutritional elasticities can be grouped into two categories based on the procedure used for the estimation of nutrition elasticity. The first set of studies estimates the intake of nutrition in households from survey data, and uses it in a regression with prices of food

Nutrition Economics. DOI: http://dx.doi.org/10.1016/B978-0-12-800878-2.00007-4

commodities and income on the right hand side. In general, the prices used for such "direct" estimations are unit prices derived by dividing expenditures by total quantities purchased (Timmer, 1981; Alderman and Timmer, 1980).

The second set of studies estimate nutrition elasticities indirectly using the price and income elasticities of food commodities. The nutrient elasticities are the weighted sum of these direct elasticities, the weights being the nutritional share of the foods for each nutrient (Pitt, 1983). The choice of the above procedures has been debated extensively in the literature. Pitt (1983, p. 10) argued for the indirect approach, equating the direct procedure to single equation regression techniques since it requires separate regression having calories as a dependent variable with prices, income, and other exogenous variables on the right hand side. However, Behrman and Deolalikar (1987) argued for the direct approach, at least when only income is used on the right hand side of the regression equations. According to them, the indirect procedure overestimates the true changes in nutrition due to income and prices.

While there could be differences in nutritional elasticities due to differences in the data and statistical methods used, the differences in procedures for calculating them introduces additional discrepancies in the available estimates.

Nutritional elasticities have been calculated for a number of countries. We briefly review them here to provide the reader with exposure to the range of estimates of nutritional elasticities. Pitt (1983), in one of the earliest studies using the indirect method, calculated elasticities for six nutrients for Bangladesh. In his study, the calorie and protein elasticities with respect to income were similar. Radhakrishna (1984) reported similar calorie elasticities for India. Behrman and Deolalikar (1987), using the indirect approach, calculated income elasticities that are very small compared to the direct approach by Sharma and Dillon (1987) with the same Indian household data. The nutrition elasticities reported for Indonesia were all based on the direct approach with large variations.

Wolfe and Behrman (1983) used the direct approach to study households in Managua, Nicaragua, and found small income elasticity estimates (0.058 for calorie and 0.29 for protein). Both of studies conducted with Brazilian data followed the direct approach (Williamson-Gray, 1982; Ward and Sanders, 1980). Pinstrup-Andersen and Caicedo (1978), using the indirect method, found the income elasticity of calorie and protein to be 0.51 and 0.65, respectively. Sahn (1988), and Gavan and Chandrasekara (1979), using different approaches, reported two different income elasticities for calories. Sahn's (1988) estimation is closer to many of the studies mentioned above. A price elasticity of calorie of −0.24 was reported by Strauss (1982) for Sierra Leone. Though any general relation between the estimates cannot be stated, income elasticity of protein in all the studies is higher than that of calories (except Pitt, 1983). In general, in these studies the elasticities using direct methods seem to be lower than those using the indirect approach.

RECENT STUDIES FROM ASIA

Recent studies that use more exhaustive data and methodology from China indicate that the income elasticities for most nutrients are much smaller than those reported earlier (Tian and Yu, 2013; Deng, 2009). Earlier studies (Alderman and Gertler, 1997; Behrman 1988) indicate that the nutrient intake of females has a more negative price elasticity than that of males. The lower income elasticity for boys is attributed to favoritism toward male children.

However, in an interesting study using the China Health and Nutrition Survey, Mangyo (2008) notes that income and price elasticities are dependent on the amount of food intake, and the manner in which intake affects the marginal utilities of the participants. For example, if elasticities are positively related to status, then there is evidence of gender bias, or even a bias against the elderly. Most importantly, Mangyo (2008) shows that the nutrient-intake elasticity for prime-age men is higher, because the marginal utility of intake falls relatively slowly for this group than for other demographic groups, as the entire household resource goes up. This implies that even with government targeting of food, males and prime-age men are likely to get higher absolute and relative levels of food.

Several studies have noted the nutritional deficiency prevalent among different groups. For example, Reddy (2010) notes the regional disparities in Andhra Pradesh, which is one of the largest states in India. Reddy (2010) examines the disparities between the coastal and noncoastal regions. In all the regions, Reddy (2010) notes that consumption of food items was less than the requirement. The share of high value commodities such as vegetables, fruits, milk, and meat products in total food expenditure is much higher in coastal compared to noncoastal regions. This difference is attributed to food habits. About 30–45% of the population was undernourished across regions. The incidence of nutritional deficiency is more prevalent among the landless, scheduled caste, scheduled tribes, and the poor. Further, within the coastal region, disparities between landless and large landholders in nutrition status is much higher than those in the noncoastal region. Overall, the study demonstrates that regional differences in nutritional deficiencies are mostly determined by food habits and income levels.

Skoufias et al. (2011) examine the impact of the 2008 food crisis using data from two cross-sectional household surveys in Indonesia. These surveys were carried out before and soon after the 1997 – 98 food crisis. The researchers note that the summary estimate of the income elasticity of starchy staple ratio is constant during the crisis. However, they also note that looking at just the summary estimates hides important information about specific nutrients. Consequently, for key micronutrients, such as iron, calcium, and vitamin B1, they find that the income elasticity is significantly higher in the crisis year compared with a normal year. Finally, their estimate of income elasticity for vitamin C is close to zero. The researchers suggest that cash transfer programs may be effective during crises to protect the consumption of many essential micronutrients. They also recommend this program to be jointly undertaken with specific nutritional supplementation programs.

Molini and Nube (2007) examine the nutritional status of males and females using data on calorie intake and anthropometric measures from Vietnam. They find strong evidence of gender discrimination on food access within households. Moline and Nube (2007) find that for the Vietnamese data, the anthropometric measures tend to improve less quickly than the calorie intake. Further, males tend to benefit more than females from economic improvements, and this feature is strongly evidenced in rural areas and in the poorest sections of the population. They also compute the BMI-growth elasticity for males and females, and find that the elasticity coefficient for males is double that of females.

The choice of the correct database to arrive at reliable estimates of nutrition elasticities is an important consideration. Gibson and Scott (2011) use evidence from Vietnam to stress this issue. The typical practice for researchers in this field is to use household data, which contain unit values (ratios of household expenditure to quantity purchased), or community prices (obtained from vendors in local markets) as proxies for market prices. They estimate a 14-food demand system separately using unit values or community prices, and arrive at calorific elasticity values with respect to rice prices, finding that the elasticity value is more than twice as large when community prices are used rather than unit values. Consequently, this study from the Vietnamese data informs us that the nutritional effects of rice price increases may be sensitive to data choices. This is an important finding, because economists typically use household survey data to estimate nutritional elasticities.

In a more recent study on Vietnam, Gibson and Kim (2013) question the relevance of estimates, because the standard demand models ignore quality substitution, in the face of increasing prices. Gibson and Kim (2013) explicitly incorporate quality substitution possibilities in an eight-food demand system estimation. They find that a 10% increase in the relative price of rice reduces household calorie consumption by less than 2%. However, they caution that this elasticity estimate would be wrongly estimated to be more than twice as large if quality substitution is ignored.

The implication from the Gibson and Kim (2013) estimation is that the consumers in Vietnam have considerable scope for protecting calorie consumption in the face of higher rice prices by downgrading the quality of the foods that they consume. Importantly, downgrading quality does not imply a decline in nutrition, because the factors that determine food quality are not dependent on factors that determine nutrition. That is, undernourished consumers focus less on food's nutritive value, but are more easily drawn toward food attributes, such as taste, appearance, odor, degree of processing, and variety and status. Unfortunately, higher quality food is more costly, yet not more nutritious. For example, in Vietnam, Gibson and Kim (2013) note that in urban areas, the price premium associated with whiteness, fragrance, and stickiness for high quality rice is 45%, and not for any difference in calories, protein, micronutrients, impurities, condition of kernels, or proportion of broken grains.

Given the above results and observations, it is critical for policymakers in Vietnam to diversify rice production in terms of varieties and quality. Unfortunately, current policy in Vietnam is distorted toward production of high quality rice, a policy purely driven by export revenues. This is not such a bad policy if lower quality rice is diverted from exports, and supports the domestic markets. Unfortunately, this is not the case. Rather, a floor is placed under the quality distribution by removing low quality rice from the market, and diverting its use as pig, poultry, or cattle feed. Under such circumstances, undernourished households may become nutritionally vulnerable to rice price increases.

Gibson and Kim (2013) use Vietnam as an important case study because published estimates for Vietnam indicate a large, negative, elasticity of calories with respect to rice prices, which has prompted policy makers to interfere with the markets. But, as Gibson and Kim (2013) point out, this policy response is based on evidence that does not correctly account for quality substitution. If quality and variety considerations are incorporated into data collection and analysis, it appears that there may be fewer conflicts between nutritional goals and export revenues.

Villa et al. (2011a, b) examine the various factors that contribute to the dietary diversity among pastoralist households in East Africa. The researchers note that income elasticities of nutrient demand do not just depend upon the total income of the household, but rather to different sources of income. They find that dietary quality instrumented via dietary diversity is critically determined by income sources, due to market failures, intra-household bargaining, and mental accounting. Their studies also indicate that income elasticities of dietary diversity depend upon household circumstances. Interestingly, they find that household heads bear disproportionately more of the nutritional burden when the household income is below mean, and enjoy more of the nutritional gains when household income is above mean. Finally, their results indicate that adult daughters are better-off than other household members in their dietary diversity, and the sons are worse off, with little difference between male heads and their wives.

Ecker and Qaim (2011) also adopt a demand systems estimation approach to estimate income and price elasticities of food demand and nutrient consumption, for household data from Malawi. They provide several policy simulations, and show that income-related policies are better suited than price policies to improve nutrition. Further, their simulations also show that price subsidies for maize can improve calorie and mineral consumption, but at the same time, the subsidies can worsen vitamin consumption in urban areas. Another study for Malawi, a country with a high level of chronic malnutrition challenges, is by Chiwaula and Kaluwa (2008), which also looks at household data, with a focus on infant foods and infant care. They find that infant care is unitary income elastic, and price inelastic. Interestingly, demand for two porridge types responds similarly to ingredients' price changes, while adult meals substitute the porridges. They provide several policy proposals, such as eradicating income poverty, diversification of household production, diet, infant food, and target recipients of infant nutritional information. Widening vitamin A fortification options and targeting mothers of infants are also related proposals.

Tsegai and Kormawa (2009) apply the Almost Ideal Demand System (AIDS) model to Nigerian data to estimate the demand for cassava, and its elasticity with respect to prices of other root and tuber crops. Their results indicate that the demand for cassava is price-inelastic, while the expenditure elasticity is found to be positive, though inelastic. They conclude from this result that cassava is a normal good and a necessity, and should be noted for its dietary strengths.

Tiffin and Dawson (2002) use cointegration techniques and examine the demand for calories in Zimbabwe. They analyze the long-run relationship between per capita calorie intake, per capita income, and food prices. Their cointegration results show that bidirectional causality exists between calories and income. Income growth reduces inadequate calorie intake, and improvements in nutritional status increase income, thereby supporting the efficiency of wages hypothesis.

Handa and King (2003) produce an interesting study on Jamaica, which relates international financial policies to nutritional adequacy. They note that in Sep. 1991, Jamaica liberalized its exchange rate as part of its Structural Adjustment Program (SAP). The sudden exogenous shock affected the prices of rice and flour, which are major imports into Jamaica. Consequently, the sudden shortage of food affected the nutritional status of children. For instance, Handa and King (2003) show that post-liberalization, the weights of children fell significantly. They find the elasticity of weight-for-height z-score with respect to food price inflation to be very high at -0.86. Further, during the rapid economic reform in 1991, this elasticity rose to -1.24, indicating a large response of weight for height to food price inflation. Jegasothy and Duval (2003) examine the food and nutrition needs of households in urban and rural Samoa. The authors estimate price and expenditure elasticities for seven food groups. From their results, they recommend an approach that combines a food – health policy that leads to improvements in agriculture, and also nutritional awareness.

Skoufias et al. (2011) examine the relationship between calories, food quality, and household per capita expenditure of poor households from rural Mexico. Skoufias et al. (2011) investigate the impact of a nutritional program in Mexico, called Programa de Apoyo Alimentario (PAL). The expenditure elasticities of calories are largely dependent on substitution between and within major food groups. The incidence of substitution within cereals is very high for poor households, and this factor alone explains about 59% of the income elasticity for food quality. Skoufias et al. (2011) note that cash transfers will have a positive impact on food diversity and nutritional status.

Bertail and Caillavet (2008) use an AIDS model for France to estimate income elasticities of fruits and vegetables. They show that price and income as policy tools are useful only for demographic clusters with slightly higher income. The demographic cluster with the lowest income remains insensitive to economic variables. Allais et al. (2010) also estimate a complete demand system using French scanner data, and calculate the impact of a fat-tax across different income groups. Their nutrient elasticity estimates show that a fat-tax has a small and ambiguous effect on nutrients purchased by French households. The tax also has a marginal effect on body weight in the short run, but a larger effect in the long run. The tax is also highly regressive, but very efficient in terms of revenue generation.

Lundh (2013) investigates the urban – rural wage gap in Sweden during the period 1914–20. Interestingly, the study connects the wage gap to the standard of living, using household surveys for five worker groups. The analysis includes household real earnings, household expenditure, and also the nutritional value of food between the worker groups. The author finds the urban – rural wage gaps to be small or moderate, and this is attributed to the higher cost of living in urban areas, and to the practice of payments in kind and home production in rural areas. Differences in nutritional value of food consumption are due to differences in earnings, working loads, and conditions. These results thus modify the picture usually given in the literature on urban – rural wage gaps and income elasticity of food items.

Similar observations also hold true for Italy, where the much of the available evidence presents a bleak view, showed that the nutritional status of the population failed to keep up with per capita income growth rates. Vecchi and Coppola (2006) investigate how nutritional status responded to economic growth in Italy during 1861–1911. They find that the incidence of under-nutrition decreased by at least 15% between 1881 and 1901. Their calculations indicate that the income elasticity of calories in 1901 was in the range of 0.3–0.6, varying inversely with income. These results contradict earlier findings, and indicate that the early phase of Italian industrialization was beneficial to the bulk of the population, particularly more so for the poorer sections.

Smed et al. (2007) estimate a demand system and elasticities for households in Denmark. They simulate the effect of taxes for their nutritional effects, across households in different age groups and social classes. The effect of prices is stronger for saturated fats, fiber, and sugar, particularly among lower social classes. The youngest decrease their demand for saturated fat the most, while the middle-aged respond heavily to the price of sugar. The authors note that while taxes may be effective in improving diets on average, the design of the tax, and the targeting of vulnerable groups with special needs should be done with care. The researchers also note potential tradeoffs from a tax on a single nutrient, as the tax may have an inadvertent effect on the demand for other food components. Hence, policy interventions should carefully combine taxes and subsidies based on nutritional information (Box 7.1).

BOX 7.1 WERE AMERICAN AND BRITISH WORKERS HEALTHIER IN THE 1880S?

Logan (2009) challenges the long-standing notion that American and British industrial workers in the late-19th century were two to four times as wealthy as those in developing countries today. This implies that the American and British industrial workers enjoyed a better standard of living than those living in developing countries. Logan (2009) also examines the demand for calories, and calculates calorie expenditure elasticities based on the 1888 Cost of Living Survey information. The results indicate that yesterday's wealthy workers were hungrier than today's poor, since the calorie expenditure elasticities based on the 1888 data are greater than those from recent data on developing countries. Logan (2009) attributes this finding to an extraordinary improvement in nutritional well-being among the poor in the last century that has not been captured by the traditional income estimates.

In a similar study conducted for the United States using the same 1888 Cost of Living data, Logan (2006) tests the hypothesis that households during this time period substituted away from carbohydrates and fiber, and moved toward proteins and fat as their incomes rose. The reason why this study is important is because it tests the ideas of anthropometric historians, who assert that there was increased nutrient intake without any nutritional substitution. Logan (2006) calculates the income elasticities of all nutritional entities including fiber, proteins, fat, and sugar. Contrary to the nutritional substitution hypothesis supported by historians, Logan (2006) finds that the income elasticity of fiber is very close to those of protein, fat, and sugar. The calorific Engel curve also indicates that the shares of carbohydrates, fat, and sugar in the diet vary with household income, but the shares of protein and fiber do not. Most importantly, Logan (2006) finds that the share of protein from animal sources increases with household income. These findings lead us to believe that the diets of late-19th century industrial workers were surprisingly balanced by modern standards.

THE UNITED STATES

In the early 1990s, the issue of nutrition labeling became a contested topic in the United States. Will nutrition labeling have a significant effect on consumer demand for nutrition quality? To answer this question, Mojduszka et al. (2001) estimate a consumer demand function for prepared frozen meals for data from 1993 to 98. The study finds that economic variables such as price, income, advertising, and price reductions have a significant effect on demand. However, concerns and knowledge about nutrition and health do not affect demand significantly. In other words, mandatory nutrition labeling has no effect on consumer preferences and purchasing patterns within the prepared frozen meals category.

The role of fiscal policy, particularly food taxes on food intake and nutrition, is also an important topic in the United States. In a recent paper, Harding and Lovenheim (2014) examine the role of taxes on soda, sugar-sweetened beverages, packaged meals, and snacks, and nutrient taxes on fat, salt, and sugar. Harding and Lovenheim (2014) analyze a huge data set of 2002–07, estimate a demand system, and compute the various expenditure elasticities. Using the estimation results, they simulate the role of taxes and find that a 20% nutrient tax has a significantly larger impact on nutrition than an equivalent product tax, due to the fact that these are broader-based taxes. From their analysis, we find that for the United States, a sugar tax can be a very effective tool to induce healthier and more nutritious eating among consumers.

Nutritional inadequacy and poverty have generated a lot of interest among US policy makers in recent years. In an extremely detailed research undertaking, Davis and You (2013) examine a basic question that is crucial in the context of the United States: is home food production worth the effort? The reason why this question is important is because, as Davis and You (2013) point out, most of the results on home food production ignore the effect of time, and this limitation distorts the nutritional adequacy estimates used in policy formulation, including the USDA Thrifty Food Plan (TFP) and SNAP.

Davis and You (2013) make a fundamental observation: food at home consumption cannot occur if there is no food at home production. In order to correctly model the home food production process, Davis and You (2013) estimate a home meal production function, which includes food and labor time as inputs. The median returns to scale and the elasticity of substitution between money and time is in the 1.2–1.9 range and 0.33–0.56 range, respectively. These estimates indicate the difficulty in substituting money for time in home meal production.

Davis and You (2013) also define and estimate a "home meal poverty rate" (HMPR), which is the percentage of the sample that produces fewer meals at home than is consistent with dietary guidelines. They find that the HMPR is about 85%, which suggests that mere cash transfers or monetary policies are not likely to be effective options for food – health concerns.

Davis and You (2013) point out that the preoccupation with cash transfers in the United States is because of the importance given to food prices, as a major determinant of the country's nutrition profile. Since most of the research indicates that food demand is inelastic, there has been a push toward policies that focus on taxes and subsidies. The research results of Davis and You (2013) inform us that in a developed country like the United States, time, and not money, is a more important factor in determining nutritional targets. They suggest policies that target time management through efficient education and improved access (Box 7.2).

BOX 7.2 ARE INDIANS LAZY?

Recently, research on food consumption, calorific intake, and the microeconomics of nutrition-adequacy has generated an interesting debate in India. India presents an excellent case study, because the economy reveals an interesting paradox that relates economic growth and food intake. In an exhaustive study, based on the National Sample Survey (NSS) data and National Nutrition Monitoring Bureau (NNMB) surveys from India, Deaton and Dreze (2009), point out the paradox, based on the following observations:

1. There has been a sustained decline in per capita calorie consumption during the last 25 years.
2. This decline is larger among better-off sections of the population, and close to zero for the bottom quantile of the per capita expenditure scale.
3. The decline in calories extends to proteins and other nutrients.
4. Fat consumption has steadily increased during this period.
5. The decline in nutritional adequacy has been during a period in which per capita incomes have increased, and food prices have been stable relative to other goods.
6. In contrast to per capita *total* expenditure, there has been no *real* increase in per capita food expenditure.
7. This period also shows a decline in fertility, which implies that Indian households had a lower number of children at the end of the period. Hence, the decline in calorie "per equivalent adult" is even larger than the decline in the traditional "calorie per person."

The above observations lead Deaton and Dreze (2009) to conclude that per capita calorie consumption is lower today at any *given* level of per capita household expenditure. In other words, the decline in calorie consumption suggests a steady downward drift of calorie Engel curves (or, the plots of per capita calorie consumption against per capita expenditure).

Deaton and Dreze (2009) also offer an explanation for this downward drift in calorific consumption that is due to better health, as well as to lower activity levels. However, the evidence offered for this explanation is mostly based on fragmentary evidence regarding major expansions in the availability of safe water, vaccination rates, transport facilities, and the ownership of various effort-saving durables.

(*Continued*)

BOX 7.2 (CONTINUED)

If calorie requirements rise sharply with activity levels, then a decline in average calorie intake must be attributed to fairly moderate reductions in activity levels, or an increase in laziness. However, this hypothesis, as Deaton and Dreze (2009) point out, is speculative, since there is no data on activity levels.

The findings and suggestions provided by Deaton and Dreze (2009) have attracted a lot of attention among researchers. For instance, if calorie reductions are purely driven by laziness or improvements in health conditions, then this implies that there are no calorie deficits in the Indian population. However, Deaton and Dreze (2009) spend a lot of time in their paper demonstrating that "nothing could be further from the truth." Indeed, they show how India experiences some of the worst anthropometric outcomes.

Finally, the reason why these findings are important is because it is puzzling that a country as poor and malnourished as India should, in periods of growing prosperity, decrease real food consumption by actually cutting back on its calorie intake.

Deaton and Dreze's results and observations have generated a heated debate among nutrition economists in India. In particular, Utsa Patnaik (2010) has called into question the value of the price index measurements and data construction methods used by Deaton and Dreze (2009). In their response, Deaton and Dreze (2010) reaffirm their findings and justify their methods. Overall, the paradox of rising undernourishment in a rapidly growing economy has received a lot of attention and critical looks in recent years in India (Palmer-Jones and Sen, 2001; Meenakshi and Viswanathan, 2005; Radhakrishna, 2005; Ray and Lancaster, 2005; Ray, 2007; Sen, 2005; Suryanarayan and Silva, 2007). A critical policy implication is that we may miss people who are food insecure in nonpoor households.

Gaiha et al. (2013) have recently reexamined Deaton and Dreze's hypothesis. They apply the standard demand theory framework, and estimate shifts in demand and food price elasticities. The researchers use food prices and expenditure in their demand estimation for calories, protein, and fats separately for rural and urban households. Their results show that there are significant price and expenditure effects. They also find that there could be other important factors that are omitted from the demand specification, which could lower consumption. Health improvements and lower physical activities are plausible explanatory factors, as shown in their demand function for calories. However, these factors have to be viewed as complementary to demand-based estimates.

Equally important in this debate is the "food budget squeeze hypothesis," advanced by Basole and Basu (2015), who show that in the Indian case, that the cost of meeting nonfood essentials has increased so fast that it has squeezed the food budget, leaving insufficient purchasing power for food, which leads to lower calorie consumption.

Finally, results of studies estimating nutritional elasticities were also reviewed for their procedures and differing results due to disparities in them. The indirect procedure followed is given below.

ESTIMATION OF NUTRITION ELASTICITIES

There are in general two common approaches to calculate the parameter estimates on how changes in income and prices will change the intake of nutrients. The first one is to estimate food demand systems using one of the demand systems (LES or AIDS) and then convert the expenditure and price elasticities obtained from these to nutrient elasticities with respect to expenditure at the same level of aggregation. Murty and Radhakrishna (1982), Strauss (1982), and Pitt (1983) have used this method in their studies. The second approach is to estimate the reduced form equations of demand for nutrients directly, using expenditure and other demographic variables as independent variables (Levinson, 1974; Timmer and Alderman, 1979; Ward and Sanders, 1980; Wolfe and Behrman, 1983; Pitt and Rosenzweig, 1985). A review of these studies was presented in the previous chapter. Table 7.1 summarizes estimates of calorie and protein elasticities from selected studies from the early literature (see Exercise 6 at the end of this chapter).

Table 7.1 Studies Relating to Nutritional Elasticities

Country	Author	Procedure for Nutrition Elasticities	Elasticities				Remarks
			Price		Income		
			Calorie	Protein	Calorie	Protein	
Bangladesh	Pitt (1983)	Indirect	−0.53	−0.42	0.82	0.79	Elasticities with respect to rice price
Brazil	Williamson-Gray (1982)	Direct	−0.016	—	0.18	—	Elasticities with respect to rice price
	Ward and Sanders (1980)	Direct	—	—	0.33	0.50	For Fortaleza, Brazil region (366 hshlds); results of 2-stage least squares
Columbia	Pinstrup-Andersen and Caicedo (1978)	Indirect	—	—	0.51	0.65	For Cali region 220 households and 22 food commodities
India	Behram and Deolalikar (1987)	Direct	—	—	0.17	0.06	Survey data for pooled households. Using linear log model
	Sharma and Dillon (1984)	Indirect	—	—	0.57	—	Same data above
	Radhakrishna (1984)	Ind. rural/ urban	−0.53 −0.31	— —	0.88 0.88	— —	Elasticities with respect to prices and total expenditure of middle income group
Indonesia	Chernichowsky and Meesook (1988)	Direct	−0.35	−0.278	0.54	0.68	With respect to rice price
	Pitt & Rosenweig (1985)	Direct	−0.07	−0.13	0.007	0.012	With respect to farm profits and grain price
	Timmer (1981)	Direct	−0.64	—	0.48	—	
	Alderman and Timmer (1980)	Direct/rural urban	−0.71 −0.94	— —	0.50 0.39	— —	Protein elasticities are not reported
Nicaragua	Wolfe and Behram (1983)	Direct			0.058	0.29	1167 households in Managua, 1977
Sierra Leone	Strauss (1982)	Direct	−0.24	—	—	—	
Sri Lanka	Sahn (1988)	Indirect	—	—	0.62	—	Relates to rural middle income households
	Gavan and Chandrashekara (1979)	Direct	—	—	0.24	—	

Source: Babu, S.C., 1989. Challenges facing agriculture in southern Africa. In: A Conference Report: Inter-Conference Symposium of the International Association of Agricultural Economists, Badplass, South Africa, 10-16, August 1998.

In earlier studies, researchers have proceeded with one of the above approaches without explaining their choice. However, Pitt (1983, p. 110) argued that using the demand system approach is clearly preferable to the direct procedure, since all the parameters of the true calorie expenditure relationship are completely identified from the individual demand equations. Also, the issues relating to the correctness and difficulties of deriving price indices for nutrients in the second approach have caused the studies following this approach to drop the prices from their equations (Behrman and Deolalikar, 1987).

FOOD PRICE – NUTRIENT ELASTICITIES

Food price – nutrient elasticities provide information on the response of nutrient intake to any change in food prices. These can be derived from the matrix of direct and cross-price elasticities calculated from the estimated parameters of the demand systems.

The elements of the matrix of food price – nutrient elasticities, $\varnothing_{nj}$ are given by:

$$\varnothing_{nj} = \frac{\sum_i a_{ni} e_{ij} e(Y_i)}{\sum_i a_{ni} E(Y_i)} \tag{7.1}$$

where the $\varnothing_{nj}'s$ are uncompensated food price – nutrient elasticities, a_{ni} is the quantity of nutrient n per unit of food i, e_{ij} is cross-price elasticity, and $E(Y_i)$ is mean consumption of food commodity i. The compensated food price elasticity of nutrients $\varnothing_{nj}^*$ is calculated in the same way, replacing e_{ij} with the compensated food price elasticity of nutrients e_{ij}^* in Eq. (7.1).

NUTRIENT EXPENDITURE (INCOME) ELASTICITIES

The expenditure elasticities of nutrients are the indicators of changes in nutrition intake due to incremental changes in expenditure. They are derived from the expenditure elasticities calculated from the estimated parameters of the demand system and are given by:

$$\varnothing_{nx} = \frac{\sum_i a_{ni} e_{ix} e(Y_i)}{\sum_i a_{ni} E(Y_i)} \tag{7.2}$$

Where $\varnothing_{nx}$ is the expenditure elasticity of nutrient n with respect to the expenditure x, and e_{ix} is the expenditure elasticity.

A NOTE ON LIMITATIONS IN DERIVING NUTRITION ELASTICITIES

Whether a direct or indirect approach was followed to derive nutrition elasticities, there are certain difficulties in the derivation and use of nutrient coefficients. Some of the limitations in the use of nutrient coefficients are pointed out, though overcoming all of them may be difficult given the limitations of data in applied studies.

The use of average nutrient coefficients for aggregated food commodities may lead to erroneous calculations of some nutrients. For example, combining rice and ragi (finger millets) in the grains group, as in Behrman and Deolalikar (1987), may result in an overvalued estimation of carotene, iron, calcium, and thiamine. Ragi has 344 mg of calcium, 6.4 mg of iron, 42 mg of carotene, and 0.27 mg of thiamine, while rice contains 9, 2.8, 2, and 0.06 mg, respectively, of these nutrients. Even if rice is considered as a single food commodity, factors such as the quality of rice varieties (some are rich in certain minerals), the use of parboiled or raw rice for cooking, and the use of hand pounded or milled rice (which changes the nutrient content of minerals and vitamins), make it difficult to use a single nutrient coefficient for rice. For example, milling rice reduces protein by 9.3%, phosphorous by 16%, carotene by %, thiamine by 71%, riboflavin by 63%, and ascorbic acid by % (Gopalan et al., 1977).

The food items which are not considered for a particular study could make a big difference if a specific nutrient is of importance. For example, if one is interested in knowing how much increase in the intake of carotene would result due to a subsidized milk program on a targeted household that suffers from vitamin A deficiency identified by the incidence of "night blindness," use of just milk and rice may not be appropriate. Other food items that are consumed by households such as vegetables, greens, and pulses, which contain high levels of vitamin A, should also be considered for such an analysis. Timmer (1981), and Alderman and Timmer (1980) use just rice and cassava for Indonesia to study nutrition elasticities.

In most of the food expenditure surveys, the categories under "other foods" are refreshments and foods taken outside the home which form a substantial portion of food expenditure. Since quantification of individual food commodities is not possible in such cases, this food expenditure is usually not considered for nutrition calculations. The nutrition coefficients are generally obtained from the nutritive values of foods (see Gopalan et al. (1977) for the nutrient content of Indian foods) and are derived based on the laboratory analysis conducted at the country level.

CAN SUBSIDIES IMPROVE NUTRITION?

Suppose we identify an extremely poor group of people in an economy, which is also particularly undernourished. Further, assume we have information about their staple diet. How might one improve this group's nutritional outcomes?

A generally accepted idea is to provide a subsidy, which is equivalent to reducing the price of the staple food product. A price reduction is likely to increase the group's consumption of the staple food product, and generate better health outcomes given other determinates remained at the same level.

Food subsidies as a policy tool to address inadequate consumption of calories and to address hunger reduction has been extensively researched (Pinstrup-Andersen, 1988). However, while the proposed policy measure is completely sound and practical, actual evidence from subsidies on nutrition intake has been elusive. Consider, for instance, the findings from an influential study by Jensen and Miller (2011), where a large subsidy scheme was launched for poor households, located in two provinces in China. The found the food subsidy had a negative impact on nutrition and in addition, the food subsidy encouraged higher consumption of goods with nonnutritional values.

The above findings are, however, consistent with the theory of consumer behavior, because as food prices decrease there is an increase in the real income of the consumers, who can technically substitute away even to nonfood items. Hence, Jensen and Miller (2011) question the value of any subsidy-like scheme that attempts to improve nutritional outcomes because, by and large, most of these attempts have large income effects. A particularly good reason why the income effects may allow for substitution is because the selected households may have already achieved subsistence levels of those staple food items.

Similar findings are also presented in another extremely interesting study on this topic for India by Kaushal and Muchomba (2015). Using three rounds of the National Sample Survey data from the Targeted Public Distribution System (TPDS), including information on staple diet preferences, Kaushal and Muchomba (2015) examine the impact of price subsidies on the intake of calories from wheat and rice. Their findings are very similar to those of Jensen and Miller (2011):

- The increase in real income has no effect on nutritional intake[1];
- There is an increase in expenditures on nonfood items; and
- There is an increase in the consumption of sugar, sugar substitutes, and edible oil.

In contrast to Jensen and Miller (2011) and Kaushal and Muchomba (2015), studies in the American context have shown positive reinforcement of healthy eating from subsidies. For example, Dong and Lin (2009), and Chang et al. (2015) show that subsidy-based programs such as SNAP and WIC positively affect the intake of fruits and vegetables, good sources of micronutrients.

The issue of a low intake of fruits and vegetables is a serious concern in the United States as well. Dong and Lin (2009) point out that although the daily recommended intakes are 1.8 cups of fruits and 2.6 cups of vegetables, most Americans consume only 1.03 cups of fruits and 1.58 cups of vegetables. Most Americans are significantly short of the requirement, and Dong and Lin (2009) estimate that a 10% subsidy would encourage low-income Americans to increase the consumption of fruits by 2.1–5.2%, and vegetables by 2.1–4.9%. Further, the cost of such a subsidy is estimated to be roughly $580 billion, for both of these products. Unfortunately, the study also notes that even with this subsidy, most Americans would still fall short of the requirement.

Chang et al. (2015) extend the above study to examine the impact of SNAP and WIC and note that, contingent on household and demographic characteristics, these programs significantly improve the intakes of fruits and vegetables.Kaushal and Muchomba (2015) explain the reason for the differences in these findings based on the following graph:

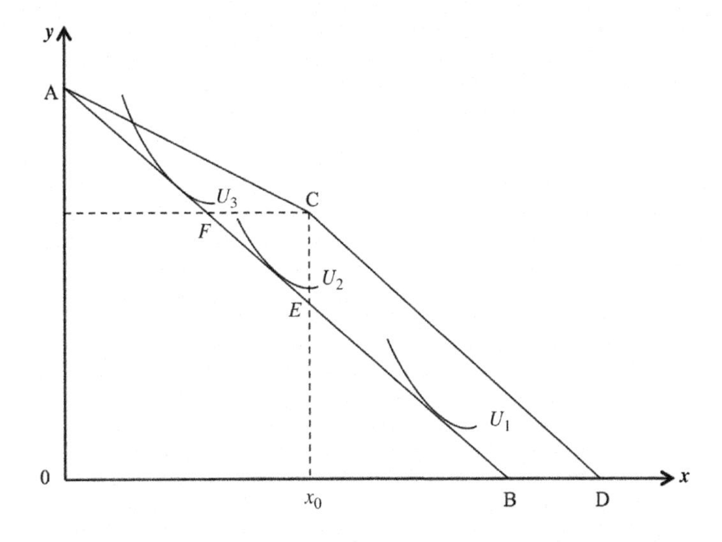

[1]A 100% increase in subsidy translates into a 0.54% increase in income, which increased calorie intake from wheat and rice by 2% to 4%, *lowered* intake from coarse grains by 40% to 75%, and *increased* sugar and sugar products by 21% to 28% (see Kaushal and Muchomba, 2015, p. 37).

Consider a typical consumer utility maximization problem with two goods: x and y, with prices p and \$1, and a budget constraint of $I = y + p\,x$, where I represents total income. Let AB represent the initial budget constraint. If the subsidy is given for good x, then the price of x falls, and AB shifts to the right.

If the government allows a food price subsidy such that the consumer can buy up to a maximum of x_0 units at a lower price cp, where $c < 1$, then the new budget constraint is given by ACD, where the budget constraint is rewritten as:

$$I = y + cp\,x_0 + p(x - x_0)$$

Before the subsidy, the consumer's equilibrium could be anywhere along the original budget line AB. However, after the subsidy, the new equilibrium will be located either on the segment AC or CD, depending upon where the original equilibrium was positioned. The issue becomes critical for the consumer with the utility function U_3, because it is possible, in this case, that the income effect may be smaller than the substitution effect to affect a net reduction in the consumption of x. As Kaushal and Muchomba (2015) point out, the final effect of the subsidy is an empirical question and cannot be determined a priori.

The empirical approaches by Kaushal and Muchomba (2015), Jensen and Miller (2011), and Chang et al. (2015) are also different from each other, and provide us with an idea of the alternative specifications researchers use in these contexts. For instance, Jensen and Miller use the following regression:

$$\%\Delta\ \text{Nutrient} = \alpha + \beta\ \%\Delta\ \text{Subsidy} + \text{disturbance}$$

where the dependent variable is the percentage change in a household's consumption of a particular nutrient at time t, and this is regressed on the percentage change in the price of the staple food at time t. Overall, Jensen and Miller (2011) are unable to reject the null, or $\beta = 0$, for many of the chosen nutrients, such as calcium, potassium, and iron.

Kaushal and Muchomba (2015) organize their empirical estimation in three stages, which can be captured by the following equations:

A logit equation for capturing the participation constraint:

$$\ln\left[\frac{\text{pr}(BPL\ \text{card} = 1)}{Pr(BPL\ \text{card} = 0)}\right] = X_i\beta + \pi_j$$

An equation representing the subsidy amount:

$$\text{Subsidy} = X_i\beta + \beta_c(\text{pr Card} * \text{Post}) + \delta_0\ \text{pr Card} + \text{Fixed Effects}$$

and finally, an equation capturing nutritional effects:

$$\text{Nutrition Intake} = X_i\phi + \varphi\ \text{Subsidy} + \varphi_0\ \text{pr Card} + \text{Dummies}$$

The first equation is a logit model that captures the probability that a household i in district j at time t has a ration card issued by the government of India. The independent variables here capture a variety of household characteristics.

The predicted probability from the logit model is substituted in the second equation (pr Card), to capture the effect of the household's subsidy amount. The predicted probability is also interacted with Post, which is a dummy variable that captures the effect of the time period when the subsidy program was expanded.

The Nutrition Intake equation finally summarizes the effect of all the predetermined variables on different nutrient intakes for a household in the sample. Kaushal and Muchomba (2015) examine a whole range of nutrients from a variety of foods including wheat and rice, which are the subsidized items. The OLS and Instrumental variables estimates indicate that an increase in subsidy increases expenditures on nonfood items.

Chang et al. (2015) use an ordered probit model for predicting the likelihood of consuming fruits and vegetables, based on their receipts of SNAP or WIC and other explanatory factors. Hence, a variety of econometric specifications and approaches are used to study the effect of subsidies on nutrition. In many instances, the functional form, the chosen estimator and methods depend upon the availability of data, and the manner in which the subsidy programs are implemented. By and large, it is not immediately obvious that a subsidy on healthy foods will lead to higher intakes of good calories. We consider this issue further in Chapter 14, Economic Analysis of Obesity and Impact on Quality of Life: Application of Nonparametric Methods, when we discuss the importance of a soda-tax on reducing obesity in the United States. We now turn toward implementation of alternative empirical methods in STATA.

IMPLEMENTATION IN STATA

We continue with the representation in the last chapter and use that framework to model the impact of a subsidy. Table 7.2 has information on the consumption of fruits (Q_1), dairy products (Q_2), and junk food (Q_3), with their prices, for 27 households. We also have a binary variable indicating whether the household is in the rural ($R = 1$), or in the urban area ($R = 0$). We also have information on the number of children (kids) and household income (m).

Table 7.2 Consumption, Price, Kids, and Income for 27 Households

Obs	Q_1	Q_2	Q_3	P_1	P_2	P_3	R	Kids	m
1	129.6	0.875	0.975	0.736	0.76	0.374	1	4	6036
2	131.2	0.864	0.975	0.769	0.796	0.378	1	4	6113
3	137	0.869	0.971	0.848	0.824	0.387	1	3	6271
4	141.5	0.868	0.965	0.86	0.835	0.393	1	3	6378
5	148.7	0.864	0.962	0.901	0.851	0.4	1	2	6727
6	155.8	0.899	0.939	0.894	0.869	0.401	1	2	7027
7	164.8	0.92	0.921	0.87	0.902	0.405	1	2	7280
8	170.9	0.95	0.93	0.9	0.95	0.413	1	3	7513
9	183.3	0.964	0.958	0.928	0.996	0.431	1	3	7728
10	195.7	0.997	0.974	0.931	1.077	0.444	1	4	7891
11	207.3	1.006	1.006	0.943	1.235	0.457	1	4	8134
12	218.2	1.013	1.05	1.002	1.327	0.477	1	5	8322
13	226.7	1.026	1.04	1.005	1.384	0.485	1	3	8562
14	237.8	1.131	1.041	1.076	1.398	0.496	1	3	9042

(Continued)

Table 7.2 Consumption, Price, Kids, and Income for 27 Households *Continued*

Obs	Q_1	Q_2	Q_3	P_1	P_2	P_3	R	Kids	m
15	225.7	1.549	1.105	1.126	1.43	0.534	0	2	8867
16	232.3	1.658	1.206	1.364	1.536	0.589	0	4	8944
17	241.6	1.729	1.287	1.579	1.692	0.625	0	3	9175
18	249.1	1.832	1.359	1.728	1.774	0.657	0	3	9381
19	261.2	1.913	1.468	1.765	1.828	0.699	0	2	9735
20	248.8	2.606	1.59	1.91	1.953	0.751	0	2	9829
21	226.7	3.641	1.723	1.981	2.466	0.822	0	4	9722
22	225.5	4.059	1.832	2.469	3.07	0.887	0	4	9769
23	228.7	3.844	1.906	2.864	3.41	0.93	0	3	9725
24	239.5	3.714	1.956	3.197	3.576	0.971	0	3	9930
25	244.6	3.657	2.015	3.657	3.802	0.968	0	4	10421
26	245.7	3.688	2.082	3.697	3.978	0.975	0	4	10563
27	269.3	2.871	2.17	3.532	4.214	0.983	0	3	10780

EXAMPLE 1: THE EFFECT OF A SUBSIDY

We can estimate the Quadratic AIDS (QUAIDS) model for this data, and check if the uncompensated elasticities vary across the rural – urban households. We can also check if all the expenditure elasticities are positive. We produce these results below.

Suppose we want to provide a subsidy that promotes fruit and dairy consumption. Suppose we know that total calories per unit of fruit, dairy, and junk food are 0.4, 0.5, and 0.6, respectively. Then total calories $C = 0.4\ Q_1 + 0.5\ Q_2 + 0.6\ Q_3$.

What will be the effect of a 1% price reduction in the price of fruit, on total calorie consumption?

The percent change in calorific consumption from this is given by:

$$s_1 \epsilon^u_{11} + s_2 \epsilon^u_{12} + s_3 \epsilon^u_{13},$$

where ϵ^u_{1j} is the uncompensated elasticity of good j when P_1 changes by 1%, and s_i represents the calorific share to total calories for good i (example $s_1 = \frac{0.4\ Q1}{C}$).

We use the `gen` commands to create C, si, and `estat uncompensated ue*` in STATA to generate the necessary values for each observation.

The uncompensated elasticities are saved as ue_1_1, ue_1_2, and ue_1_3. Finally, the percentage change in calories is generated using:

```
s1*ue_1_1 + s2*ue_2_1 + s3*ue_3_1
```

We can also compute the mean of this entity. We illustrate these commands and the STATA output screens below.

We first use gen to generate total expenditure, and define the relative share of each good. The four sequential commands in STATA are:

```
gen expfd = p1*q1 +p2*q2 + p3*q3

gen w1 = p1*q1/expfd
gen w2 = p2*q2/expfd
gen w3 = p3*q3/expfd
```

We then implement the quaids command to estimate the QUAIDS model, similar to the process from the last chapter, and the STATA input and output results are:

```
quaids w1-w3, anot(10) prices (p1-p3) expenditure(expfd) nolog
```

```
Quadratic AIDS model

Number of obs            =        27
Number of demographics   =         0
Alpha_0                  =        10
Log-likelihood           = 330.93232
```

	Coef.	Std. Err.	z	P>|z|	[95% Conf.	Interval]
alpha						
alpha_1	1.141389	.0997319	11.44	0.000	.9459183	1.33686
alpha_2	-.2344438	.093194	-2.52	0.012	-.4171007	-.051787
alpha_3	.0930545	.0116734	7.97	0.000	.070175	.1159339
beta						
beta_1	.0744037	.0409941	1.81	0.070	-.0059432	.1547506
beta_2	-.1123671	.0383384	-2.93	0.003	-.187509	-.0372253
beta_3	.0379634	.0047634	7.97	0.000	.0286274	.0472995
gamma						
gamma_1_1	.0081077	.0166209	0.49	0.626	-.0244686	.040684
gamma_2_1	-.0137494	.0188513	-0.73	0.466	-.0506973	.0231984
gamma_3_1	.0056418	.0031688	1.78	0.075	-.0005689	.0118525
gamma_2_2	.0237449	.0209252	1.13	0.256	-.0172677	.0647576
gamma_3_2	-.0099955	.0028892	-3.46	0.001	-.0156582	-.0043328
gamma_3_3	.0043537	.00083	5.25	0.000	.0027271	.0059804
lambda						
lambda_1	.0088134	.0042405	2.08	0.038	.0005022	.0171246
lambda_2	-.0128274	.0039687	-3.23	0.001	-.0206059	-.0050489
lambda_3	.004014	.0004851	8.28	0.000	.0030633	.0049648

As mentioned in the last chapter, the uncompensated elasticities are computed for every observation, and the means are produced below. All own elasticities are negative, and we note that only goods 1 and 2 are substitutes, while the other two combinations represent complementarity in this data.

```
. estat uncompensated ue*

. summarize ue*
```

Variable	Obs	Mean	Std. Dev.	Min	Max
ue_1_1	27	-.9949049	.0029014	-.998989	-.9892969
ue_1_2	27	.0058247	.001355	.0041285	.008466
ue_1_3	27	-.0008297	.0004088	-.0016227	-.0003147
ue_2_1	27	-.7099307	.7202711	-1.868265	.1628326
ue_2_2	27	-1.772059	.468742	-2.509744	-1.200282
ue_2_3	27	-.0130663	.079715	-.1070366	.1331494
ue_3_1	27	-.217953	.5015302	-1.009156	.5805875
ue_3_2	27	-.1115348	.2361806	-.4499804	.2680736
ue_3_3	27	-.5663684	.1271834	-.8183446	-.4067908

In the next set of commands, we calculate the share of calories, s_i, the calorific share to total calories for good i (example $s_1 = \frac{0.4Q1}{C}$). Finally, we compute the impact of a subsidy on a 1% reduction in the price of fruit, by defining `frt_sub`

```
gen c = 0.4*q1 + 0.5*q2 + 0.6*q3
gen s1 = 0.4*q1/c
gen s2 = 0.5*q2/c
gen s3 =0.6*q3/c
gen frt_sub = s1*ue_1_1 + s2*ue_2_1 + s3*ue_3_1
```

summarize s1 s2 s3 frt_sub

Variable	Obs	Mean	Std. Dev.	Min	Max
s1	27	.9802481	.0068899	.9664766	.987643
s2	27	.0104464	.0052419	.0055872	.0217457
s3	27	.0093055	.0019418	.0064853	.0123228
frt_sub	27	-.982659	.0059454	-.9900663	-.9689668

The result indicates that a 1% decline in the price of food increases calorie consumption by 0.98%. We can also check if the effect is greater for rural households (see Exercise 1), and the impact of demographics on the size of the fruit subsidy (Exercise 2). Finally, we can also compare the impact of a 1% price increase on junk foods, and see if a policy like a soda-tax will be effective in reducing obesity.

EXAMPLE 2: BASED ON JENSEN-MILLER (2011)

In this example we examine the following data with 30 observations for 15 households (H), where each family is observed for two time periods (time). We have information on whether the household unit has a ration card (Card = 1 if yes; 0 if not), the quantity of nutrient consumed (q), the market price (p1) of the good and the price paid by the household (p2), the size of the household (N), and the percentage of females in the household (F). We use this information to test for three different models:

obs	time	H	Card	q	p1	p2	N	F
1	1	1	0	53.4	1	0.66	3	0.66
2	2	1	0	54.2	1.25	0.83	4	0.5
3	1	2	0	49.4	1	0.66	5	0.6
4	2	2	0	50.2	1.25	0.83	3	0.66
5	1	3	0	49.9	1	0.66	4	0.75
6	2	3	0	50.7	1.25	0.83	5	0.6
7	1	4	0	46.4	1	0.66	3	0.66
8	2	4	0	47.2	1.25	0.83	3	0.33
9	1	5	0	53.4	1	0.66	3	0.33
10	2	5	0	53.4	1.25	0.83	4	0.75
11	1	6	0	54.7	1	0.66	4	0.75
12	2	6	0	49.4	1.25	0.83	4	0.5
13	1	7	0	53.4	1	0.66	5	0.6
14	2	7	0	54.2	1.25	0.83	5	0.6
15	1	8	1	50.9	1	0.66	3	0.66
16	2	8	1	51.7	1.25	0.83	3	0.66
17	1	9	1	53.4	1	0.66	3	0.33
18	2	9	1	54.7	1.25	0.83	3	0.33
19	1	10	1	49.4	1	0.66	3	0.33
20	2	10	1	50.7	1.25	0.83	4	0.5
21	1	11	1	49.9	1	0.66	5	0.4
22	2	11	1	51.2	1.25	0.83	3	0.66
23	1	12	1	46.4	1	0.66	4	0.5
24	2	12	1	47.7	1.25	0.83	5	0.4
25	1	13	1	53.4	1	0.66	3	0.66
26	2	13	1	54.2	1.25	0.83	3	0.66
27	1	14	1	54.7	1	0.66	3	0.33
28	2	14	1	50.7	1.25	0.83	4	0.75
29	1	15	1	53.4	1	0.66	4	0.75
30	2	15	1	54.2	1.25	0.83	3	0.66

In the first few commands, we follow Jensen and Miller (2011) and use the arc-elasticity formula to compute the percentage changes in the subsidy amount and nutrient intake.

```
• gen sub = p2 - p1      /* we define a new variable called sub */
• gen lags1 = sub[_n-1]      /* we get the lag value of sub */
• gen lagN = q[_n-1]      /* we get the lag value of q */
• gen sn = sub - lags1      /* computing the numerator for arc elasticity */
• gen sd = 0.5*(sub + lags1)      /* the denominator for arc elasticity */
• gen perchsub = (sn/sd)*100      /* %Δ Subsidyis calculated */
• gen nN = q - lagN
• gen nD = 0.5*(q + lagN)
• gen perchNut = (nN/nD)*100      /* %Δ Nutrient is calculated */
```

We wish to estimate three versions of the model:

$$\%\Delta \text{ Nutrient} = \alpha + \beta \, \%\Delta \text{ Subsidy} + \text{disturbance}$$

We use the option `quietly` to run three models, and summarize the results below using STATA's `estout`:

```
• quietly regress perchNut perchsub
• estimates store m1, title(Model 1)
• quietly regress perchNut perchsub n f
• estimates store m2, title(Model 2)
• quietly regress perchNut perchsub n f card
• estimates store m3, title(Model 3)
• estout m1 m2 m3, cells(b(star fmt(3)) se(par fmt(2))) legend label varlabels
  (_cons constant) stats(r2 df_r, fmt(3 0) label(R-sqr dfres))
```

After each regression, we store the output under different titles, and then use the options under `estout` to place a star to denote significance, use three decimal places for the estimates and two decimal places for the standard error, use the variable names as legends, and produce R-squared and degrees of freedom. STATAs summary output version is given as:

	Model 1	Model 2	Model 3
	b/se	b/se	b/se
perchsub	0.014	0.015	0.015
	(0.05)	(0.06)	(0.06)
N		-0.647	-0.754
		(1.50)	(1.60)
F		-1.590	-1.832
		(7.87)	(8.10)
Card			-0.560
			(2.55)
constant	0.042	3.340	4.184
	(1.13)	(6.78)	(7.90)
R-sqr	0.002	0.012	0.014
dfres	27	25	24

* p<0.05, ** p<0.01, *** p<0.001

Although the sign of β is positive in every case, none of the models produce any significant results, and have very poor R-squared values. The results may be recalibrated using different sets of independent variables (see Exercise 3).

EXAMPLE 3: BASED ON KAUSHAL AND MUCHOMBA (2015)

We extend the above example to include the household's income (`Inc`), age of the head of the household (`Age`), and whether the household possesses a bicycle (`Bicycle` = 1 if yes; 0 if not).

obs	time	H	Card	q	p1	p2	N	F	Inc	age	Bicycle
1	1	1	0	53.4	1	0.66	3	0.66	1.1	45	1
2	2	1	0	54.2	1.25	0.83	4	0.5	1.5	45	1
3	1	2	0	49.4	1	0.66	5	0.6	1.3	43	1
4	2	2	0	50.2	1.25	0.83	3	0.66	1.7	43	0
5	1	3	0	49.9	1	0.66	4	0.75	1.5	37	0
6	2	3	0	50.7	1.25	0.83	5	0.6	1.9	37	0
7	1	4	0	46.4	1	0.66	3	0.66	1.7	46	1
8	2	4	0	47.2	1.25	0.83	3	0.33	2.1	46	0
9	1	5	0	53.4	1	0.66	3	0.33	1.9	34	1

(Continued)

Continued

obs	time	H	Card	q	p1	p2	N	F	Inc	age	Bicycle
10	2	5	0	53.4	1.25	0.83	4	0.75	2.3	34	1
11	1	6	0	54.7	1	0.66	4	0.75	2.1	56	0
12	2	6	0	49.4	1.25	0.83	4	0.5	2.5	57	0
13	1	7	0	53.4	1	0.66	5	0.6	2.3	42	1
14	2	7	0	54.2	1.25	0.83	5	0.6	2.7	43	1
15	1	8	1	50.9	1	0.66	3	0.66	1.02	46	0
16	2	8	1	51.7	1.25	0.83	3	0.66	1.42	47	0
17	1	9	1	53.4	1	0.66	3	0.33	1.22	44	0
18	2	9	1	54.7	1.25	0.83	3	0.33	1.62	45	1
19	1	10	1	49.4	1	0.66	3	0.33	1.42	38	1
20	2	10	1	50.7	1.25	0.83	4	0.5	1.82	39	0
21	1	11	1	49.9	1	0.66	5	0.4	1.62	47	0
22	2	11	1	51.2	1.25	0.83	3	0.66	2.02	47	0
23	1	12	1	46.4	1	0.66	4	0.5	1.82	35	1
24	2	12	1	47.7	1.25	0.83	5	0.4	2.22	35	1
25	1	13	1	53.4	1	0.66	3	0.66	2.02	57	0
26	2	13	1	54.2	1.25	0.83	3	0.66	2.42	58	0
27	1	14	1	54.7	1	0.66	3	0.33	2.22	43	1
28	2	14	1	50.7	1.25	0.83	4	0.75	1.96	44	1
29	1	15	1	53.4	1	0.66	4	0.75	0.94	47	1
30	2	15	1	54.2	1.25	0.83	3	0.66	1.34	48	0

First, we estimate a logit function and compute the predicted probabilities of having a ration card for each household using `logit` in STATA:

```
. logit card age bicycle inc, nolog

Logistic regression                             Number of obs   =         30
                                                LR chi2(3)      =       2.41
                                                Prob > chi2     =     0.4922
Log likelihood = -19.523888                     Pseudo R2       =     0.0581

------------------------------------------------------------------------------
        card |      Coef.   Std. Err.      z    P>|z|     [95% Conf. Interval]
-------------+----------------------------------------------------------------
         age |   .0344645    .0655129     0.53   0.599    -.0939384    .1628673
     bicycle |  -.3362617    .8568965    -0.39   0.695    -2.015748    1.343225
         inc |  -1.133942    .8920355    -1.27   0.204    -2.882299    .6144158
       _cons |   .8127834    3.316005     0.25   0.806    -5.686467    7.312033
------------------------------------------------------------------------------
```

The logit model is not a best fitting model in this case, as none of our variables are significant. However, to best illustrate the method, we proceed by generating the predicted probabilities (`pcard`) for each observation using the `predict pcard`.

Next we use the `ivregress` procedure in STATA to generate the IV estimates, using the predicted values of the subsidy variable as an instrument, in the nutrition equation. We first derive a subsidy amount for each observation, using the definition provided by Kaushal and Muchomba (2015):

$$\text{Subsidy}_i = q\,(p_1 - p_2)/n$$

```
. ivregress 2sls q age f inc n time (sub = pcard bicycle)

Instrumental variables (2SLS) regression               Number of obs =        30
                                                       Wald chi2(6)  =      7.32
                                                       Prob > chi2   =    0.2919
                                                       R-squared     =    0.4709
                                                       Root MSE      =    1.8425

------------------------------------------------------------------------------
           q |      Coef.   Std. Err.      z    P>|z|     [95% Conf. Interval]
-------------+----------------------------------------------------------------
         sub |   9.767989   7.075962     1.38   0.167    -4.100641    23.63662
         age |  -.0741175   .1315552    -0.56   0.573    -.3319609    .1837258
           f |   3.591102   2.888484     1.24   0.214    -2.070222    9.252427
         inc |   .9446723   1.009038     0.94   0.349    -1.033006    2.922351
           n |   12.78356   9.539571     1.34   0.180    -5.913659    31.48077
        time |  -11.49782   8.369243    -1.37   0.169    -27.90124    4.905596
       _cons |  -32.85679   58.42414    -0.56   0.574     -147.366    81.65242
------------------------------------------------------------------------------
Instrumented:  sub
Instruments:   age f inc n time pcard bicycle
```

The IV estimator indicates that nutrition increases with subsidy, but our estimates are not significant. Consequently, we cannot reject the null that the subsidy has no effect on nutritional intake. Perhaps other specifications may yield better results (Exercise 4).

EXAMPLE 4: BASED ON CHANG ET AL. (2015)

Chang et al. (2015) estimate an ordered logit model where the latent variable (Y^*) is the preference for nutrient, such that:

$$Y^* = \beta X + \text{disturbance}$$

where the disturbance term has the usual properties. In our example, the observable equivalent of Y^* can be constructed by a censoring of nutrient intake as follows:

$y = 1$ if $q < 50$, $y = 2$ if $50 < q < 51$ and $y = 3$ if $q > 3$, where q is the amount of nutrient consumed by household i, which can be derived in STATA using the **gen** command:

```
. generate y = .
(30 missing values generated)

. replace y = 1 if (q < 50)
(9 real changes made)

. replace y = 3 if (q > 50)
(21 real changes made)

. replace y = 2 if (q >= 50) & (q < 51)
(5 real changes made)
```

The ordered logit estimation is achieved by **ologit**:

```
. ologit y sub card n f inc age bicycle

Iteration 0:   log likelihood = -29.852291
Iteration 1:   log likelihood = -22.636231
Iteration 2:   log likelihood = -22.231789
Iteration 3:   log likelihood = -22.227032
Iteration 4:   log likelihood = -22.227029
Iteration 5:   log likelihood = -22.227029

Ordered logistic regression                       Number of obs   =         30
                                                  LR chi2(7)      =      15.25
                                                  Prob > chi2     =     0.0329
Log likelihood = -22.227029                       Pseudo R2       =     0.2554

------------------------------------------------------------------------------
           y |      Coef.   Std. Err.      z    P>|z|     [95% Conf. Interval]
-------------+----------------------------------------------------------------
         sub |   2.363084   .9875328     2.39   0.017     .4275554    4.298613
        card |   .3299308    .920349     0.36   0.720     -1.47392    2.133782
           n |    2.18694    1.34236     1.63   0.103    -.4440381    4.817917
           f |   3.039109   3.397692     0.89   0.371    -3.620245    9.698463
         inc |  -.4744724   1.192905    -0.40   0.691    -2.812523    1.863578
         age |    .078078   .0812107     0.96   0.336    -.0810922    .2372481
     bicycle |   1.572625   .9656897     1.63   0.103    -.3200922    3.465342
-------------+----------------------------------------------------------------
       /cut1 |   25.04199   10.68661                      4.096623    45.98736
       /cut2 |   26.15139   10.77653                      5.029776      47.273
------------------------------------------------------------------------------
```

The iteration stops at Step 5, without any problem of convergence, since the procedure uses Maximum Likelihood methods. The log-likelihood test and the LR chi2(7) is 15.25, indicating that the model overall is qualitatively significant, since all the variables are not zero. The coefficients indicate the change in the log-odds scale of the response variable (y), for a unit change in each of the independent variables. Interestingly, in this model, the variable *sub* representing the subsidy amount is positive and significant ($z = 2.39$), and if the households were to get a one unit increase in subsidy, then the log-odds of the household going to a higher group increase by 2.36. We can also compute the proportional odds ratio model using STATA (Exercise 5). This model captures the odds of going to a higher category, given the cumulative odds of other categories.

CONCLUSIONS

The major themes in this chapter and in the previous chapter were to use consumer demand theory to study the demand behavior of households toward food commodities, so as to help devise policies to achieve proper nutrition and reduction of malnutrition. This chapter extends the consumer theory models from the previous chapter to construct nutritional elasticities using prices and incomes as key variables.

We have also looked at recent microeconometric evidence regarding the impact of food subsidies, and note that the issue is an empirical one; it is not always obvious that a subsidy will encourage healthy nutrient intakes. We provide a review of three recent studies from the literature covering India, China, and the United States. We also provide examples to implement these ideas using STATA, particularly with respect to estimating IV estimators, ordered probit, and the QUAIDS models.

An important application of these issues in many countries is the role of subsidies in promoting good eating habits. Similarly, the role of "soda-taxes" and "fat-taxes" has also gained sufficient traction in this context. We use STATA procedures to illustrate the usefulness of these methods. For instance, our STATA examples help in calibrating different nutritional elasticities, and also derive the impact of a specific subsidy amount on the consumption of different food items. The STATA examples also illustrate the application of the relevant econometric methodology pertaining to the type of data collected. The framework developed here can help policy makers to quickly determine the effect of a sizeable subsidy on consumption among the poorer sections in the community.

Based on the information from this chapter, is it possible to predict, which of these two effects is likely to be more significant on nutritional outcomes—a tax on soft drinks or a subsidy on milk? The exercises at the end of this chapter help in approaching such issues.

In addition, this chapter highlights the recent importance given to the analysis of demand for nutrients by academic researchers in development policy making. There is a large body of research that looks at such analyses and derives implications for policies to address malnutrition. However, the application of data, methods of estimation, and results have differed for policy making. This chapter is an attempt to expose the readers to such issues.

EXERCISES

1. Use the data from Example 1 and replicate the QUAIDS model. After the above procedure, type the following commands to produce the uncompensated elasticities for rural and urban households.

```
.estat uncompensated if R ==1, atmeans
.matrix uprural = r(uncompelas)
.estat uncompensated if R==0, atmeans
.matrix upurban = r(uncompelas)
```

   Use the `matrix list` command and check if the elasticities are different across these groups.
2. Implement the role of demographics for this data following the steps indicated in the last chapter: first create the log of income using `gen lm = log(m)`. Then estimate using:

```
quaids w1-w3, anot(10) prices (p1-p3) expenditure(expfd) nolog demographics (kids r lm)
```

   Compute the impact of a subsidy on fruit and examine the empirical issue that "no-subsidy effect" holds under this specification.
3. Use the data from Example 3 to examine the Jensen and Miller (2011) model:

$$\%\Delta \text{ Nutrient} = \alpha + \beta \; \%\Delta \text{ Subsidy} + \text{disturbance}$$

   Use the `gen` command and redefine *sub* following Kaushal and Muchomba (2015):

```
gen sub = q*(p1 - p2)/n
```

   Use the arc-elasticity computations to compute the two variables following Example 2, and include *inc*, *age*, and *bicycle* in your regression. Does this specification reject the null hypothesis of a no-subsidy effect?
4. Reexamine the data associated with Example 3, to the `ivregress` procedure, in which income (*inc*) is added as an instrument. Can you reject the null hypothesis of a zero-subsidy effect under this specification?
5. Use Example 4 and the associated data, and apply the following command in STATA:

```
ologit y sub card n f inc age bicycle, or
```

   Note the difference in output. Is the coefficient of *sub* still positive and significant?

   In this case the `'or'` option allows us to interpret the change in levels cumulatively. The coefficients in odds, compares the households who are in groups greater than *2 or 3* with those in lower groups, based on the observed household level. For a one unit increase in *sub*, the odds of high *y* versus the combined middle and low *y* groups are K times greater, given the other variables are held constant. K is the estimated coefficient in STATA using the `'or'` option. Find K for this data.
6. Consider Table 7.1 and the selected earlier studies that estimated nutrient elasticities for calorie and protein. Develop a comparative table for micronutrients using studies based on demand for micronutrients published in the last 10 years. Write your commentary on the estimates. Why does it matter if the estimates differ by a few decimals? What implications do such differences have for food and nutrition policies?

PART

DETERMINANTS OF NUTRITIONAL STATUS AND CAUSAL ANALYSIS

D

CHAPTER

SOCIO ECONOMIC DETERMINANTS OF NUTRITION: APPLICATION OF QUANTILE REGRESSION

8

As you are aware, diarrhoeal diseases, which are also associated with a high risk of malnutrition and stunting, take a huge socio-economic toll on society.

Stephen Kampyongo, Zambia Minister of Local Government and Housing (*Zambia Daily Mail*, November 21, 2015)

INTRODUCTION

The nutritional status of individuals is often depicted as the final outcome of the process in which various factors interact. While food consumption which can provide adequate nutrients is a key ingredient to produce nutritional outcomes, various other socio economic determinants play their role in this "production" process (Smith and Haddad, 2015; Linnemayr et al., 2008). In the conceptual framework presented in Chapter 3, A Conceptual Framework for Investing in Nutrition: Issues, Challenges and Analytical Approaches, we discussed several basic, underlying, and immediate determinants of nutritional security. These determinants and their roles vary according to geopolitical, socio economic, demographic, and other climatic contexts. In Chapter 4, Microeconomic Nutrition Policy, we discussed the econometric formulations that are used to explain the magnitude and nature of the role such determinants play in nutritional outcomes. In this chapter, we present studies that have analyzed such determinants from policy perspectives, and provide specific analytical examples for practitioners to identify the role of these determinants in the formulation of nutritional policy and program interventions. The magnitude and the nature of the nutritional contribution of these socio economic factors will vary, depending on their interaction. We demonstrate the method of quantile regression to capture the policy and program effect of different socio economic groups.

SAFE WATER, SANITATION, MOTHERS' EDUCATION, CHILDREN'S DIETS, AND NUTRITION

Socio economic factors that determine nutritional outcomes go beyond food security variables. Recently, Smith and Haddad (2015) revisited their earlier work to show the contributions of various socio economic determinants toward reducing child stunting in developing countries from 1970 to 2010. They found that just three factors—safe water, sanitation, and secondary school education of the mother—contributed to 63% of the total reduction in stunting.

Nutrition Economics. DOI: http://dx.doi.org/10.1016/B978-0-12-800878-2.00008-6

Food-related variables contributed to only one third of the stunting reductions. Other recent studies have also shown that nonfood factors, such as the wealth of the household, women's human capital, and health care can have a significant influence on the stunting levels of children in developing countries (Headey et al., 2015). Monteiro et al. (2009) studied the reasons for the reduction in child under-nutrition in Brazil during the years 1996–2007. They found up to 40% of the reductions could be attributed to safe water, sanitation, health care, and maternal education. Studies have specifically looked at maternal education, stature, and status as determinants of child nutrition (Hernandez-Diaz et al., 1999; Wamani etal, 2004). Child spacing and feeding practices have been investigated (Dewey and Cohen, 2007; Varela-Silva et al., 2009; Kumar et al., 2006; Mamiro et al., 2005).

The quality of child care has been of interest as well (Blau and Hagy, 1998; Cleveland and Krashinsky, 1998; Morris, 1999). Other studies that looked at socio economic factors in general include: Willey et al., 2009; Marjan et al., 2002; Delpeuch et al., 2000; Abubakar et al., 2012; Reyes et al., 2004. Below we review specific studies in detail to motivate methodological discussions on estimating the contribution of various factors to nutritional outcomes in all forms.

In the context of a developed country, Variyam et al. (1999a,b) examine the hypothesis that a mother's stock of nutritional knowledge and diet–health awareness influence her children's diets. As seen in Chapter 4, Microeconomic Nutrition Policy, the positive effect of education on health and nutrition outcomes is a prediction from household production theory. Household production theory posits that the ability of the more educated to acquire and process a greater amount of health information than the less educated leads to greater allocative efficiency.

Variyam et al. (1999a,b) examine the process through which a mother's health and nutrition knowledge enters into good dietary intake and improved child welfare. They demonstrate that the role of the mother's education in the child's nutritional status is not as simple or straightforward as household theory makes it out to be. A reduced-form single-equation that relates a child's health status to the mother's health knowledge may overstate the effect of education, if the estimation procedure does not control for endogeneity issues surrounding education and knowledge. Indeed, Variaym et al. (1999b) show that once the information effect of schooling is taken into account, fat density does not significantly differ by education level. For example, Variaym et al. (1999a) control for endogeneity by explicitly modeling knowledge as a function of several demographic characteristics. Using data from the USDAs 1989–91 Continuing Survey of Food Intakes by Individuals (CSFII), and the companion Diet and Health Knowledge Survey (DHKS), the researchers find that:

- The mothers' health and nutrition knowledge have substantial effects on the diets of preschool children. A higher level of maternal knowledge translates into significantly lower intakes of fat, cholesterol, and sodium, and higher intakes of fiber among preschoolers.
- Maternal knowledge has no significant influence on calcium and iron intakes for children belonging to the same age groups.
- The influence of the mothers' knowledge on nutritional intakes diminishes for children who are 6 years or older. Fat, sodium, calcium, and iron intakes are not significantly affected by the mothers' knowledge. The knowledge effect on fiber is smaller for this group when compared with that on preschoolers.
- School-age girls have higher energy intakes from fat compared to boys.

- For preschoolers, the effect of the mother's employment status occurs mainly through nutrition knowledge, while for school-age children the effect occurs through time available for food preparation at home.
- School-age children whose mothers work part-time have significantly lower intakes of fat and saturated fat when compared to those with mothers who work full-time.
- Policies must target mothers with both young children and also school-age children.

The way in which information filters into a household's nutrient allocation also provides key insights into policy. Variyam et al. (1998, 2002) point out that the benefits of health and nutrition information may not be the same across different groups. For instance, a group with better education and a higher income may benefit more from programs such as the Nutrition Labeling and Education Act. The standard OLS estimates may not be able to capture these group differences, because the econometric methodology is restricted to analyze the data at the mean values. Consequently, Variyam et al. (2002) employ quantile regression techniques to examine this issue.

Why are quantile regression estimates better than OLS estimates? Variyam et al. (2002) provide the answer by demonstrating that, while the OLS estimates show that income has a beneficial effect on cholesterol intake, the quantile estimates show that much of the beneficial effect was located at the upper quantiles. The same is also true for intakes of saturated fat and total fat.

Similarly, while education is positively correlated with better diets, the beneficial effects are much greater for the upper quantiles, which are the riskiest group. Variyam et al. (2002) also show that for this riskier quantile group, there are black–white differences in cholesterol and fiber intakes. For instance, black men consume less saturated fat, but more cholesterol and less fiber than white men. Similar differences in intakes also exist between black and white women. Further, overall results show that the Hispanic segment of the US adult population has a better nutrient intake than non-Hispanic (also see Box 8.1).

BOX 8.1 DOES ACCULTURATION SOFTEN HISPANIC YOUTHS?

In the context of social customs and background, Mazur et al. (2003) ask an interesting question: given the evidence that adolescents in lower-income groups are twice as likely to be overweight when compared to those in higher-incomes, and given that Hispanics are disproportionately poor when compared to non-Hispanic white groups, how then is it possible for Hispanics to exhibit more nutritious diets? The researchers attribute this tendency among Hispanics to acculturation. The term "acculturation" refers to the process where individuals learn and modify their behavior as they adapt themselves to another culture. The modifications in the behavioral traits can include changes to values, norms, lifestyle, food, and diet. If low-income Hispanic youth are not obese or overweight, then Mazur et al. (2003) posit that low acculturation of this group to the mainstream must be related to low food insufficiency in this group.[a]

The researchers use data on 2985 Hispanic youths aged 4–16 from the third National Health and Nutrition Examination Survey (1988–94) to test whether acculturation influences food choices and insufficiency, alongside other determinants including income levels, maternal education, and employment status. The researchers collected information on nutrient intakes through interviews, and computed the consumption of calcium, folate, vitamin A, iron, zinc, and saturated fat. This information was combined with information on food insufficiency, based on the responses to a questionnaire. The language spoken at home by the head of the household was a proxy for acculturation. The researchers included a host of explanatory variables such as parents' country of birth, age and sex of the children, location, household headship, occupation of the head of the household, poverty index of the household, and social integration which was captured by the duration of residence in the current dwelling.

(Continued)

BOX 8.1 (CONTINUED)

The statistical results from regression estimates clearly indicate, as in previous studies, that household socio economic characteristics are important determinants of diet and food sufficiency among Hispanic youths. In particular, the study's main findings are that, holding other things constant, limited acculturation is associated with lower food insufficiency among poor Hispanic youths. Spanish language use by parents at home, which signals low acculturation, is associated with lower intakes of macronutrients, with diets having a lower percentage of energy from fat, and with less food insufficiency. The study also has other important results that complement previous literature: income levels shape dietary and food sufficiency outcomes, since lower-income households are at a greater risk of food insufficiency.

- Female-headed households overall have much lower average annual incomes than did other Hispanic households, and hence are at greater risk of food insufficiency.
- Children of agricultural workers had problems of food insufficiency, because agricultural income is often low and derived from part-time employment.
- Youths in households receiving food stamps experience both food insufficiency and inadequate intakes of vitamin A.
- Through directly comparable analysis of several key components of the diet and of food insufficiency, our findings underscore the importance of improving educational opportunities, job placement, and wages for members of lower-income households.

The policy implications from this study indicate that nonfood interventions have to be innovative, explore opportunities that include culturally specific programs, and target specific groups. Policies must focus on education, which is also a key driver of food choice. This is because Hispanic youths whose parents had low levels of education had a higher percentage of energy from saturated fat, for given levels of acculturation. Since culture-based buffering seems to be an important factor in determining the lives of Hispanic youths, policies must identify pathways to capitalize on the positive aspects of culture, because the benefits of culture-based protection diminish at higher levels of acculturation.

[a]Interestingly, first-generation Mexican–American immigrant adults consumed less fat and more fiber, and had a lower prevalence of overweight than did immigrants from subsequent generations (Mazur et al. (2003, p. 1120) and their references).

Finally, education and the process of information acquisition on better nutrition cannot be adequately overemphasized. Crutchfield, Kuchler, and Variyam (2001) show that in the United States, the benefits of information via labeling in the poultry industry can generate health benefits amounting to $62 to $125 million annually.

Aturupane et al. (2008) use data from Sri Lanka's Demographic and Health Survey, and stress the importance of quantile estimation in the context of socio economic background and nutrition intake. The researchers argue that OLS estimates can be misleading in predicting the effects of determinants at the lower end of the distributions of weight and height. For example, while OLS estimates show that on average girls are not nutritionally disadvantaged relative to boys, quantile estimates reveal that among children at the highest risk of malnutrition, girls are disadvantaged relative to boys.

Similarly, while OLS estimates show a strong association between expenditure per capita and nutritional improvement on average, the quantile estimates indicate that this association is not a significant determinant of child height or weight at the lower end of the distribution. Further, parental education, access to electricity, and the availability of piped water have larger effects on child weight and height at the upper quantiles than at the lower quantiles. Consequently, much in line with previous findings, the researchers call for targeted policy interventions.

ARE NUTRITIONAL OUTCOMES AGE-SPECIFIC?

If increases in income levels significantly reduce malnutrition, then income transfer programs should be able to achieve nutritional objectives easily. However, the solution is not that simple, for two reasons. First, the magnitude of the responsiveness of a 1% change in income on anthropometric indicators varies across studies, and second, the number of studies that link income effects on nutritional status including calories are few and far between. Sahn and Alderman (1997) take up this issue and examine the importance of age-specific effects on the demand for nutrients in Maputo, Mozambique. An important reason as to why some basic indicators, such as the mother's education, do not show any significant relation to improved anthropometric measures is the problem of aggregation that arises when all age-specific cohorts are lumped together in the estimation. Sahn and Alderman (1997) demonstrate that there are significant age differences in the determinants of height-for-age which do not get properly accounted for by models that fail to control for cohort-specific effects.

Using data from the Maputo Integrated Household Survey covering 1816 households, the authors estimate three sets of reduced-form models that take into account the fact that per capita expenditures are endogenous. They use a production function approach where calorie intake, birth weight, and health clinic visits are endogenous choices. Regional prices, distance to the clinic, mother's fluency with Portuguese, mother's place of birth, and child's place of birth were used as instruments. Data was also collected on child's gender and age, the presence of a father, and the availability of sanitation facilities.

Controlling for age-specific effects, the researchers find, adds a new dimension to policy analysis. For example, the results indicate that the impact of higher incomes on height-for-age is significant only for children 2 years of age and older. For younger children, the mother's education is a key driver of height-for-age. Policy proposals must be designed in such a way as to improve child care practices for mothers of children who are younger than 2 years. In particular, if a poor family has children in the highest risk age group, a transfer program that focuses on nutritional improvement would be more effective than a regular income transfer. Welfare improving schemes that focus on immunizations, prenatal care, weighing programs, and growth monitoring will provide the best results, alongside income transfers.

DOUBLE BURDEN AND SOCIO ECONOMIC DETERMINANTS IN INDONESIAN HOUSEHOLDS

Double burden arises when obesity and under-nutrition occur at the same time. The dual burden phenomenon is usually observed in the country-level data. However, in a recent study Roemling and Qaim (2013) find evidence of dual burden from Indonesian household panel data. In fact, they find overweight and underweight individuals living in the same households, which suggests nutritional inequality in intra-household resource allocation. In particular, Roemling and Qaim (2013) focus on mother–child pairs, because the most common and paradoxical combination within households is the simultaneous presence of underweight children and overweight mothers.[1] Their

[1]Jehn and Brewis (2009) also observe this paradoxical feature in US data among lower and middle income groups.

study tracks the dynamics of this phenomenon using panel data spanning 1993–2007, and finds that 17% of all households in Indonesia can be classified under this category. However, households do not stay in this category for long. The dual burden is a transitory phenomenon in Indonesia. While the dual burden phenomena started with the upper income quantiles, of late it has been observed mostly within the lower quantiles. Unfortunately, most households escape the double burden category and move directly into the overweight category. This means that most individuals who are likely to be obese tomorrow are today's underweight children. Location in urban centers, the number of children in households, and overall household consumption spending are the top socio economic drivers of nutritional inequality. Like previous studies in the literature, Roemling and Qaim (2013) find that the mothers' educational level significantly improves nutritional outcomes within households.

POLICY ISSUES FOR REDUCING UNDER-NUTRITION: AN EXAMPLE

Do large numbers of Americans eat food away from home (FAFH), and how does this behavior affect nutritional outcomes? Liu et al. (2013) examine this important behavior among Americans, using a large dataset from the 2008–09 Consumer Expenditure Surveys, which provide information on FAFH expenditure for over 11,000 households. The researchers examine many sub groups within the data: households with a husband–wife with and without children, households with single-parents with children, and single-member households. Using data on expenditure on breakfast, lunch, and dinner, Liu et al. (2013) find that most households consume lunch away from home. Further, total per capita expenditure is highest for dinners.

Most interestingly and importantly for policy purposes, the researchers uncover that FAFH now takes up half of the food dollars for American households.[2] The researchers attribute the rise in FAFH to many economic and social factors, such as household income, household size and composition, characteristics of the household heads such as their working hours, age, education level, race, ethnicity, and region of residence. US food including the Supplemental Nutrition Assistance Program (SNAP) also plays a role in FAFH behavior. Since FAFH may not always be a healthy option, policy makers are concerned that the rise in FAFH may lead to high medical and health coverage costs in the future. Understanding the drivers of FAFH by type of meal is crucial to mitigate future health costs, and Liu et al. (2013) uncover key socio economic determinants of this trend. The researchers find that FAFH is positively related to income.

Overall, the key determinant of FAFH turns out to be the allocation of time. The researchers find that, within the sample of husband–wife households without children, the probability of FAFH increases with the work hours of both husband and wife. This is also true with the work hours of the reference person among husband–wife households with children. In single-person households, work hours play an even more important role, increasing not only the probability of FAFH, but also conditional expenditures on FAFH breakfast.

[2]FAFH was about $433.5 billion in 2010, and was about 41.3% of total food expenditure, compared to 32.0% in 1980 (see Liu et al., 2013).

Most importantly, the researchers find that FAFH is strongly related to education among all types of households. This finding has implications for nutritional education. The authors suggest that policy makers can identify segments of the population that should be targeted for nutritional education. The target population should be single men and household members with busy working schedules, as well as those with college or higher education. These groups should be informed about the relatively higher levels of sodium, cholesterol, and saturated fats in FAFH meals, and given recommendations about healthy FAFH choices such as fruits, vegetables, milk, and oils, and educational messages about moderating consumption of fats, added sugars, and alcohol. The results also show that SNAPs effects on FAFH are negligible.

An important aspect that makes nutritional policies and programs difficult is the endogeneity between nutrition and economic growth. While economic growth can reduce nutrition, it is also possible that nutrition is essential for human capital, which is an important input in the aggregate production function and productivity growth. Hence, when we examine the determinants of nutrition and its effect on growth, we have to account for the endogeneity underlying these two variables. Linnemayr et al. (2008) allow for the potential of community and socio economic determinants and program interventions in Senegal. The nutrition intervention programs in Senegal are effective, as the programs help compensate the risks young mothers face. This is an important aspect of the problem, because children born to young mothers are highly prone to receiving low anthropometric scores. The programs also alleviated the problems faced by mothers with low social status. Nutrition programs that concentrated on the stresses of mothers were developed by the government and NGOs, and have helped children overcome disadvantages at birth. Linnemayr et al. (2008) is a good example of how socio economic determinants can be very useful pointers to a successful public policy in nutrition.

GENERATING ANALYTICAL RESULTS FOR POLICY DISCUSSIONS

The seminal work of Koenker and Bassett (1978) on quantile regression analysis has had a major influence in econometric estimation, spanning several research areas, and leading to many extensions.[3] For example, Variyam et al. (2002) note that the fat intake for men at the 90th percentile is considerably higher than the recommended level, and the cholesterol intake at the 7th percentile for women is 273 milligrams, much below the recommended level. Essentially, researchers have begun to look at the standard relation between the outcome variable y and the independent variables x at different points in the distribution of y.

The standard regression estimation provides estimates at the average of the distribution $E(y|x)$. Quantile regressions provide information about the intergroup difference along the whole distribution of y, rather than only at the average locations. From a policy angle, it is important to know where the between-group difference is located, and as an example, if too much fat intake takes place at the upper end of the distribution more than at the lower end of the intake distribution, then policy makers may be able to make informed decisions, rather than pursuing a broad-based approach.

[3]See Koenker and Hallock (2001) for a review of the developments and further references.

Let $y = x'\beta + e$ represent a standard linear model. Then the qth quantile regression estimator $\hat{\beta}_q$ is obtained by minimizing the following loss function with respect to the true β_q:

$$Q(\beta_q) = \sum_{i:y>x'\beta}^{N} q|y - x'\beta_q| + \sum_{i:y<x'\beta}^{N} (1-q)|y - x'\beta_q|$$

where q is a parameter which lies between (0,1), and where $Q(\beta_q)$ achieves symmetry when $q = 0.5$, and becomes asymmetrical when q approaches 0 or 1. Cameron and Trivedi (2010, p. 212) note that the quantile regression estimates are robust to the presence of outliers in the data, and can capture special features of the data at the extremes. Further, the technique has become widely applicable to many research fields including nutrition economics.

QUANTILE ESTIMATION USING STATA

Quantile regression estimates on socio economic determinants of nutrition intakes have become very popular in recent studies. We use the approach from Variyam et al. (2002) to motivate the example in this section. We present a simple data set and illustrate the usefulness of this regression approach, and implement the estimation using STATA. In the table below, we have observations on the macronutrient intakes of 34 households:

Obs	Energy	Fat	Chol	Fiber	H	Income	Age	Ed	W
1	3829	22	368	27	3	2580	26	12	1
2	3895	26	276	27	3	1830	29	12	1
3	3955	19	276	31.5	2	3430	31	10	1
4	3984	30	460	18	5	2120	32	11	0
5	3995	30	368	31.5	4	1980	35	12	0
6	4010	18	184	36	4	3600	36	10	0
7	4060	18	184	45	3	3330	42	11	1
8	4082	19	276	31.5	3	3400	44	12	1
9	4172	24	184	18	3	2690	47	12	0
10	4816	20	276	40.5	3	3250	51	13	0
11	4890	18	368	36	3	3690	26	15	1
12	4934	18	392	13.5	3	3470	29	12	0
13	5079	24	368	22.5	4	2280	31	12	1
14	5104	22	184	18	4	3220	32	10	1
15	5886	16	184	36	4	3600	35	11	1
16	5899	18	460	22.5	5	2410	36	12	0
17	6165	15	276	31.5	5	3720	42	10	0
18	6229	23	368	13.5	5	2370	44	11	0

Continued

Obs	Energy	Fat	Chol	Fiber	H	Income	Age	Ed	W
19	6295	23	276	22.5	3	2070	47	12	1
20	7827	15	368	36	3	4080	51	12	1
21	8129	21	368	22.5	3	2750	32	13	0
22	8814	21	368	36	6	4060	35	15	0
23	9690	17	460	27	5	2830	36	16	1
24	9735	25	368	22.5	3	2650	42	17	0
25	10,371	16	276	31.5	3	4030	44	17	1
26	10,372	16	276	31.5	3	3880	31	15	1
27	11,385	14	276	36	4	4330	32	16	1
28	11,497	12	276	31.5	4	4840	35	14	0
29	11,995	17	460	22.5	4	3170	36	15	0
30	12,990	14	276	31.5	3	3420	42	16	0
31	13,466	14	276	31.5	3	3830	44	17	1
32	13,594	12	276	22.5	3	4720	47	17	1
33	14,500	14	184	31.5	4	3900	51	18	0
34	15,906	21	276	27	4	4290	49	20	1

Data Description *We have information on:*
Energy: *Total energy consumed, measured in calories*
Fat: *Total saturated fat intake, measured in grams*
Chol: *Total cholesterol intake, measured in milligrams*
Fiber: *Total fiber intake, measured in grams*
H: *Size of the household*
Income: *Monthly income of the household, measured in dollars*
Ed: *Years of education of the head of the household*
W: *A binary variable representing race (W = 1 for whites, and W = 0 for non whites)*

We first estimate the following linear regression model:

$$\text{Energy} = a + b\,H + c\ \text{Income} + d\ \text{Ed} + e\ W + \text{disturbance term}$$

Our goal is to see which of the independent variables significantly affect the total nutrient intake, as measured in total calories consumed. We exploit the `regress` command in STATA:

```
regress energy h income age w, vce(robust)
```

where `vce(robust)` is the robust estimate of variance that takes into account some of the assumptions underlying the standard OLS model concerning the dependence of the independent variables with the disturbance terms.

The input line and STATA output are given below for the above regression:

```
. regress energy h income age ed w, vce(robust)

Linear regression                                      Number of obs =      34
                                                       F(  5,     28) =   67.29
                                                       Prob > F      =  0.0000
                                                       R-squared     =  0.8388
                                                       Root MSE      =  1616.2

------------------------------------------------------------------------------
             |               Robust
      energy |      Coef.   Std. Err.      t    P>|t|     [95% Conf. Interval]
-------------+----------------------------------------------------------------
           h |   404.9245   357.1885     1.13   0.267    -326.7429    1136.592
      income |   1.122582   .2939234     3.82   0.001     .5205068    1.724656
         age |    48.5694   50.65411     0.96   0.346    -55.19085    152.3296
          ed |    1029.98   102.2836    10.07   0.000     820.4616    1239.498
           w |  -360.8429   607.6877    -0.59   0.557    -1605.635     883.949
       _cons |  -13019.43   1981.946    -6.57   0.000    -17079.27   -8959.601
------------------------------------------------------------------------------
```

Note that both *ed* and *income* are significant variables and are directly related to energy intakes. The `qreg` option in STATA produces regressions for different quantiles. The estimation of the median regression is produced by the default `qreg h income age ed w` command:

```
. qreg energy h income age ed w
Iteration  1:  WLS sum of weighted deviations =  19377.793

Iteration  1: sum of abs. weighted deviations =  18927.284

Median regression                                    Number of obs =        34
  Raw sum of deviations    52020 (about 6165)
  Min sum of deviations 18927.28                     Pseudo R2     =    0.6362

------------------------------------------------------------------------------
      energy |      Coef.   Std. Err.      t    P>|t|     [95% Conf. Interval]
-------------+----------------------------------------------------------------
           h |   376.3618   429.6761     0.88   0.389    -503.7897    1256.513
      income |   1.282743   .4995349     2.57   0.016     .2594917    2.305993
         age |    47.5377   50.06492     0.95   0.350    -55.01563     150.091
          ed |   941.0422   151.0845     6.23   0.000     631.5597    1250.525
           w |  -174.5219   766.5413    -0.23   0.822    -1744.711    1395.667
       _cons |  -12078.08    2964.16    -4.07   0.000    -18149.89   -6006.275
------------------------------------------------------------------------------
```

The estimates of the quantile regression are very close to the linear model for *h*, *income*, and *age*. Once again, both *ed* and *income* are significant variables in the quantile regression. STATA allows for simultaneous estimation for different quantiles. Suppose we want to estimate the regressions for different quantiles, say $q = 0.25$, 0.50, and 0.75. This can be achieved through the **sqreg** command, as this procedure produces the variance–covariance matrix needed for hypothesis testing, as shown in Box below.

```
. sqreg energy h income age ed w, q(.25,.5,.75) reps(50)
(fitting base model)

Bootstrap replications (50)
----+--- 1 ---+--- 2 ---+--- 3 ---+--- 4 ---+--- 5
..................................................    50

Simultaneous quantile regression                    Number of obs =         34
  bootstrap(50) SEs                                 .25 Pseudo R2 =     0.5058
                                                    .50 Pseudo R2 =     0.6362
                                                    .75 Pseudo R2 =     0.7008

------------------------------------------------------------------------------
             |              Bootstrap
      energy |      Coef.   Std. Err.      t    P>|t|     [95% Conf. Interval]
-------------+----------------------------------------------------------------
q25          |
           h |   820.1326    622.527     1.32   0.198    -455.0561    2095.321
      income |   1.153049   .6872737     1.68   0.105    -.2547673    2.560866
         age |   18.88277   74.18404     0.25   0.801    -133.0763    170.8419
          ed |   935.9112    229.035     4.09   0.000     466.7543    1405.068
           w |   132.2481   974.4069     0.14   0.893    -1863.734     2128.23
       _cons |   -13460.4   4565.829    -2.95   0.006    -22813.08   -4107.722
-------------+----------------------------------------------------------------
q50          |
           h |   376.3618    575.605     0.65   0.519    -802.7117    1555.435
      income |   1.282743   .4716682     2.72   0.011      .316574    2.248911
         age |    47.5377   58.98472     0.81   0.427    -73.28702    168.3624
          ed |   941.0422   142.3929     6.61   0.000     649.3636    1232.721
           w |  -174.5219   794.3127    -0.22   0.828    -1801.598    1452.554
       _cons |  -12078.08   3275.858    -3.69   0.001    -18788.37    -5367.79
-------------+----------------------------------------------------------------
q75          |
           h |  -328.8738   472.5766    -0.70   0.492    -1296.903    639.1555
      income |   1.175301    .505829     2.32   0.028     .1391569    2.211444
         age |   70.87121   64.62571     1.10   0.282    -61.50856     203.251
          ed |   1027.882   126.3172     8.14   0.000     769.1328    1286.631
           w |  -1069.982   603.7009    -1.77   0.087    -2306.607    166.6433
       _cons |  -9746.797   2358.423    -4.13   0.000    -14577.81   -4915.787
------------------------------------------------------------------------------
```

The `reps(50)` command restricts the iterations to 50 replications within the STATA bootstrap estimation procedures. The coefficients vary across quantiles, and the results of the median regression are repeated when $q = 0.50$. The highly statistically significant variables (*income* and *ed*) have a much greater impact on *energy* at higher conditional quantiles of energy intake. OLS coefficients differ for the upper and lower quantiles. As Varyiam et al. (2002) point out, the quantile estimates depicting the quantile regression estimates have the capacity to capture the slope coefficients at different points in the distribution. This feature is particularly useful if the underlying data exhibits heteroscedasticity. Researchers often examine this feature by performing hypothesis tests on the equality of the slope coefficients derived from the different conditional quantiles.

Suppose we wish to test the equality of the coefficient of *income* from the quantile regressions with $q = 0.25$, $q = 0.50$, and $q = 0.75$. We can use the `test` command in STATA, which provides a Wald test, under the null that the coefficients are equal:

```
. test [q25=q50=q75]: income

 ( 1)  [q25]income - [q50]income = 0
 ( 2)  [q25]income - [q75]income = 0

       F(  2,    28) =    0.03
            Prob > F =    0.9693
```

The F-value indicates that the null hypothesis of equal slope coefficients cannot be rejected for our data. We can also test the equality of the *ed* coefficients:

```
. test [q25=q50=q75]: ed

 ( 1)  [q25]ed - [q50]ed = 0
 ( 2)  [q25]ed - [q75]ed = 0

       F(  2,    28) =    0.28
            Prob > F =    0.7565
```

Again, the null hypothesis of equality of slope coefficients cannot be rejected for this data. Cameron and Trivedi (2010, p. 226) note that the test of $C_{0.25,\ 2} = C_{0.75,\ 2}$ is similar to a test of heteroscedasticity independent of the functional form of the heteroscedasticity. We can use the **test** command in STATA in a similar fashion as before:

```
. test [q25]income = [q75]income

 ( 1)  [q25]income - [q75]income = 0

       F(  1,    28) =    0.00
            Prob > F =    0.9793
```

We reject the null hypothesis, which indicates that the variable *income* does not affect the location and the scale of *energy*. We leave the other tests of heteroscedasticity as an exercise for the reader.

Researchers often display quantile regression results graphically so as to compare the quantile estimates with the OLS coefficients and along with OLS confidence intervals. The `grqreg, cons ci ols olsci scale(0.75)` command in STATA provides the visualization:

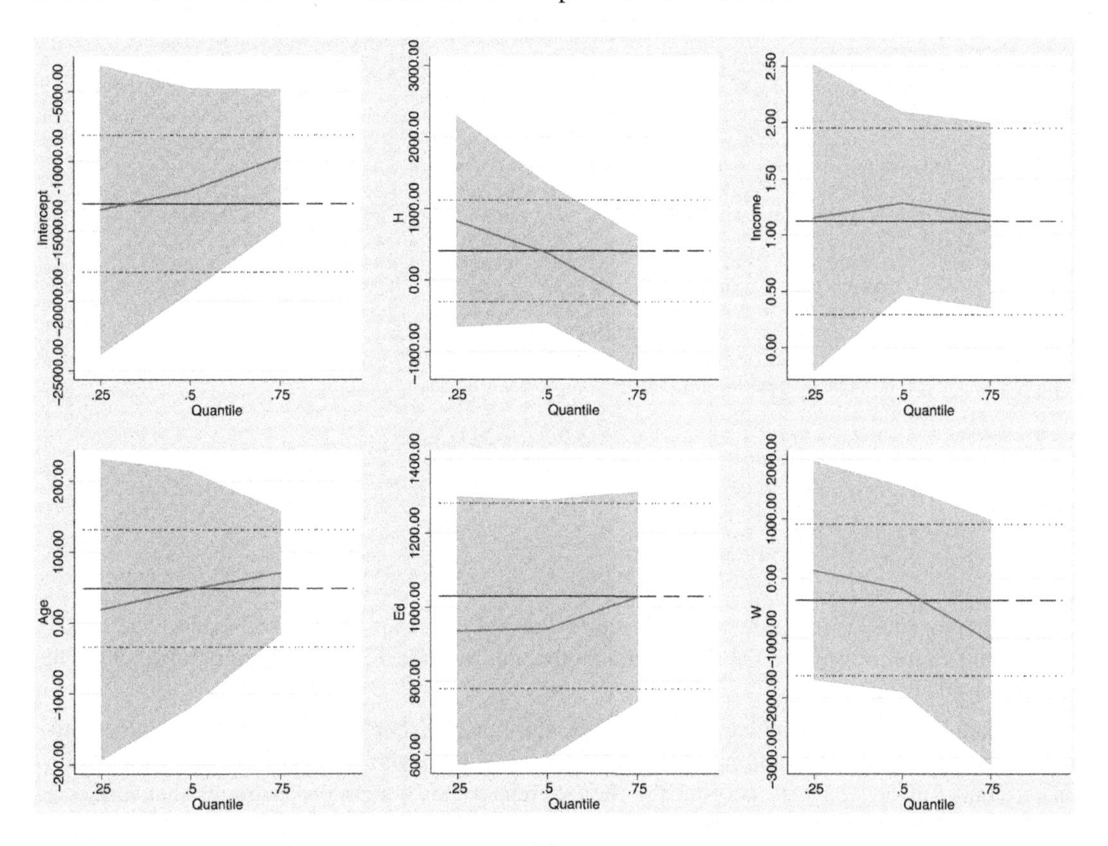

The `grqreg` procedure produces the graphs from STATA. The intercept term can also be included in the graph with the option `cons`. In addition, `ci ols olsci` produces the 95% confidence interval, the OLS coefficients, and their confidence intervals. The option `scale(1.1)` increases the size of the titles in the graphs along the axes.

In the graphs above, the horizontal lines are the OLS estimates which are constants, and hence, do not vary with the location on the x-axis; the confidence intervals also appear as dashed lines in each graph. The top right-hand plot shows that the coefficient of *income* is positive over the entire range, and has a larger effect in the median quantile. Similarly, the coefficient of *ed* is much smaller than the OLS estimate, for almost all of the quantiles, and is closer to the OLS coefficient around the upper quantiles. However, both the coefficients are within the confidence intervals of the OLS estimates, and hence, are not significantly different from the OLS coefficients.

We can also estimate a linear model and test for heteroscedasticity directly in STATA:

```
. reg energy h income age ed w

      Source |       SS       df       MS              Number of obs =      34
-------------+------------------------------           F(  5,    28) =   29.14
       Model |  380599127     5  76119825.3           Prob > F      =  0.0000
    Residual |  73134575.2    28  2611949.11           R-squared     =  0.8388
-------------+------------------------------           Adj R-squared =  0.8100
       Total |  453733702    33  13749506.1           Root MSE      =  1616.2

------------------------------------------------------------------------------
      energy |      Coef.   Std. Err.      t    P>|t|     [95% Conf. Interval]
-------------+----------------------------------------------------------------
           h |   404.9245   347.4829     1.17   0.254     -306.862    1116.711
      income |   1.122582   .4039784     2.78   0.010     .2950693    1.950094
         age |    48.5694   40.48795     1.20   0.240    -34.36641    131.5052
          ed |    1029.98   122.1834     8.43   0.000     779.6987    1280.261
           w |  -360.8429   619.9089    -0.58   0.565    -1630.669     908.983
       _cons |  -13019.43   2397.143    -5.43   0.000    -17929.76   -8109.108
------------------------------------------------------------------------------

. estat hettest h income age ed w,iid

Breusch-Pagan / Cook-Weisberg test for heteroskedasticity
         Ho: Constant variance
         Variables: h income age ed w

         chi2(5)      =     4.28
         Prob > chi2  =   0.5105
```

The `reg` procedure is used for the linear regression and the `estat hottest` is used to test for the presence of heteroscedasticity. The results tell us the null hypothesis of homoscedasticity cannot be rejected for this data.

The main emphasis in this chapter is to relate nutritional outcomes to socio economic determinants such as safe water, sanitation, culture, mother's education, and access to social capital. Policy makers almost always have to account for the inherent socio economic constraints that influence the effectiveness of specific interventions.

The STATA analysis of quantile regression employed in this chapter is very useful in estimating the effect of public policy, because the econometric methodology controls for the quantile group under consideration. The examples from STATA show that OLS estimates can be misleading in predicting the effects of determinants at the lower end of the distributions (say for weight and height of specific sub groups).

The implications of the research in this chapter and the STATA exercises is to stress the importance of targeted policy interventions. Consider a policy question that wishes to address the issue of eating away from home. Will such behavior affect nutritional outcomes of all households in a similar manner? Answers to such questions are empirical in nature, and the exercises in this chapter and the corresponding examples from the previous sections provide the necessary framework for understanding these complex and sometimes complicated matters.

CONCLUSIONS

Over the past three decades, country level studies have shown that nutritional status can be influenced by various socio economic factors (Scrimshaw et al., 1997; Delpench et al., 2000; Willey et al., 2009; Abubakar et al., 2012; Hien et al., 2008; Abuya et al., 2011; Semba et al., 2008; Arifeen et al., 2001; Ramli et al., 2009; Cohen et al., 1995; Dekker et al., 2010). Understanding the nature and magnitude of the contribution of these socio economic variables can help in the formulation of policies and programs that influence these factors, which in turn can improve the nutritional status of the population. Analysis of such factors helps policy makers to go beyond food-based interventions to affect changes in eating habits, other nutrition, and health-related behavior changes. We explore more of these factors in the rest of the chapters in this book.

EXERCISES

1. Develop a conceptual framework for your study country to identify the socio economic variables that are critical in determining children's nutritional status. Support this framework with relevant literature from the country.
2. Consider the data given in the text regarding nutrient intake. Variyam et al. (2002) point out that as income and age increases, the consumption of cholesterol decreases. Use the **sqreg** procedure and test the hypothesis for $q = 0.25$, 0.50, and 0.75. Are the slope coefficients the same across the quantiles?
3. Variyam et al. (2002) point out that as income and age increases, the consumption of fiber increases. Use the `sqreg` procedure and test the hypothesis for $q = 0.25$, 0.50, and 0.75. Are the slope coefficients the same across the quantiles?
4. Consider the data below with observations on energy and macronutrient intake for 36 households:

obs	Energy	BMI	Income	Ed	Age	H	G	w
1	2000.40	7.628	27.883	15	85	3	1	1
2	2001.70	7.539	11.628	8	83	1	1	0
3	2002.00	9.030	38.974	13	77	2	0	0
4	2002.00	8.126	36.861	14	76	2	1	0
5	2002.30	6.430	34.12	12	69	2	0	0
6	2808.30	7.780	126.317	17	78	2	0	1
7	2808.30	9.092	24.678	17	76	2	1	1
8	7008.40	7.710	14.562	12	65	2	1	1
9	7030.50	8.142	20.965	6	65	6	1	1
10	7031.80	7.836	105.461	17	65	2	0	0
11	7332.90	8.078	48.814	12	65	2	1	0
12	7337.70	8.637	0.069	15	65	3	0	0
13	7346.40	9.635	38.37	12	65	1	0	1
14	7359.00	7.415	67.299	17	65	2	1	0

(Continued)

Continued

obs	Energy	BMI	Income	Ed	Age	H	G	w
15	7367.10	8.046	36.114	12	65	2	1	0
16	7679.90	5.375	45.901	12	65	1	0	0
17	7683.40	7.188	23.853	13	65	1	1	1
18	7686.90	7.824	11.146	16	65	1	1	0
19	8001.50	11.263	5.736	12	81	1	0	0
20	8006.60	8.284	12.61	8	71	2	0	1
21	8010.20	6.057	6.955	11	78	1	0	1
22	8023.10	6.780	12.137	12	76	1	1	0
23	8028.60	10.217	0	12	67	2	0	1
24	8039.50	9.989	20.876	8	89	3	1	1
25	8046.50	9.668	31.111	16	74	3	0	1
26	8046.50	6.510	31.111	12	75	3	1	0
27	8062.10	8.513	12.231	12	78	1	1	1
28	8066.21	7.697	11	13	70	1	1	0
29	9832.80	9.269	17.961	16	70	1	1	1
30	9833.10	8.571	4.95	6	83	2	0	1
31	9833.50	10.079	13.545	9	72	1	1	0
32	9834.60	9.909	11.463	13	80	1	1	0
33	9834.70	8.615	8.5	3	67	2	1	1
34	9834.70	7.170	8.5	4	65	2	0	1

In the table above, *energy* indicates total energy in calorie consumption, *BMI* represents the *i*th individual's Body Mass Index. *Income*, *Ed*, *Age*, and *W* are as defined before. *G* represents the gender of the head of the household (1 = Female, 0 = Male). Perform sqreg and tests of equality of slope cofficients using `sqreg energy h income age ed w G, (.25, .5, .75) reps (50)`. In a different regression test, find out whether *BMI* also has the same features as *energy* to the selected independent variables.

CHAPTER

INTRA-HOUSEHOLD ALLOCATION AND GENDER BIAS IN NUTRITION: APPLICATION OF HECKMAN TWO-STEP PROCEDURE

9

Women receive hand hoes, and men receive tractors in DRC – Tractors rot behind their houses and women continue to use hoes.

—Mrs. Monique Kande (Executive Director, Institut Panafricain de Droits de la Femme et de la Bonne Gouvernance, Democratic Republic of Congo)

INTRODUCTION

Women, both as mothers and care givers, play a critical role in determining the nutritional status of the household. Thus, a major factor in explaining under-nutrition levels in developing countries is the status of women in terms of how much they have control over the resources which could be spent on proper nutrition for the members of their household. Studying the status of women and its implication for nutrition programs and policies requires addressing several critical questions (Kabeer, 1994; Fafchamps and Quisumbing, 1997; Quisumbing, 2012; Smith and Haddad, 2015; Brown et al., 2009).

How does studying the internal dynamics of decision-making help explain the women's and the nutrition status of the household members? Why is there gender bias in some societies, where women and female children get a lesser share of the nutrition compared to their male counterparts? Addressing these questions requires study of food security and nutrition challenges with a "Gender" lens (Babu et al., 1993; Brody et al., 2014; Brown, 2015). In applying a gender lens to nutrition economics, an important step is to understand the intra-household decision-making process (Brown et al., 2009). In this chapter we will review the models of intra-household resource allocation, take a critical look at the policy studies that have applied them, and use implications of such models to understand the gender bias in nutritional status from an applied policy perspective.

The nutritional status of a society is dependent on and aggregated from the nutritional status of its individual members. Among other factors, how resources are allocated among the individuals within a household will determine the nutritional status of the members. In designing nutritional interventions, it is important to understand how the benefits of these interventions accrue to various members of households. Household decision-making varies depending on the nature of the society and its organization, and differs from culture to culture. In some societies, e.g., women control the resources within households and make key decisions; in some men dominate the decision-making process; and in some other societies the decision is made jointly through negotiation and

Nutrition Economics. DOI: http://dx.doi.org/10.1016/B978-0-12-800878-2.00009-8

bargaining. Nutrition policy interventions could be made most effective in reaching the targeted members of the household through better understanding of the internal dynamics of decision-making within the households (Haddad et al., 1997; Haddad and Kanbur, 1990).

To begin with, if the individual levels of nutrition status and poverty are not measured, the levels of malnutrition and poverty could be grossly underestimated (Quisumbing et al., 1995; Quisumbing and Maluccio, 2003; Smith et al., 2003; Black et al., 2008). In terms of intra-household resource allocation to improve nutrition and health, studies have shown that providing resources to female members of households can significantly improve the welfare outcomes of children (Lundber, Pollak, and Wales, 1996). Targeted nutrition interventions could also have an adverse effect if the intra-household dynamics of resource allocation are not fully understood (Ruel and Alderman, 2013).

In this chapter, we analyze the factors that affect intra-household allocation of food and nutrients in order to develop cost-effective nutrition interventions. We address several issues that have received wide attention among the development community in recent decades. We ask how the members of the household distribute the resources available between men and women. We analyze the process that households use to redistribute the resources. We also ask how such resource allocation processes and decisions affect the welfare outcomes of women and children differently in terms of nutritional status, education, and health.

Intra-household allocation of resources largely depends on two broad processes: the resource generation process and the resource distribution process (Browning et al., 2014; Fafchamps and Quisumbing, 1997; Chiappori, 1988, 1992). The resource generation process depends on how various members of the household allocate their labor to production activities within and outside the home. The resource distribution process, on the other hand, depends on the pattern of consumption such as food and nonfood commodities and other investments in human capital development. We explore these processes in detail below and follow closely the developments in existing literature (Haddad et al., 1997; Doss, 1996; Browning and Chiappori, 1998; Alderman et al., 1995; Apps, 2003; Xu, 2004; Mangyo, 2005).

In general, in much of the economic analysis, the decision-making unit is considered to be the household. Models that are called "Unitary Models" treat a household as a single unit and study the welfare of the household and the process by which consumption and employment decisions are made. This is based on the assumption that individuals are alike in the economic unit called a family or household. Individuals within a household earn and spend incomes in a collective manner. They save and borrow for all the members together. They invest in various forms of resources and in accumulation of health and education jointly.

Household activities such as preparation of the meals, providing child care, and performing household chores are undertaken collectively. Decisions related to where to live and what jobs to choose are jointly decided upon by the household members in a unitary model of the household economy. Yet, these models have come under criticism, since they assume all the members of the household have the same tastes and preferences, which often is not realistic. This aspect of differing preferences has profound implications for the nutritional status of the members of the household and how nutrition programs and policies reach various members of the household (Alderman et al., 1995; Haddad et al., 1997).

The study of intra-household allocation of nutrients begins with relaxing the assumption that all members of the household have the same tastes and preferences, and relaxing the requirement that the household is the decision-making unit. The relaxation of these assumptions results in the study of the nature of the welfare of the individual members of the household when their preferences are in fact different. Such enquiries also help the analyst to address the issues related to the distribution of resources among the members of the household. Then, the maximization of welfare of the household members depends on the process of collective decision-making among the members of the household (Doss, 2001; Charman, 2008).

In Chapter 4, Microeconomic Nutrition Policy, on the microeconomics of nutrition, we showed that the utility of the household depends on the prices households face for the bundle of goods they purchase, and of course on the income they have to buy this bundle of goods. When the household members are assumed to have different tastes and preferences, household welfare will not only depend on prices and income, but also on how the household members decide on what to buy and for whom. The members of the household are then engaged in the process of negotiating and bargaining among themselves to come up with a collective decision. The models that analyze the welfare of households, taking into account the bargaining power of each household member, are called "collective models." In collective models, the responsibilities of each member of the household are assigned within and outside home. The outcome of these models will depend on who is assigned to work on what type of responsibilities, and the division of labor between the male and female members and younger and older members of the households. In agrarian societies in developing countries, for example, women spend a lot of their time in production activities which can compromise their time spent on child care (Malapit et al., 2013; Nelson, 2015). This has implications for the institutional well-being of mothers and children (World Bank, 2007; Brown, 2015; SPRING, 2014; Quisumbing and Meizen-Dick, 2012).

Analysis of the welfare of the members of the household, however, requires information on the individual members of the households. Individual levels of information on time and resources allocation, consumption patterns, food security, and nutritional status, for example, can help in the study of how the household's resources are allocated among the members of the household, and how the responsibilities are shared among the members of the household. Through the individual level of information one can study the intra-household dynamics of the household. Based on such individual levels of information, analysts are able to study the relationship between household incomes and expenditures, further helping in understanding the implications of such allocation of resources among the household members on their welfare outcomes, such as nutritional intake, health, and other individual level socioeconomic variables. Understanding such intra-household decisions can help in designing program interventions that improve the efficiency and equity in resource allocation (Okali, 2011; Fischer and Qaim, 2012).

ECONOMICS OF INTRA-HOUSEHOLD BEHAVIOR[1]

In studying household behavior, some preliminary concepts need clarification and attention. We begin with the concept of preference of individuals within the household. Let us assume that we

[1]The theoretical exposition of intra-household dynamics presented in this chapter is based largely on the review of several well-known publications including Browning et al. (2014), Haddad et al. (1997), Chiappori (1988), Fafchamps and Quisumbing (1997), Browning and Chiappori (1998), Quisumbing and Maluccio (2000), Haddad and Kanbur (1995), Lundber et al. (1996), Dercon and Krishnan (2000), Udry (1997), and Alderman et al. (1995). We keep the same notations of these authors to help readers to explore this literature further.

have one male (m) and one female (f) within the household. These two individuals consume public goods (Q) and private goods (q). Both Q and q are the vectors with N and n goods, respectively. The male and female members of the household share the private good such that $q_i = q_t^m + q_t^f$. Thus, the household members should allocate resources to N numbers of public good, and "n" numbers of q^m private good to male, and q^f private good to female members.

As seen in Chapter 4, Microeconomic Nutrition Policy, on the microeconomics of nutrition, the consumption of Q_1, q_1, and q_2 through allocation is made by increasing the utility of both male and female members of the households U^m (Q, q^m, q^f), and U^f (Q, q^m, q^f). It should be noted that since this is joint decision-making, the choice of one member of the household (e.g., q^f) will affect the utility level U^m, and vice versa. Given that the total income of the household is finite, the choice has to be made through negotiation among the family members on who will get what commodities, and in what quantities. Thus, the allocation among the members of the household will depend on how they negotiate and bargain, and the power they have in bargaining for their fair share of resources. When a particular member does not have the required power, his or her capacity to negotiate and bargain for their share of resources is reduced.

Formally, the utility relationship of the members of the household's male (m), and female (f), is normally written as follows (Chiappori, 1988; Browning and Chiappori, 1998):

$$U^m\left(Q, q^m, q^f\right) = U^m(Q, q^m) + \varphi^m U^f(Qq^f) \text{ (The utility of the male member of the household);}$$
$$U^f\left(Q, q^m, q^f\right) = U^f(Q, q^m) + \varphi^f U^f(Qq^m) \text{ (The utility of the female member of the household).}$$

In this formalization, the utility of one member is affected by the private consumption of the other members. This formulation is generally referred to in the literature as a special case of caring in which, for example, the male member cares about the allocation of resources of the female member only through the utility that the female member derives from her consumption of goods. The shares φ^f and φ^m are nonnegative, and less than unity. This implies that one member cares about the other member's consumption, but not as much as their own consumption. This formulation with caring for each other avoids challenges in accounting for externalities arising from the consumption of q^m and q^f to female and male members, respectively.

Application of collective models for analysis of nutritional outcomes could provide useful insights for program interventions and policy designs aimed toward nutritional improvements. One of the real world challenges in improving the nutritional status of the child is to decide which of the parents should allocate more time to child care. Given that child care involves spending time at home with the child, and assuming both male and female parent can contribute equally to child care, and can influence the nutritional strategy of the child equally, the collective decision-making involves finding out who will spend time with the child. Given that each parent will have a different wage rate in the market, the household decision normally involves sending the parent who can earn higher wages into the labor market, and allowing the parent with lesser income earning capability to stay at home to care for the child.

Using the above logic, the nutritional status of a child in a household could be through an output production process that uses purchased goods such as food, water, sanitation, and other health inputs q_N, and the time the parents' spend in order to care for the child T^m for the male parent, and T^f for the female parent. The child nutrition (N) production process could be denoted using the following relationship $N = f\left(q_N, T^m T^f\right)$; this intra-household production problem could be solved for the best allocation of resources toward child nutrition, given the time and income constraints of the household.

The household decision-making process often depends on the status of the members of the households, and how much power they have in bargaining toward achieving their goals. This bargaining power may vary depending on the culture, educational status, employment and income, and caring ability relative to others in the family. The factors that do not affect the preferences of the household members and budget constraints, but still influence the intra-household decision-making process, are called distribution factors. The fear of one member leaving the household can affect the bargaining power of the member who depends on the other member for survival. Such distributional factors are hard to measure, and yet could play an important role in the intra-household dynamics and decision-making processes. The household's total income, a combination of the contribution of the male (Y_m) and female (Y_f) members could be a distribution factor. This is because a higher relative contribution of male income in the total income could have an influence on the decision-making process, resulting in different outcomes than when the male and female members are contributing to the income pool in equal proportions.

Another useful concept in the study of intra-household dynamics is the "transferable utility" within the household. Household members could transfer their utility to others by redistributing the utility among them. In societies where women consume last and sacrifice food for the husbands and boys by neglecting a female child is a good example of transferable utility. We will apply this concept later in the chapter in the context of analyzing food and nutrition interventions for their distributive effects on the members of the household.

MODELS OF INTRA-HOUSEHOLD DECISION-MAKING

In this section we take a look at the literature on intra-household decision-making to understand various models that have been developed to explain the decision-making process within households. Three types of models are described in the literature (Haddad et al., 1997; Quisumbing and Maluccio, 2000); the unitary model, the noncooperative model, and the cooperative model, we describe each in turn.

The unitary model is based on the assumption that every member of the household has his or her own utility function. It also assumes that there is a household head who cares for the family, but acts as a benevolent dictator and makes decisions related to income distribution within the household.

The individual utility of each member of the household (V^i) is a function of the private good consumed by the individual (q^i) and the public good he or she consumes (Q). The household aggregate utility function U (Q, q) is maximized subject to income constraints ($P^{'}Q + P^{'}q \leq Y$), given the price ($P^{'}$) of the vector of public goods, and price ($p^{'}$) of the vector of private goods.

Solving this utility maximization problem requires the assumption of differentiability and concavity of preferences (see chapters: Microeconomic Nutrition Policy; and Macroeconomic Aspects of Nutrition Policy, on the microeconomics of nutrition) and will give the domain for Q and q, which depends on (P, p, Y), $Q = Q$ (P, p, Y), and $q = q$ (P, p, Y). These demand functions conform to the regular properties of the demand functions described in Chapter 5, Macroeconomic Aspects of Nutrition Policy, and are not affected by the distribution of income which is pooled among members of the households.

The second set of models that help describe intra-household decision-making are non-cooperative models. These models assume no binding agreement between the members of the households. Each individual member in the household acts on their own. In the noncooperative model, each member of the household maximizes his or her utility. Their utility maximization problems are as follows: the male member of the household maximizes $U^m(Q^m + Q^f, q^m)$ subject to $(PQ_m + pq^m = Y_m)$, whereas the female member of the household maximizes $U^f(Q^m + Q^f, q^f)$ subject to $(PQ^f + pq^f = Y_f)$. Q^f and Q^m are the contributions of the female and male members of the household, respectively, to the public good.

When both members of the household contribute to the public good, the demand functions of individuals for private goods (q) and public goods (p) are as follows:

$$q^{*m} = f(P, p, Y_m + Y_f);$$

$$q^{*f} = f(P, p, Y_m + Y_f); \text{ and}$$

$$Q = f(P, p, Y_m + Y_f).$$

Once again, the demand functions depend on the prices of private and public goods, and the pooled income of the individuals. These outcomes are not affected by the distribution of total income among the members of the households. The noncooperative model has also been applied to situations in which only one member of the household contributes to resource allocation.

The price ratio of the public good and the private good (P/p) should be greater than or equal to the ratio of marginal utility derived from these goods. In this type of model, redistribution of income among the members of the household will result in a new set of market demands. This is because, with redistributed income, the member who gets additional income may increase their demand for a specific commodity (which could result in increased or decreased nutrition outcomes), even when the other member may not alter the demand set with new levels of income.

The optimal demand for public and private goods in this case could be written as:

$$Q^* = f(P, p, Y_m + Y_f);$$

$$q^* = q^*_m + Y_{f/p} = f(P, p, Y_m) + Y_{f/p}.$$

The third group of models that help describe the allocation of resources within households is called the cooperative model. Note that in the noncooperative models, inefficient outcomes are possible and acceptable. Such outcomes can be problematic when nutrition interventions are constrained by social norms, such as women eating last in the household, and can lead to inefficient outcomes. Further, social norms that may also call for commitment of one member of the household without the context of the other toward long-term nutrition intervention, can reduce efficient outcomes.

In cooperative models outcomes are maximized through Pareto efficiency allocation. This type of allocation maximizes the utility of one member of the household holding the utility of the others. For example, in a household with two members, one male and one female, the utility of the male member is given by U^m (Q, q^m, q^f), and the utility of the female member is given by U^f (Q, q^m, q^f). This utility maximization problem to achieve Pareto efficiency is given by max U^m (Q, q^m, q^f) subject to $P'Q(P'(q^m, q^f) \leq$ Y, and U^f $(Q, q^m, q^f) \geq U^f$.

The Pareto solution frontier is a collection of all efficient allocation solutions $U^m = (P, p, Y, U^f)$ with varying levels of U^f. In the context of who in the household has more "say" in the allocation of resources, the above model could be modified by introducing coefficients that reflect this "power." These coefficients are called Pareto weights, and are assigned to each member of the household. For the two-member household discussed above, λ_m and λ_f are assigned as nonnegative numbers and are normalized such that $\lambda_m + \lambda_f = 1$. Then the model presented above could be rewritten as maximizing the joint utility of both male and female members weighted by their Pareto weights: $\lambda_f U^f(Q, q^m, q^f) + \lambda_m U^m(Q, q^m, q^f)$.

In this version of the model, it should be noted that the Pareto weights themselves are a function of the price of the goods, income, and other factors, such as the distribution factor (z) discussed earlier. The male or female member of the household would have absolute power when the weight of the other person is equal to 0. Further, when the weight of one person increases and that of the other person decreases, the utility of the person with the higher Pareto weight will increase, given that the total weight ($\lambda_m + \lambda_f$) is still equal to 1.

The implication of the model to nutritional outcomes is that when one can identify the distribution factor (z) that favors the Pareto weights of an individual which in turn can favor that person's utility outcomes, interventions could be designed to work in favor of that person by providing incentives that will increase the distribution factor z. The Pareto optimum outcomes thus obtained need not be long-lasting, since it may not be acceptable to both members and they will continue to search for a better solution from their own perspective. In reality, the final welfare outcome is based on the resource allocation that is obtained through continued negotiations and bargaining among the members of the household.

The household utility (U^H) maximization with the Pareto weights could be written as $U^H(Q, q, \lambda_m(P, q, Y, z))$. Incorporating the Pareto weights to individual utility functions, the utility maximizing problem becomes, $\max\{\lambda_f(P, p, Y, z)U^f(Q, q^m, q^f) + ((1 - \lambda_f)(P, p, Y, z))U^m(Q, q^m, q^f)\}$ subject to $(q^m, q^f = q)$. Thus, in this formalization of the model, prices and income enter the decision-making process through the Pareto weights, which are normally developed as an index in applied studies. Also, in this model formulation the distribution factor plays a critical role in deciding the value of the Pareto weights. Changes in prices and income can result in various levels of Pareto efficient allocation, as they will affect the Pareto weights.

A fourth type of model incorporates bargaining in the process of decision-making within the households. They are considered as an improvement over the cooperative models, as they give more specific predictions on the powers of decision-making. In the bargaining models, each individual is assumed to have a "threat point" V^i, in addition to the assumption of the individual utility function. The "threat point" indicates the level of utility that an individual can obtain, when there is no consensus with the decision-making partner.

A major assumption made in the bargaining models is that the utility levels of the individual at the end of the decision-making process is Pareto efficient, and that the point representing the utility levels (U^{m*}, U^{f*}) are on the utility possibility frontier. The role of threat points in the bargaining models is to influence the location of the final utility levels, and the process of influence is guided by the theory of bargaining. The threat point set $V(V^m, V^f)$ is expected to be in the Pareto efficient allocation, to ensure the individuals in a household reach an agreement.

Several factors affect the level of threat point of the individuals in the household. In some cultures the fear of divorce and the social shame attributed to it could work against one or both members of the household. In cultures where the male members of the household own the property and the property could only be inherited by males, female members of the households may be left with no bargaining power, resulting in very low threat points in intra-household distribution. Thus, the behavior of the household members is influenced by the level of threat points.

In what follows we review selected recent applied studies that have used these models to explain nutritional outcomes, and further show how policy analysis and applications could be undertaken using household survey data.

ARE GIRLS EATING MORE? A REVIEW OF SELECTED STUDIES

Promoting gender equality is listed as an important concern in the third Millennium Development Goal, and it is also one of the major goals for the newly established Social Development Goals listed in Chapter 2, Global Nutrition Challenges and Targets: A Development and Policy Perspective. Dercon and Singh (2012) note that while in recent years gender inequality has declined, particularly in education, gender bias still persists in other dimensions. Dercon and Singh (2012) present evidence of gender bias using data from four countries: Ethiopia, India, Peru, and Vietnam. Dercon and Singh (2012) provide an interesting approach and framework, which indicates gender bias across different cultures and socioeconomic contexts in developing countries.

Dercon and Singh (2012) use a very unique data set from the Young Lives Cohort study. The data covers about 12,000 children over three rounds of data collection: in 2002, 2006, and 2009. Further, the data also covers both rural and urban areas; Dercon and Singh (2012) supplement the data with other related information, and generate a 13-indicator multidimensional scale to evaluate their research questions.

Dercon and Singh (2012) use three anthropometric measures to evaluate nutritional status. Most importantly, the researchers find that there is a striking pro-female bias in nutritional status in all four countries.

The height-for-age, weight-for-age, and the BMI-for-age all reveal a significant pro-girl bias in all four countries. Indeed, the researchers, along with Dasgupta (2012) and Udry (1997), point out that the gender bias against girls, commonly found in the literature, is no longer evident with recent results and data sets. This line of research enquiry also provides us with a lot of related results for policy:

- Maternal education is a key driver in reducing gender bias in nutrition and educational attainment.
- Children from poorer households, both boys and girls, are less likely to be enrolled in schools.

It is difficult to find a common set of causes of gender bias across different countries. Gender biases are often context-dependent and go beyond the dimensions limited to standard data collection. For example, in India, while gender inequality has fallen substantially in school enrollment, there has been an unequal and sharp increase in pro-boy enrollment in private English-medium schools. Effective policy requires an understanding of specific institutional contexts, and an understanding of specific biases.

Further, evidence of gender bias has been difficult to find, primarily due to insufficient data, and as Dercon and Singh (2012) point out, the available data up to this point have been large surveys, which view these matters rather narrowly. Another important problem, as Mitra and Rammohan (2011) point out, is that of data truncation that arises due to self-selection. That is, Mitra and Rammohan (2011) observe, that even though the problem of "missing women" is well recorded, there is very little evidence to support gender bias in health or nutritional status.

Mitra and Rammohan (2011) point out, correctly, that this lack of evidence on nutritional gender bias could be attributed to data truncation. That is, girls who would have been otherwise discriminated against have died as victims of female infanticide. Hence, there is a problem of self-selection that creates a downward bias on the number of girls born alive, who hence are more valued, relatively, and thereby well-fed. Consequently, one must control for sample selection, to properly account for nutritional differences across genders.

Using a sample of 16,652 children from the National Family Health Survey (NFHS) data, Maitra and Rammohan (2011) reject sample truncation and self-selection. As in Decron and Singh (2012), in this study as well, there is no evidence of gender bias in survival outcomes or in short-term nutritional weight-for-height z-score (WHZ). However, Maitra and Rammohan (2011) do find better height-for-age (HAZ) outcomes for boys relative to girls, indicating better long-term nutrition for boys.

Further, Maitra and Rammohan (2011) also demonstrate families in poorer quantiles are most severely affected in survival and nutrition for all children. Additionally, the researchers also note large regional disparities in all outcomes relating to infant mortality, HAZ, and WHZ. This is an important finding, which complements the well-known North–South divide in India. That is, the north-western states such as Uttar Pradesh, Bihar, Punjab, and Haryana have high female mortality, compared to the south-eastern states of Andhra Pradesh, Kerala, Karnataka, Tamilnadu, and also West Bengal. For instance, according to the researchers' estimate, a child from Kerala, in the south, has a 5% higher probability of survival, relative to a child born in Uttar Pradesh, in the north.

While there is no evidence of nutritional differences across gender at the overall aggregate data level, Maitra and Rammohan (2011) are able to observe substantial gender differences, based on regional classification. WHZs for example, are lower for boys in south-eastern states, whereas they are better for boys in Haryana, Punjab, and Assam, located in the north. Similarly, Haryana also exhibits lower HAZ for girls. These differences are important for policy, because these differences occur even in an affluent state like Punjab. Similar to almost all studies in the literature, Indian data exhibits the importance of maternal education, and the adverse effects on survival and nutrition in the lower wealth quantile groups (Box 9.1).

BOX 9.1 GENDER-BIAS IN 19TH CENTURY ENGLAND

Horrell and Oxley (2015) observe that factory labor by children of 9–12 hours per day in 19th century England was excessive and stunting, and it was particularly sordid for girls relative to boys. The authors examine data on 16,402 children working in northern English textile factories in 1837, and show that by modern height standards, there was substantial gender-bias.

The authors demonstrate the gender difference after accounting for occupational sorting, differential susceptibility to disease, and poorer nutrition for girls. Disproportionate stunting from the effects of nutritional deprivation, and type and amount of work undertaken, are also included in the analysis. These are important considerations, because the main culprit behind disproportionate stunting in girls is household work.

It is possible that the observed stunting among girls can be due to occupational sorting, if healthier boys, but not girls, self-sorted themselves into factory occupations. However, the study shows that it was the less strong boys who sorted into factory work, because alternative employment in mining was even more physically demanding.

The authors argue that if such sorting occurred among boys, then it must be less likely among girls, who faced fewer job alternatives. Hence, occupational sorting offers little to explain the inferior heights of factory girls. Next, were girls more susceptible than boys to poor sanitation and associated disease? The authors' findings with data indicate that differential resistance to disease is not the main driver of the gender disparity in heights.

Did girls eat less than boys? The regression results from the study indicate no relationship between the percentage of females in the household and calorie or protein availability. Thus, there is no indication of any substantial difference in food intake of these factory girls and boys. Similarly, the question about the effects of nutritional deprivation can also be dismissed, because the daily difference in energy needs is just five calories, which cannot explain the large difference in stunting.

Consequently, after ruling out these possibilities, the authors posit that girls had less genuine leisure than boys, since girls were expected to contribute to household labor in addition to paid work. Although they brought equal economic value as their brothers, these girls had lower net nutritional intake, the main culprits being factory work and unpaid household work, or what is termed as "double burden." The double burden arises when girls are in both factory work and unpaid domestic labor, which increased physical demands on girls, with no compensating nutrition. This observed double burden constitutes gender bias.

POLICY APPLICATIONS OF STUDIES ON INTRA-HOUSEHOLD RESOURCE ALLOCATION

Public policy is often designed in terms of in-kind transfers, grants, supplemental aid, and unconditional and conditional cash transfers. The nutritional program under Supplemental Nutrition Assistance Program (SNAP) known as the Woman, Infants and Children (WIC), and the AID to Families with Dependent Children (AFDC) are conditional in-kind transfers in the United States. Public policy is now more geared toward providing conditional in-kind transfer contributions to women, because there is growing evidence that indicates that allocations made to women are more effective.

Cash transfers have also raised important questions for policy makers. Should governments adopt cash transfer programs, and will these transfers achieve the desired results? Should the transfers be conditional or unconditional? Duflo (2003) examines an interesting data set from South Africa to answer several important policy concerns. The South African data set is unique because it provides information about a pension program which is universal and noncontributory. Does the South African pension program improve children's nutritional status? In order to answer this question, Duflo (2003) compares the weight-for-height measure of children from households without any eligible members with those from households with an eligible man, and finally, with those in households with eligible women.

Duflo (2003) finds that pensions received by men do not influence the nutritional status of boys or girls. Pensions received by women also do not affect the outcomes for boys, but increase the weight-for-height for girls by 1.19 standard deviations. The above finding tells us that cash transfers work in special ways, depending upon the gender of the recipient. This result is also observed for comparisons dealing with height-for-age. Duflo (2003) notes that the weight-for-height measure responds quickly to changes in a household's income. However, a child's height is a result of past and current nutrition. This feature allows Duflo (2003) to formulate the following hypothesis: older children living in pension eligible households should be smaller than those in noneligible households, while for younger children, the difference in height between the two households should be smaller. Once again, Duflo's results indicate that pensions received by men do not influence the height of either boys or girls. Pensions received by women do not influence boys' heights, but increase girls' heights by 1.16 standard deviations.

Most interestingly, Duflo (2003) also considers the household's composition, and estimates the effect of pension eligibility of the paternal and maternal grandmother and grandfather. Once again, the eligibility of the mother's mother had the strongest effect on the nutritional outcomes of girls. Duflo (2003) thus points out that the efficiency of cash transfer programs depends upon the way they are administered. As to why grandmothers prefer girls is still an open question.

Duflo's results are also important for Brazil's nutrition policy programs. Thomas (1990) examines Brazilian data and finds that families do not pool their unearned income, or transfers, toward equal allocation. Rather, mothers prefer to direct resources toward daughters, and fathers toward sons.

The Brazilian data is used to test the equality of the slope coefficients between maternal and paternal regressions of unearned income on nutritional outcomes. Thomas (1990) rejects the null hypothesis of equality, and shows that these results are consistent with differential intra-household preferences. Indeed, this line of research shows that resources in the hands of women have a bigger impact on a family's health.

Duflo (2003) also presents a related result with respect to savings: the propensity to save out of a man's pension is much lower than the propensity to save out of a woman's pension income. Taken together, these results indicate that households are not a collective unit, or a single entity. Individual preferences and bargaining power are crucial determinants of intra-household allocations.

Dercon and Krishnan (2000) also examine the issue of consumption smoothing in the presence of exogenous shocks, for an interesting data set from rural Ethiopia. The researchers develop an intertemporal model of households to capture nutritional smoothing over time. If the households act as cooperating units, then risk sharing is likely to spread out individual shocks. The sharing rule will not affect individual outcomes. However, if individual preferences, bargaining power, and dictatorial allocations exist, then the individual outcomes are likely to be skewed in some fashion.

Dercon and Krishnan (2000) examine the difference in outcomes for males and females, using a Quetelet index, which reveals huge variations across gender. The index is particularly biased against women in poorer households, and for those in the southern part of Ethiopia.

Dercon and Krishnan (2000) note that there is no evidence of risk pooling in poor households, if there are illness shocks to women. The differences in the ages between husband and wife, and the rules governing divorce settlements, determine intra-household allocations. A wife's allocations were larger if a household in the south had a landholding. This result once again contradicts the notion of the household as a unitary entity.

In-kind transfers have also been examined for similar concerns. Jacoby (2000) uses data from a school feeding programmer (SFP) in the Philippines to test for an intra-household Flypaper Effect. If the household behaved as a cohesive unit, then altruistic parents will adjust calorific allocations in response to the program. This implies that SFPs should have no major impact on child nutrition. Using a difference-in-difference estimator, Jacoby (2000) shows that there is a Flypaper Effect, and that the SFP improves calorific intakes and health outcomes in the Philippines. From a public policy point of view, in-kind transfers in the form of SFPs are likely to be successful in terms of children's health outcomes.[2]

GENERATING ANALYTICAL RESULTS

A substantial portion of this line of literature exploits using parametric estimation of binary outcome models with selection. The Heckman selection model provides a two-step estimator that is usually applied to nutrition and health economics. In all of these applications, the outcome variable, say y_i, takes two values, such that:

$$y = \begin{cases} 1 & \text{with probability } p \\ 0 & \text{with probability } (1-p) \end{cases}$$

A distribution function describes the behavior of p through different functional forms. Logit and probit models are used very extensively in the literature to capture the distribution functions of the outcome variable, which is the selection equation. A regression model is developed by introducing exogenous variables and parameters, such that:

$$p_i = Pr(y_i = 1|\boldsymbol{x}) = F(\boldsymbol{x}_i'\beta)$$

where $F(.)$ is the cumulative distribution function, usually modeled as:

$$\text{Logit}: F(\boldsymbol{x}_i'\beta) = \frac{e^{\boldsymbol{x}_i'\beta}}{1 + e^{\boldsymbol{x}_i'\beta}}$$

Or:

$$\text{Probit}: \phi(\boldsymbol{x}_i'\beta) = \int_{-\infty}^{\boldsymbol{x}_i'\beta} \phi(z)dz$$

where $\phi(z)$ is the standard normal cumulative distribution function. The estimation of the selection equation is usually achieved through maximum likelihood methods, which generate the parameter $\hat{\beta}$, by maximizing the likelihood function:

$$\sum_{i=1}^{n} \left[y_i \ln F(\boldsymbol{x}_i'\beta) + (1 - y_i) \ln (1 - F(\boldsymbol{x}_i'\beta)) \right]$$

[2]We explore this issue in Chapter 11, Methods of Program Evaluation: An Analytical Review and Implementation Strategies, under program evaluation, and in greater detail in Chapter 13, Economics of School Nutrition: An Application of Regression Discontinuity, which deals exclusively with school feeding programs.

Consider the following system of two equations:

$$y_j = \boldsymbol{x}_j\beta + u_{1j} \tag{9.1}$$

and

$$z_j\beta + u_{2j} > 0 \tag{9.2}$$

where (9.1) is a regression equation, and (9.2) is a selection equation, with disturbance terms u_{lj} and u_{2j}. The two-step procedure begins with the estimation of the selection equation. Consider the probit estimates of (9.2):

$$\Pr(y_j > 0)z_j) = \Phi(z_j\gamma)$$

The estimates are used to compute the hazard ratio, the Inverse Mills ratio, or λ_j for each observation j as follows:

$$\lambda_i = \frac{\phi(z_j\hat{\gamma})}{\Phi(z_j\hat{\gamma})}$$

where ϕ is the density function form normal distribution.

The regression Eq. (9.1) is now estimated in the second step, after including λ_j. This yields another parameter $\beta_\lambda$associated with λ_j. In addition, the disturbance terms have the following assumptions: $u_1 \sim N(0,\sigma)$ and $u_2 \sim N(0,1)$ and $corr(u_1,u_2) = \rho$. These can also be calculated after estimation:

$$\hat{\rho} = \frac{\beta_\lambda}{\hat{\sigma}}$$

where

$$\hat{\sigma}^2 = \frac{e'e + \beta_\lambda^2 \sum_{j=1}^{n}(\beta_\lambda(\beta_\lambda + \hat{\gamma}z_j)}{n}$$

IMPLEMENTATION IN STATA

In this section, we develop an example using STATA, to illustrate the estimation of nutritional linkages to child mortality, after controlling for selection. Our example is motivated by issues and methods presented by Maitra and Rammohan (2011). For illustrative purposes, consider the following information on 35 households:

obs	*si*	*whz*	I_1	I_2	C_1	C_2	H_1	H_2	H_3	H_4
1	0	0.705	3	0	1	0	7.61	10.02	0	0
2	0	0.705	3	1	0	0	8.28	10.02	1	0
3	0	0.640	2	0	1	0	6.82	9.43	0	0
4	0	0.705	2	1	0	0	6.91	10.46	1	0
5	0	0.705	2	0	0	0	7.92	10.02	0	0

(Continued)

Continued

obs	*si*	*whz*	I_1	I_2	C_1	C_2	H_1	H_2	H_3	H_4
6	0	0.705	3	1	0	0	7.39	10.22	1	0
7	0	0.640	2	1	0	0	7.84	9.77	0	0
8	0	0.705	2	1	0	0	6.87	10.02	0	0
9	0	0.705	1	0	0	1	7.42	10.46	1	0
10	0	0.705	1	1	0	0	7.31	9.77	0	0
11	0	0.675	3	1	0	0	7.48	9.77	0	0
12	0	0.675	2	1	0	0	7.89	9.43	1	0
13	0	0.705	2	0	1	0	6.80	8.29	0	1
14	0	0.728	3	0	1	0	7.92	10.22	0	0
15	0	0.705	3	0	1	0	7.59	10.02	1	0
16	1	0.705	2	0	1	0	7.57	10.02	0	1
17	1	0.705	3	0	0	0	7.19	10.22	0	0
18	1	0.705	3	1	0	0	7.34	9.77	0	1
19	1	0.705	3	0	1	0	7.80	9.77	0	0
20	1	0.675	3	0	1	0	8.09	10.22	0	0
21	1	0.675	3	1	0	0	7.77	10.46	0	0
22	1	0.675	3	1	0	0	7.24	9.77	0	1
23	1	0.705	2	0	1	0	8.12	10.22	0	0
24	1	0.728	2	1	0	0	7.70	10.02	0	0
25	1	0.675	5	0	1	0	8.75	10.46	1	0
26	1	0.750	2	1	0	0	8.00	10.82	0	0
27	1	0.705	3	1	0	1	9.02	9.77	0	0
28	1	0.640	1	0	0	0	7.07	9.43	1	0
29	1	0.675	3	0	1	0	7.83	9.77	0	1
30	1	0.705	3	0	1	0	7.53	10.02	1	1
31	1	0.675	3	0	0	0	7.12	10.02	0	0
32	1	0.728	1	1	0	0	8.05	10.22	0	0
33	1	0.705	3	0	1	0	7.41	10.02	1	1
34	1	0.705	1	0	0	0	8.28	10.02	1	0
35	1	0.599	1	0	1	0	7.09	8.92	0	1

The variables used for this analysis presented in the first row are:
si = A binary variable indicating that the child in household i *is "Alive" at the time of the survey implemented 60 months after the child was born.*
whz = The i*th child's WHZ score, which measures the child's weight according to height.*

This indicator has been used to monitor the growth of children, and is typically regarded as a measure of short-term rather than long-term health status. Children with low weight-for-height are considered *wasted.*

Child-specific variables:

obs = Observation for the ith child in the sample.
I_1 = Birth-order of the child in the household.
I_2 = Gender of the child, a binary variable (female = 1; male = 0).

Health inputs:

C_1 = Whether the child was born in a private hospital (1 = yes; 0 = no).
C_2 = Whether the household has access to private toilets (1 = yes; 0 = no).

Household characteristics:

H_1 = Number of years of mother's education.
H_2 = Log of the household's total income earned over the last 10 years.
H_3 = Whether the mother works outside home (1 = yes; 0 = no).
H_4 = Whether the mother has heard of oral rehydration salts (ORS) (1 = yes; 0 = no), where this variable represents the mother's knowledge of ORS.

The use of ORS is described by UNICEF as the best way to combat dehydration caused by diarrhea (Maitra and Rammohan, 2011).

In order to test whether a nutritional outcome such as WHZ is affected by the survival probability of the child, we have to estimate two equations, one that captures the survival probability, and another that connects the exogenous variables to WHZ. We begin with the **summarize** command and the output from STATA:

```
. summarize
```

Variable	Obs	Mean	Std. Dev.	Min	Max
obs	35	18	10.24695	1	35
si	35	.5714286	.5020964	0	1
whz	35	.6928	.0294626	.599	.75
i1	35	2.4	.8811757	1	5
i2	35	.4285714	.5020964	0	1
c1	35	.4	.4970501	0	1
c2	35	.0571429	.2355041	0	1
h1	35	7.629143	.5200813	6.8	9.02
h2	35	9.938857	.4592562	8.29	10.82
h3	35	.3142857	.4710082	0	1
h4	35	.2285714	.426043	0	1

For purposes of comparison, we first examine the following regression:

$$\text{Whz} = a + b_1H_1 + b_2H_2 + b_3H_3 + b_4H_4 + b_5C_2$$

The regress command and the output from STATA are given below:

```
. regress whz h1 h2 h3 h4 c2

      Source |       SS       df       MS              Number of obs =      35
-------------+------------------------------           F(  5,    29) =    2.07
       Model |  .007758247     5  .001551649           Prob > F      =  0.0983
    Residual |  .021755348    29  .000750184           R-squared     =  0.2629
-------------+------------------------------           Adj R-squared =  0.1358
       Total |  .029513595    34  .000868047           Root MSE      =  .02739

------------------------------------------------------------------------------
         whz |      Coef.   Std. Err.      t    P>|t|     [95% Conf. Interval]
-------------+----------------------------------------------------------------
          h1 |   .0042675   .0102409     0.42   0.680    -.0166776    .0252125
          h2 |   .0328928   .0120616     2.73   0.011     .0082241    .0575615
          h3 |  -.0043918    .010153    -0.43   0.669    -.0251569    .0163733
          h4 |   .0061894   .0124525     0.50   0.623    -.0192789    .0316577
          c2 |   .0064857   .0209226     0.31   0.759    -.0363059    .0492772
       _cons |   .3329209   .1232089     2.70   0.011     .0809304    .5849114
```

Besides the total household income (H_2) and the constant terms, none of the other variables are significant. We now incorporate the selection constraint using a probit model for survival. The procedure involves two steps. In the first step, a probit equation is fitted to capture the likelihood of survival. The inverse Mills ratio is computed for each observation, and then in the second step, the same regression equation is estimated with the selection term (or the inverse Mills ratio) as an additional independent variable.

We implement the first-stage estimation using the `probit` command in STATA:

```
. probit si i1 i2 c1 h1

Iteration 0:   log likelihood = -23.901784
Iteration 1:   log likelihood = -21.663419
Iteration 2:   log likelihood = -21.656977
Iteration 3:   log likelihood = -21.656976

Probit regression                                 Number of obs   =         35
                                                  LR chi2(4)      =       4.49
                                                  Prob > chi2     =     0.3438
Log likelihood = -21.656976                       Pseudo R2       =     0.0939

------------------------------------------------------------------------------
          si |      Coef.   Std. Err.      z    P>|z|     [95% Conf. Interval]
-------------+----------------------------------------------------------------
          i1 |    .113586   .3009436     0.38   0.706    -.4762526    .7034247
          i2 |  -.7127836   .6451503    -1.10   0.269    -1.977255    .5516879
          c1 |  -.2319729   .6748081    -0.34   0.731    -1.554573    1.090627
          h1 |   .7726133   .5003812     1.54   0.123    -.2081158    1.753342
       _cons |  -5.567426   3.697164    -1.51   0.132    -12.81373    1.678881
------------------------------------------------------------------------------
```

Although the signs of the variables are interesting, the statistical z values indicate that none of the variables are significant. We now proceed to compute the inverse Mills ratio and the second-stage regression, using predict and the generate commands:

```
. predict xb
(option pr assumed; Pr(si))
. generate invmills = normalden(xb)/normal(xb)
```

Notice that we explicitly incorporate the invmills variable in the augmented regression below. Recall that this was not possible with our original regression estimation.

```
. regress whz h1 h2 h3 h4 c2 invmills

      Source |       SS       df       MS              Number of obs =      35
-------------+------------------------------           F(  6,    28) =    1.72
       Model |  .007964217     6  .00132737            Prob > F      =  0.1521
    Residual |  .021549377    28  .000769621           R-squared     =  0.2698
-------------+------------------------------           Adj R-squared =  0.1134
       Total |  .029513595    34  .000868047           Root MSE      =  .02774

------------------------------------------------------------------------------
         whz |      Coef.   Std. Err.      t    P>|t|     [95% Conf. Interval]
-------------+----------------------------------------------------------------
          h1 |   .0102516    .015537     0.66   0.515    -.0215745    .0420777
          h2 |   .0331139   .0122243     2.71   0.011     .0080735    .0581542
          h3 |   -.003837   .0103394    -0.37   0.713    -.0250163    .0173423
          h4 |   .0070228   .0127153     0.55   0.585    -.0190233    .0330689
          c2 |   .0064888   .0211919     0.31   0.762    -.0369209    .0498985
    invmills |   .0450944   .0871682     0.52   0.609    -.1334616    .2236504
       _cons |   .2631434   .1837571     1.43   0.163    -.1132659    .6395527
------------------------------------------------------------------------------
```

The output indicates that the invmills estimate is not statistically significant. This implies that selection in this model is not a critical factor. The results from the augmented regression are also very close to the original regression without the selection term, and in both the total household income (or H_2) is the only significant variable.

Cameron and Trivedi (2010, p. 442) point out that the standard errors from this procedure are not correct, because the procedure does not allow for the randomness of the intercept term. However, it is a correct procedure to adopt, as a robustness check, and then compare it to STATAs heckman procedure, which researchers use extensively to test for sample selection. The heckman procedure is based on bivariate normal distribution. STATA also has a heckman twostep procedure, which produces results based on a univariate normality assumption, and the standard errors are assumed to be more robust. We produce the command and the output from both the procedures below:

```
. heckman whz h1 h2 h3 h4 c2, select(si = i1 i2 h1 c1) nolog

Heckman selection model                         Number of obs      =        35
(regression model with sample selection)        Censored obs       =        15
                                                Uncensored obs     =        20

                                                Wald chi2(5)       =     21.67
Log likelihood =  26.69254                      Prob > chi2        =    0.0006

------------------------------------------------------------------------------
             |      Coef.   Std. Err.      z    P>|z|     [95% Conf. Interval]
-------------+----------------------------------------------------------------
whz          |
          h1 |   .0021105   .0161225     0.13   0.896    -.0294889    .0337099
          h2 |    .068477    .018065     3.79   0.000     .0330703    .1038837
          h3 |  -.0019798   .0126696    -0.16   0.876    -.0268117    .0228522
          h4 |   .0104314   .0130427     0.80   0.424    -.0151317    .0359946
          c2 |   .0361067    .033556     1.08   0.282    -.0296619    .1018753
       _cons |  -.0233771   .1819039    -0.13   0.898    -.3799022     .333148
-------------+----------------------------------------------------------------
si           |
          i1 |   .2246504   .2750975     0.82   0.414    -.3145309    .7638316
          i2 |  -.7331469   .6519621    -1.12   0.261    -2.010969    .5446754
          h1 |   .6177292   .5073436     1.22   0.223     -.376646    1.612104
          c1 |  -.2270122   .6624883    -0.34   0.732    -1.525465    1.071441
       _cons |  -4.666859   3.612487    -1.29   0.196     -11.7472    2.413485
-------------+----------------------------------------------------------------
     /athrho |   .7445551   .7164109     1.04   0.299    -.6595845    2.148695
    /lnsigma |  -3.704425   .2555772   -14.49   0.000    -4.205347   -3.203503
-------------+----------------------------------------------------------------
         rho |   .6318894   .4303594                     -.5780868    .9731571
       sigma |   .0246144   .0062909                      .0149156    .0406197
      lambda |   .0155536   .0139355                     -.0117595    .0428666
------------------------------------------------------------------------------
LR test of indep. eqns. (rho = 0):   chi2(1) =     0.70   Prob > chi2 = 0.4019
```

```
. heckman whz h1 h2 h3 h4 c2, select(si = i1 i2 h1 c1) twostep
note: two-step estimate of rho = 1.0330205 is being truncated to 1

Heckman selection model -- two-step estimates   Number of obs      =        35
(regression model with sample selection)        Censored obs       =        15
                                                Uncensored obs     =        20

                                                Wald chi2(5)       =     17.04
                                                Prob > chi2        =    0.0044

------------------------------------------------------------------------------
             |      Coef.   Std. Err.      z    P>|z|     [95% Conf. Interval]
-------------+----------------------------------------------------------------
whz          |
          h1 |   .0087739     .02528     0.35   0.729     -.040774    .0583219
          h2 |   .0649934   .0193658     3.36   0.001     .0270372    .1029497
          h3 |  -.0006315   .0141218    -0.04   0.964    -.0283097    .0270467
          h4 |   .0074846   .0146254     0.51   0.609    -.0211807    .0361498
          c2 |   .0323447   .0406018     0.80   0.426    -.0472334    .1119228
       _cons |  -.0509332   .2229704    -0.23   0.819     -.487947    .3860807
-------------+----------------------------------------------------------------
si           |
          i1 |   .1135861   .3009435     0.38   0.706    -.4762523    .7034244
          i2 |  -.7127836   .6451502    -1.10   0.269    -1.977255    .5516876
          h1 |   .7726134   .5003807     1.54   0.123    -.2081148    1.753342
          c1 |  -.2319729   .6748081    -0.34   0.731    -1.554572    1.090627
       _cons |  -5.567427    3.69716    -1.51   0.132    -12.81373    1.678873
-------------+----------------------------------------------------------------
mills        |
      lambda |   .0341831   .0438807     0.78   0.436    -.0518214    .1201877
-------------+----------------------------------------------------------------
         rho |    1.00000
       sigma |  .03418312
------------------------------------------------------------------------------
```

The estimates and the standard errors of all the parameters in both versions are roughly the same. In both sets, we see that lambda is statistically insignificant, and the values of rho indicate that the hypothesis that the two parts (children who survive versus those who do not) are independent has to be rejected. In other words, factors that determine nutritional status and survivability are not jointly determined in this example. Further, the estimates of the selection equation using `probit` do not differ very much from those produced by the `heckman` procedures. This result is similar to the observations in Maitra and Rammohan (2011, p. 105), which imply that the problem of incidental truncation is not a major issue in this example. We explore this issue in Exercise 2, at the end of this chapter.

As pointed out in this chapter and also in the analytical section, the problem of self-selection is extremely crucial in this line of research. The major themes in the chapter indicate the variety of ways in which gender-bias in nutritional allocation may or may not occur, and in most cases, the deciding factor is self-selection. The STATA exercise below using the popular Heckman two-step procedure helps to find out whether the underlying data exhibits self-selection. Policy makers will have to examine self-selection issues to avoid implementing misguided policy instruments. The

STATA exercises at the end of this chapter illustrate how we can test for the presence of self-selection by accounting for its possibility in the estimation process. Several refinements within STATA allow for further testing of incidental truncation and robustness checks. The exercises at the end of the chapter help in uncovering these matters.

CONCLUSIONS

In this chapter, we introduced a set of intra-household issues that affect the nutritional status of individuals. The status of women as mothers, care givers, and preparers of food consumed at home determine what household members eat in the context of a family (Quisumbing et al., 1995; Smith et al., 2003). The resources controlled by women, their empowerment in society, and their bargaining power at the household level compared to other members of the household affect the decisions women make, including that of investment in good nutrition (Quisumbing, 2012). Studies have highlighted the need for empowering women in the context of their role as food producers as well (Quisumbing and Meinzen-Dick, 2012; Doss, 1991; Doepke and Tertilt, 2011; Fishcher and Qaim, 2012; Kabeer, 1999; Babu et al., 1993).

Understanding the theory and applications of intra-household decision-making in the context of gender bias in nutrition is an important step toward formulating nutrition intervention policies and programs. In this chapter, we introduced this theory and reviewed recent studies that have analyzed gender bias and inequality in the context of nutrition policy making. The chapter also demonstrated the use of the Heckman selection model in STATA. Although gender bias is decreasing as girls education is improving, more research is needed to explore the role of women in empowering the nutritional status of the population. The role of improving data, methods, and the capacity to analyze such issues for improved policy making cannot be over-emphasized.

EXERCISES

1. Use the data from the example and estimate a logit equation in STATA for the likelihood of survival using `logit si i1 i2 h1 c1`. Are the estimates different from those obtained using the probit selection equation?
2. Suppose we want to test whether a mother's employment status (H_3) influences infant survivability. Udry (1997, p. 61) argues that a mere inclusion of H_3 in the selection equation is likely to miss a lot of information, because H_3 is observed only for those children who have survived. If the mother works outside the home (or $H_3 = 1$) is known only at the time of the survey; the employment status of the mother during pregnancy or one year after that is missing. Consequently, the current status of unemployment is a reasonable indicator of health status of those children who have survived, and Maitra and Rammohan (2011) address this issue. Estimate a `logit si i1 i2 h1 h3 c1` in STATA for the likelihood of survival, and then estimate a `heckman twostep` procedure with a suitable regression for WHZ. Is the inverse Mills estimate significant? Are the estimates of the selection equation from both models roughly the same? What can you conclude about incidental truncation?

3. Here is a useful exercise for students to understand the differences between the `heckman twostep` procedure and the one where we run a separate probit or logit for sample inclusion followed by a regression, which is usually known as a two-part model. The table below has information similar to our previous example. The STATA website (see http://www.stata.com/manuals13/rheckman.pdf) indicates that the `heckman twostep` procedure is a useful two-step procedure when the dependent variable in the regression model is observed in the absence of selection.

Use the `heckman whz h2 h4 c2, select(si = i1 c1) twostep` procedure. Now, type the following command in STATA: `replace yt = 0 if yt == .` which replaces all the WHZ for the "missing" children with the value 0. Reestimate the `heckman whz h2 h4 c2, select(si = i1 c1) twostep` model, and see if the selection estimation results match. Estimate the WHZ regression with the two-part procedure. What can you conclude about the assumptions underlying incidental truncation and the STATA procedures?

Whz	Si	c_1	c_2	I_1	H_2	H_4
	0	0	0	0	0.445	1
	0	0	1	0	0.642	0
	0	0	0	1	0.001	1
	0	0	1	1	0.586	1
	0	0	1	0	0.086	1
	0	1	1	0	0.773	1
	0	1	0	1	0.399	1
	0	1	0	1	0.867	1
	0	0	0	0	0.673	0
	0	0	0	0	0.911	0
	0	0	1	0	0.453	0
	0	1	1	0	0.052	0
	0	1	1	1	0.708	0
	0	0	1	1	0.573	0
	0	0	0	1	0.882	1
0.197	1	0	0	0	0.091	1
0.275	1	0	0	0	0.737	1
0.252	1	0	1	0	0.259	1
0.221	1	0	1	0	0.367	1
0.319	1	1	1	1	0.623	0
0.140	1	1	1	1	0.726	0
0.254	1	1	1	1	0.646	0
0.341	1	1	1	1	0.038	0
0.248	1	1	1	1	0.204	1
0.206	1	1	1	1	0.411	1
0.169	1	1	0	0	0.191	0
0.111	1	0	0	0	0.982	1

CHAPTER

10 ECONOMICS OF CHILD CARE, WATER, SANITATION, HYGIENE, AND HEALTH: THE APPLICATION OF THE BLINDER–OAXACA DECOMPOSITION METHOD

Nutrition is a key element in any strategy to reduce the global burden of disease. Hunger, malnutrition, obesity and unsafe food all cause disease, and better nutrition will translate into large improvements in health among all of us, irrespective of our wealth and home country.

—Dr. Gro Harlem Brundtland, former Director-General, WHO, at the World Economic Forum in 2000

INTRODUCTION

Persistent malnutrition in regions and countries where food self-sufficiency has been achieved for >30 years remains an enigma. A good example of this is the South Asia region where the green revolution in food production, which increased food production three times between 1960 and the 1980s, did very little to reduce the levels of child malnutrition. Researchers have attempted several explanations of this phenomenon. Ramalingaswami and Jonson (1995) conjectured that gender inequality in food consumption and women's empowerment are key toward achieving any improvement in child malnutrition. Throughout the last two decades, nutritionists and economists alike have moved from one variable to another trying to solve the nutrition puzzle; however, they fully recognize that nutrition is a collective outcome of all these factors (Smith and Haddad, 2003; Smith and Haddad, 2015).

In addition, as seen in Chapter 9, Intra-Household Allocation and Gender Bias in Nutrition: Application of Heckman Two-Step Procedure, certain societies where female children and women consume less compared to their male counterparts exhibit gender bias in rural nutrition, and this may vary depending on the productivity levels and seasonality in food consumption (Babu et al., 1993). Thus, translating food availability at the household level into the nutritional well-being of the members of the household depends on factors that go beyond food. Recently, factors related to water, sanitation, and health have been recognized under the common term "WASH" as areas for nutritional interventions. Access to quality water, sanitation, and health care are important, along with the care that is provided to vulnerable members of the household such as preschool children, pregnant and lactating mothers, and the elderly.

The fact that nonfood factors play a significant role in the nutritional status of society has been recognized for some time (Smith and Haddad, 2003; Smith and Haddad, 2015). These factors have been

Nutrition Economics. DOI: http://dx.doi.org/10.1016/B978-0-12-800878-2.00010-4

addressed collectively, as well as individually, in explaining the nutritional status of the population. In this chapter we look at these factors, specifically, child care, water, sanitation, and health care, and the associated policy interventions to measure, explain, and address the challenges of malnutrition.

Although some of the socioeconomic factors affecting nutritional status were addressed in Chapter 8, Socioeconomic Determinants of Nutrition: Application of Quantile Regression, the factors related to child care, water, sanitation, and hygiene require a separate chapter, as they affect the nutritional status of the population by entering the nutrition production function as complementary inputs. Clean water is needed to convert food consumed into good nutrition. Water, sanitation, and hygiene also affect the absorption of nutrients and minerals. For example, anemia, a major nutritional and public health challenge, is caused by poor absorption of iron in the body. A safe water supply, improved sanitation facilities, and education to improve hygienic behaviors can help in the prevention of environmental enteric dysfunction. These are also crucial factors that help improve the absorption of iron from the food consumed.

Lack of WASH factors causes serious damage to the health and nutritional status of the population. For example, Bremmer (2010) notes that about 1.5 million deaths are caused by diarrhea each year, and the main reason for this is the lack of clean water and sanitation facilities. Bremmer (2010) notes that any diarrhea-prevention policy must first address access to clean sanitation. In fact, Gunther and Gunther (2013), using household survey data from 40 developing countries, show that a reduction of about 8–22 deaths per 1000 children is possible if sanitation and water technologies are provided. The 2015 target set in Millennium Development Goal 4, was a reduction of 32%, and the said reductions in deaths will account for 11% of this target, amounting to 0.6–1.7 million children in developing countries. The cost of technologies amounts to roughly 80% of per capita GDP in these countries.[1] The recently agreed upon Sustainable Development Goals also address the challenge of ensuring access to clean water and sanitation for all (Wage et al., 2015).

CONCEPTUAL APPROACH

Researchers have conceptualized the role of various factors in contributing to the nutritional status of the population. One way to identify the pathways of such contributions is through causal analysis, which is commonly used in field situations. Here, we adopt one such conceptual causal pathway developed by Action Against Hunger (ACF) (2014) to exemplify how various factors could be accounted for in explaining the nutritional outcomes (Fig. 10.1). At each stage of the factor's contribution to the next stage, the factors are ranked or measured where possible, and used in the analysis. They are also used to compare the status of one community with another in the context of nutritional outcomes. For example, the hypothesis such as: does a community that has better access to health care workers have better nutritional outcomes, all other factors being the same? Three broad pathways are traced from the community level to individual nutritional outcomes. Food availability, child care, and health and sanitation pathways are measured and analyzed through the collection of data and information on specific indicators.

[1]For a sequence of papers on this issue see Jha et al. (2014).

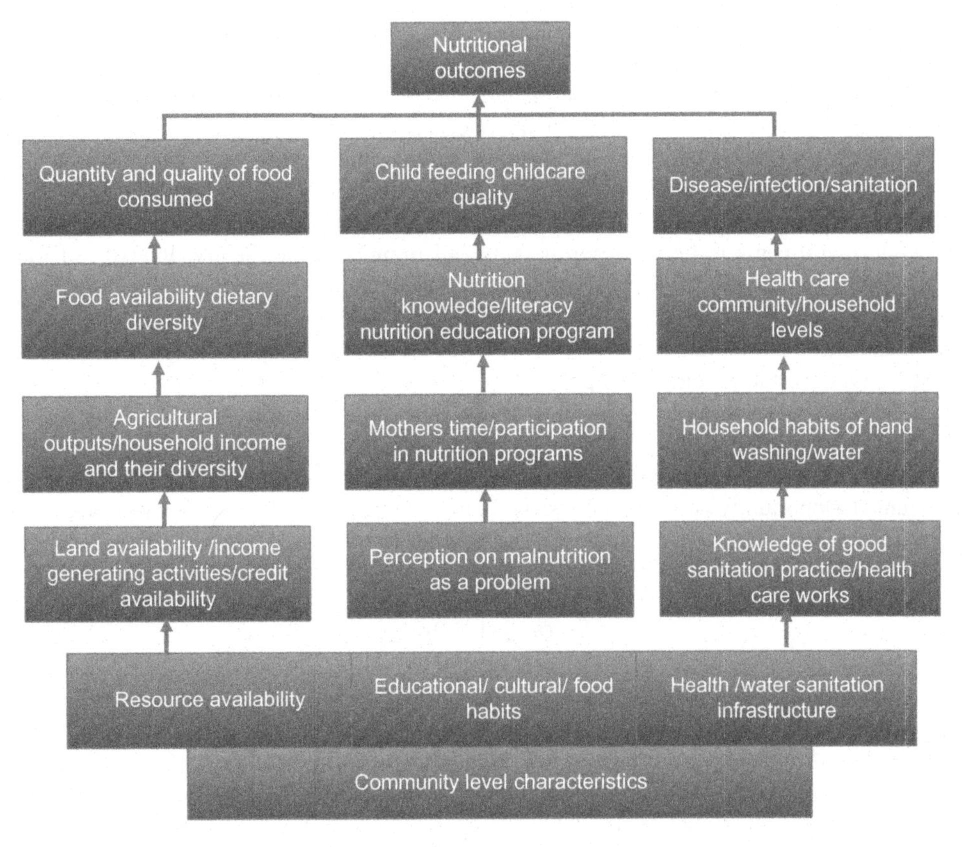

FIGURE 10.1

Hypothetical causes of malnutrition.

Adapted from ACF Nutrition Causal Analysis (ACF, 2014).

WHAT DO WE KNOW: POLICY LESSONS?

In this section we briefly review the role of various factors that determine nutritional status, and examine possible interventions to reduce malnutrition. Smith and Haddad (2015) present an exhaustive study tracking child under-nutrition across 116 countries over the time period 1970–2012. In order to make improvements in children's nutritional outcomes, countries must increase and maintain adequate food supplies, and governments must invest in agriculture with concentrated efforts, because agricultural output is highly susceptible to climate change. These environmental effects will be highly damaging to South Asia and Africa, where child under-nutrition is the highest. However, increasing food supplies is not the only necessary condition for an improvement in child nutrition. Smith and Haddad (2015) note that across all countries, food, care, and health environment are the major drivers of a reduction in stunting. National governance is also a key driver, which will be discussed in related literature later in this chapter as well. Governance issues are particularly vital for water-related programs.

In keeping with the issues discussed in this chapter, Smith and Haddad (2015) find that the best programs needed to reduce stunting should include: a greater percentage of dietary energy from nonstaples, better access to sanitation, women's education, and access to safe water. Programs that promote income growth also help in supporting accessibility to these goods.

Investments in women's education and increasing gender equality have the best short-term impact, while investments in health environments and improving the dietary diversity of food are beneficial in the long run. We present several examples in this chapter, where countries have designed and implemented different randomized design interventions to improve children's and women's health and nutritional outcomes. We particularly examine these issues in the context of child care, sanitation, and public health policies.

Singh (2011) presents a set of interesting results that connect nutritional deficiency to unequal opportunities in India. The notion of "unequal opportunities" is measured through two related indices: the Dissimilarity Index (*D*), and the Human Opportunity Index (*O*).

The Human Opportunity Index (*O*) consists of two elements: (1) how many services are available and (2) how equitably these services are distributed. Both the Dissimilarity Index (*D*), and the Human Opportunity Index (*O*) are related as follows:

$$O = \overline{p}(1 - D)$$

where $\overline{p}$ is the average probability or the access rate of a child to health services. Both the *D* index and the *O* index vary between 0 and 1. A Dissimilarity index $D = 1$ means a high unequal placement in society and opportunities are close to zero, since $O = 0$, in that case.

Singh's (2011) research uses two rounds of the National Family Health Survey (NFHS), and constructs the *D* index and the *O* index, with specific reference to immunization services and nutritional outcomes, and presents several results:

- Parental education and the family's wealth are key drivers of a child's health and nutrition.
- Overall, >50% of the children do not receive full immunization, and do not receive minimum nutrition.
- There are substantial geographic variations in the opportunity for immunization and nutrition.
- While the south and north-east perform well on both index measures, the central, west, and north-west regions of India fare very poorly.
- Overall, scheduled castes and scheduled tribes who are considered vulnerable members of Indian society have the lowest measures in both indices, with respect to full immunization and minimum nutrition.
- Children in urban areas fare much better than rural children, in terms of both elements. This is a rather disturbing finding, given that as many as 75% of children live in rural areas.

Despite many interventions launched by the government of India, regional disparities have widened. Once again, Singh's (2011) study indicates the importance of regional-specific policies that promote parental education. Interregional inequalities in opportunity are a special concern in India, particularly after the economic reforms introduced in the late 1980s. Singh's (2011) study shows that those regions like the south, with the highest *D* index and *O* index, continue to show upward movement post reforms. Further, those states in the central and eastern regions, with low measures of both indices, continue on their downward trend.

The above observations not only hold for nutrition and immunization programs, government sponsored schemes and food-subsidy programs also suffer from these disparities. Lokshin et al. (2005) point out the deficiencies in the Integrated Child Development Services (ICDS), which is one of the biggest nutrition interventions for preschool children.

Similar to the findings of Singh (2011), Lokshin et al. (2005) find that the poor northern states, with the greatest need for the program, have the lowest coverage, and the lowest budgetary allocations. Further, two of the states in this region do not spend their allotted amounts, highlighting poor governance. The level of unequal regional opportunity worsens because, as Lokshin et al. (2005) point out, the rich northern states get most of the funding, and also use the funding effectively. Recently, Jain (2015) showed how the ICDS program improves nutritional outcomes for very young children.

Unequal opportunities can exist in many forms. For example, Sethuraman (2008) provides an extensive analysis of how tribal and rural women fall way short of opportunities, which is reflected in the incidence of domestic violence, and a consequent reduction in the children's nutrition.[2]

Sethuraman's (2008) study links women's empowerment to nutritional outcomes. An interesting feature in this study is the empirical tracking of women's empowerment, through the mother's decision-making capabilities, freedom of movement, employment, and experience of domestic violence. To this, a host of socioeconomic variables are added. Sethuraman's (2008) regressions convey some crucial information for policy: women's empowerment and experience with domestic violence affect current child nutrition levels, and also affect a child's long-term growth potential.

The role of gender-equality and equal rights must accompany policy interventions; mere food subsidies cannot rectify inequalities that are household-driven. Note that these findings are in keeping with those of Jensen (2011) on fertility, Singh (2011), and Lokshin et al. (2011). Datta's (2015) study on Bihar's JEEVIKA Project is an excellent example of a successful government program that achieved these goals. Kavitha and Lal (2013) highlight the importance of sanitation on women's health.

CHILD MALNUTRITION AND RIGHTS TO WATER AND SANITATION

The UN General Assembly has recognized basic sanitation as a human right. Mcgranahan (2015) presents an exhaustive analysis covering Latin America and Asia, where the quasi-public goods problem associated with sanitation is severe. Mcgranahan (2015) argues that the "rights-based" approach is narrow in scope. Attempts where public sanitation upgrades have worked best have been due to grassroots movements, involving co-production and polycentric approaches. Mcgranahan (2015) cites two examples, which followed community-based co-production movements, where the public goods problems were overcome. The Orangi Pilot Project in Karachi, Pakistan, and the Indian National Federation of Slum Dwellers in Pune and Mumbai, are the two examples from Mcgranahan (2015), where sanitation upgrades have worked well.

[2]Also see Adhau (2011) for malnutrition issues affecting tribal societies.

The importance of access to water and its relation to child nutrition has been well documented. For instance, Zewdie and Abebaw (2013) identify the main drivers of child malnutrition in the Kombolcha districts of Eastern Hararghe in Ethiopia. They apply logit regressions to data from 249 children, and show that a significant number of them are stunted, wasted, and malnourished. The main determinants of child malnourishment turn out to be water source and sanitation facilities, among other important factors such as immunization, farm size, household size, gender, and the mother's use of antenatal care. Zewdie and Abebaw (2013) stress the role of clean and safe water access, along with proper sanitation, as requiring immediate attention. Craven and Stewart (2013) go as far as to demonstrate that resources should be directed away from HIV-related activities, and moved toward basic hygiene and nutrition, better water quality, and living standards in Africa, where recurrences of other lethal diseases are often in conjunction with HIV/AIDS.

The environment in which people live affects their nutritional status. A clean environment reduces the probability of infectious diseases. Recently, the nutritional status of the population has been connected to levels of sanitation. Defecating in open in developing countries and living with animals have been studied for their possible connection to nutritional outcomes. Heady et al. (2016) look at this possible contributor to poor nutrition. The environmental-centric disorder caused by the ingestion of poultry feces, and the diarrhea caused by this act, increase the morbidity and affect nutritional intake. Heady et al. (2016) use a multicountry survey from Ethiopia, Bangladesh, and Vietnam discussing the presence of animal feces and the cleanliness of mothers. They reported a significant connection between livestock ownership and the resultant presence of feces in household compounds, which negatively influence child growth outcomes in Bangladesh and Ethiopia, but not in Vietnam.

Open defecation is another serious challenge to the hygienic environment to maintain the good health that facilitates better absorption of nutrients. In this context, evidence and studies from India have proliferated in recent years. The reason for bringing sanitation into the context of this chapter is to show that access to clean sanitation is an important indicator in the *O* index discussed earlier. Further, provision of water and sanitation has direct links to child malnutrition. Cuesta (2007) presents evidence of this link using a longitudinal data set from the Philippines, which tracks 3289 children for three rounds. The likelihood of birth weight stunting is reduced most when households have access to flush toilets. Access to public latrines also reduces the probability of household birth weight stunting. Cuesta (2007) also finds significant effects of maternal education on lower birth weight stunting, although the effect disappears when the women's empowerment index is taken into account. Also see Lee et al. (1997) for an excellent discussion relating nutrition, mortality, and sanitation for Bangladesh and the Philippines, and the related endogeneity issues.

Besides the obvious advantages to health, improvements in sanitation can contribute to human capital formation and cognitive skills among children. Spears and Sneha (2013) use data from India's Total Sanitation Campaign (TSC), and demonstrate the increase in literacy among 6-year-old children who were exposed to pit-latrines for the first time.

Although, in all of these studies, we find that access to sanitation and water improve nutritional outcomes, we cannot ignore the fact that these factors alone cannot substitute for wider policy interventions that emphasize good health practices and provide relevant healthy goods.

The fact that sanitation and water supply act as public goods creates additional problems for their provision. Mader (2012) shows that the attempt to create these public goods via microfinance, in India and Vietnam, revealed that there was a severe under-provision of these goods, because agents act only on the basis of self-interest, and are not concerned about the collective welfare aspects of the problem. Mader (2012) questions the effectiveness of microfinance, and also the accountability of the governance systems in these countries for the provision of these public goods.

The above observation echoes an earlier World Bank Study by Ban, Das Gupta, and Rao (2008), which points out that local politicians in South India capture new technology for themselves. The study emphasizes the need to educate the public and all the stakeholders about the health benefits of overall improved sanitation.

THE ASIAN ENIGMA

India provides an excellent case study to issues related to sanitation and health outcomes. The Indian government was proactive in establishing a TSC in 1999, and researchers have produced a host of studies documenting the impact of several interventions across India. Most of the studies examine the impact using randomized control designs and evaluate the effect of different treatments. For instance, Hammer and Spears (2013), in a World Bank study for India, note the widespread prevalence of open defecation in India, and relate this to the very high rates of child stunting. In a randomized control trial in the state of Maharashtra, the study shows an increase in child height after implementing a sanitation program. The extensive research conducted by Spears and his colleagues relate this finding to what is frequently referred to as "the Asian enigma," where child stunting in India is much higher than in poorer African countries (Ramalingaswami and Jonson, 1995; Spears 2013).

Spears (2013) demonstrates that open defecation accounts for almost all of the stunting in India. The government of India, through its Total Sanitation Campaign, offers local governments monetary incentives to combat this issue. Spears (2013) presents robust econometric evidence from country-level regressions, within-country regressions, and randomized control experiments to support the stunting–defecation link.

Most importantly, Spears (2013) applies decomposition methods to largely account for the Asian enigma by comparing differences in the height-for-age in Africa and India. Recently, Duflo and others (2015) have shown that at an annual cost of $60 per household, a 30–50% reduction in diarrhea was achieved with integrated water and sanitation in rural India.

All of the studies, including Duflo et al. (2015), point to several institutional rigidities that prevent the programs from being fully successful. Lamba and Spears (2013) relate the prevalence of caste-based discrimination in winning prizes and awards implemented in a Clean Village Prize in India.

Augsburg and Rodriguez (2015) also indicate several new insights into the sanitation dynamics in India. The researchers stress that there are several key drivers of toilet ownership and acquisition that policy makers must take into account for a successful TSC. In particular, in India, social status plays a vital role in the acquisition of toilets. There are severe financial constraints that prevent lower income groups from acquiring toilet facilities. In addition, they argue that toilet interventions must use campaigns and slogans that stress status, such as "no loo, no bride," which will drive males to invest in toilets. The issue of status, shame, and dignity, along with personal hygiene, should be taken into account when policies are designed.

As Reddy and Snehalatha (2011) point out, these gender issues are crucial for women, who are particularly vulnerable to lack of sanitation facilities. Patil et al. (2013) is one of the early studies on the impact of TSC, which adopts a cluster-randomized, controlled trial set in the state of Madhya Pradesh, India. This is an extensive study that analyzes about 3000 households, including >5000 children. The program shows marginal improvements in latrine use, water quality, and water-borne infections within the treatment groups. Since this was an early study, with the intervention lasting only 6 months, the effects are not statistically significant. However, follow-up studies for India have generated a lot of information to support the health impact of public interventions.

As mentioned before, randomized trials have become a common research tool to evaluate the impact of a treatment. In the Indian context, researchers have adopted this tool to study sanitation programs. Dickinson et al. (2015) recently studied the effect of a randomized sanitation promotion program in the state of Orissa in India. The researchers note that the program increased ownership and use of latrines. Importantly, the intervention also shows an increase in children's mid-upper-arm circumference, height, and weight *z*-scores. The cost–benefit calculations also take into account the negative externalities, and this is an excellent study that legitimizes active public policy intervention promoting good sanitation.

Agoramoorthy and Hsu (2009) suggest that the government of India increases the subsidy to build new toilets, based on their study from the state of Gujarat, India. The costs of medical treatment, and the opportunity cost of lost wages, were significantly lower after the construction of 100 newly built toilets. Kumar and Vollmer (2013), using District Level Household Survey data and a propensity matching technique, show that access to improved sanitation reduces the risk of contracting diarrhea by 2.2% points. Again, using the same propensity matching technique, Begum, Ahmed, and Mansur (2011) point out that while Bangladesh has concentrated on improving the provision of clean water, the results show that this policy must be associated with improved sanitation. The study shows that the synergy between safe water and sanitation has the best health effects.

Cameron et al. (2013), in another World Bank Study conducted for Indonesia, show that when new toilets were constructed, open defecation decreased for nonpoor households by 6% points. Diarrhea declined by 30% in the treatment groups. Briceno et al. (2015), in a World Bank Study for rural Tanzania, also document the effects of a government policy promoting hand-washing and clean sanitation. The policy, run as a randomized trial, finds marginal improvements in health only when programs are combined with hand-washing and sanitation. Just as in India and Bangladesh, the authors conclude that an overall increase in hygiene is a vital part of any successful campaign.

CHILD CARE, FERTILITY, AND MALNUTRITION

Can reductions in fertility improve childhood nutrition? This question has occupied researchers ever since Gary Becker (1960) presented the famous "quantity–quality tradeoff," in the context of childhood malnutrition. Jensen (2012) presents a succinct explanation of the situation confronting researchers in this area. Suppose we consider two children, from two different families, who are identical in every respect, and where both sets of parents wish to have an additional child. Suppose only one of the children ends up with a younger sibling.

Will the child without a younger sibling be better nourished, relative to the other child who now has a sibling? This question cannot be easily verified because of many difficulties associated with data availability and interpretation. For instance, the environmental and social factors associated with infertility may also be linked to a child's health status. Consequently, to separate the effects of various factors becomes an econometric issue.

To tackle the above statistical issues, Jensen (2012) uses a unique panel data set from four states in India, which are Haryana, Punjab, Rajasthan, and Uttar Pradesh. The data captures information on fertility and socioeconomic variables for two rounds, in 2003 and 2006. Their results indicate that reductions in fertility improve children's nutritional status and demonstrate the validity of the quantity–quality tradeoff.

Just how important is Jensen's (2012) finding for India? The findings are crucial, because in rural India about 27% of births are unwanted, particularly in Bihar, where about 33% of births are unwanted. Consequently, interventions linked to fertility are likely to produce desired effects. Indeed, Jensen's (2012) results suggest that between 1979 and 2006, when India experienced a decline in fertility, there was a corresponding 23% decline in malnutrition. Fertility differences explain about half of the nutrition gap observed between urban and rural centers in India. Jensen's (2012) findings have a lot of relevance for intervention programs for India (Box 10.1).

BOX 10.1 ARE MONSOON BABIES CURSED?

Is it possible that children born during certain months are doomed to ill-health? Apparently, yes, according to Lokshin and Radyakin (2012), who point out that children born during the monsoon months in India have lower anthropometric scores compared to children born during fall–winter months.

The idea that "month-of-birth" is an important indicator of a child's health outcome has been documented and tested positively for other countries, such as Australia (Weber et al., 1998), Sweden (Kihlbom and Johansson, 2004), the United Kingdom (Phillips and Young, 2000), and for the United States (Van Hanswijck de Jonge et al., 2013). Following this line of research, Lokshin and Radyakin (2012) demonstrate large "month-of-birth" effects, which result in poor environmental conditions leading to poorer health outcomes.

Lokshin and Radyakin (2012) connect medical, climatic, and environmental data to the socioeconomic conditions of households, and demonstrate "month-of-birth" effects after controlling for several characteristics and sample selection. The researchers use three rounds of the Indian NFHS data set that cover thousands of children from over 30,000 households.

The results show anthropometric scores as the lowest for children born in summer, while improving for those born in fall and early winter. Further, the same result is true for boys and girls. Children born during the monsoon months are also more likely to be stunted than those born after the start of the monsoon.

Potential endogeneity and sample selection are common data problems in this line of research. For instance, parental behavior may influence a child's health outcomes. Parents may plan pregnancies, so as to give birth in nonmonsoon months, or overcompensate on child care, if there is a birth in a "bad" month. Since parents' beliefs, behavior, and actions are not observed, there is a chance that some of the bad health outcomes are not accounted for properly.

Similarly, household wealth and maternal health may affect the month-of-birth and, hence, health outcomes. Likewise, unwanted pregnancies or a child's strength at birth may also confound the "month-of-birth" effects. The

(Continued)

BOX 10.1 (CONTINUED)

authors address all of these econometric issues through probit and multinomial regressions. Overall, they are able to reject sample selection and endogeneity concerns. Beside the "month-of-birth" effects, the authors find:

- A higher birth order has a negative impact on the health outcomes of boys and girls.
- Health outcomes worsen with the age of a child.
- Children from wealthier homes and better-educated mothers are less likely to be stunted.
- A child's health and mother's age has an inverted U-shape relation, where peak health is when a mother is 40 years old.

Adequate information on nutrition, along with family planning guidance, can go a long way with improvements in the "month-of-birth" effects.

HEALTH CARE AND NUTRITION

Health economists have also been very interested in examining the role of sanitation and access to safe water, alongside improvements in overall hygiene, such as hand washing, on the effects of infant mortality and overall health outcomes. We have already pointed to the study by Whittington et al. (2012) regarding the need to plan public health interventions based on evidence and situations, rather than on a cost–benefit calculation. In a very interesting demand-based study, Persson (2002) calculates the compensating variation of welfare improvements on consumer surplus from two different models, and shows that welfare calculations from health improvements are sensitive to the choice of models.

Interestingly, Andres et al. (2014) also estimate health benefits from an intervention program situated in rural areas of India, during 2007 – 08. The health benefits particularly accrue to children when a household moves from an open to a fixed-point defecation system, or from open defection to an improved sanitation system. In terms of health benefits, the study notes a substantial 47% reduction in diarrhea among children, with 25% of the benefit resulting from direct intervention, and the remaining being due to external benefits. All health benefits in this study are attributed solely to improved sanitation facilities.

Hathi et al. (2014) examine some of the important questions in health economics: is open defecation, without a toilet or latrine, worse for infant mortality and child height where the population density is greater? Is poor sanitation an important mechanism by which population density influences health outcomes? The researchers use a large, international child-level data set of 172 demographic and health surveys, matched to census population density data for 1800 subnational regions, and also separately, other data from Bangladeshi districts, to explore these questions. The results indicate statistically robust interactions between sanitation and population density. The authors demonstrate that open defecation externalities are more important for child health outcomes where people live more closely together.

In a related paper Geruso and Spears (2015) show that Indian Muslims are about 25% more likely than Indian Hindus to use latrines or toilets, and negative externalities from neighborhood effects are substantially linked to infant mortality rates. This is a serious health concern in India, where roughly more than 500 million people openly defecate. In a very interesting study on family migration in rural China, Mu and de Brauw (2015) show that parental migration has no significant effect on the height of children, but it certainly improves their weight. However, improvements in height are obtainable through access to tap water in migrant households.

ECOHEALTH APPROACHES: MAHARASHTRA, INDIA, CAMEROON, AND LEBANON

An Ecohealth perspective is an excellent model that can be adopted by many countries. The approach to the subject matter is thorough, and includes a range of environmental health issues, including solid waste collection, drainage of stagnant waters, improved pedestrian walkways, reduced groundwater contamination through safer disposal of human waste, improved household hygiene, safe drinking water, monitoring of health, and provision of basic care. Most of the interventions were through promotional campaigns and education, and concentrated on safe water use and hygiene. The results show substantial reduction in rates of infection in children. Similar efforts were also undertaken in Bebnine, Lebanon, and once again, multifaceted efforts along with an understanding of Ecohealth, were noted to be the best suited approach to affect interventions concerning hygiene, water safety, and clean sanitation (Habib, 2012).

Ngnikam et al. (2012) present a success story from Yaoundé in Cameroon, where a project team from various fields assembled to tackle the serious issue of diarrhea and intestinal parasitoses in children. The team managed to put together a comprehensive project bringing in new forms of social engagement and, along with government staff, educated the stakeholders about water and sanitation. The project linked the information to regular monitoring by medical staff, so as to get information and data on children's health and progress.

In an exhaustive study using stakeholder analysis, Haddad et al. (2014), show a dramatic decline in stunting in Maharashtra, in India. The authors point out that "if stunting cannot decline in this kind of context, then it would struggle to decline in many places." In this study, all the actors worked together to produce a dramatic policy effect, even in an environment where governance was not the best, but not the worst, food security levels were average, and water and sanitation levels were vulnerable. The success was due to civil activists and media working together to produce good health outcomes, with shared governance and a commitment from the top.

PANAGARIA–GILLESPIE DEBATES AND POLICY CONSIDERATIONS

The observations and policy conclusions discussed in this chapter, and in the preceding ones, provide us with the necessary insight into recent debates on nutrition policies. Most recently, this debate involves an economist, a policy maker, and a nutrition policy researcher.

Panagaria (2013) has recently argued that the extent of child malnutrition in India is overestimated. In particular he doubts that Indian children fare much worse than their counterparts in sub-Saharan Africa.

For instance, Panagaria (2013) questions the validity of the findings on under-nutrition, because the researchers place common weight and height standards across all populations. Thus, Panagaria (2013) argues that the estimates ignore differences that may arise due to genetic, cultural, and environmental factors. Second, Panagaria (2013) also points out that India's performance is much better than many sub-Saharan African countries when it comes to mortality, longevity, and other health outcomes. Thus, he argues the implications of nutritional outcomes are misleading.

Gillespie (2013) has responded to both of these criticisms, and indicates that the researchers, including WHO, do construct their estimates carefully after taking into consideration genetic differences and other cultural diversity measures. Further, Gillespie (2013) points out that although India's mortality statistics are better than sub-Saharan Africa, they do not guarantee that India's children obtain sufficient nutrition. Gillespie (2013) further points out that India's performance with children's underweight has improved marginally, but not significantly, particularly during a time of rapid economic boom.

India is listed as one of the 17 countries with a 40% or more occurrence of stunting, which amounts to roughly 60 million children, and is 36% of the global total. More than one-third of Indian infants are underweight, and 20% are stunted, a disadvantage that carries on into later life.

As pointed out by the literature pertaining to this area of research, India's child nutritional status depends upon many factors beside the provision of a "balanced diet." Nutritional policies call for a broader response to health, water, sanitation, maternal education, and women's empowerment, along with coherent family planning mechanisms.

The idea that the mere provision of a "balanced diet" itself can be mired in political shortsightedness makes the issue of child nutrition even more crucial in India. Consider the editorial in *Economic and Political Weekly* (June 6, 2015, p. 9), where a state-sponsored nutritional meal scheme has become controversial because some religious groups do not want eggs served to children. That fact that these decisions on lower caste non-vegetarian children's diets are being made by upper caste and wealthy Hindus means the issue could be seen as highly ironic.

ANALYTICAL METHODS

This section is motivated by a question posed by Dean Spears (2013) in his extensive research on the impact of sanitation on health outcomes: can differences in sanitation coverage explain the difference in height between children in two different regions?

Pooled regressions often obscure these interesting features because of Simpson's Paradox: the slope coefficients of individual regressions may be different for different subsets, since the averages of the independent variables within these subsets may be correlated differently with the outcome variable. Consequently, Spears (2013) uses decomposition methods to explain the differences in the outcome variable that arise between the two geographical areas.

Decomposition methods can tell us how much of the inequalities, say in height, are due to inequalities in x_1 (say in calorific intake) versus inequalities in x_2 (say in mother's education). Decomposition methods are useful, because, once regressions are performed, the methods can inform us about the fraction of the difference between two groups, and this is explained by the differences in the exogenous variables.

Following Spears (2013), we illustrate the application of the Blinder–Oaxaca decomposition method. Suppose there is a difference between the means of the outcome variable, height-for-age (H) for similar cohorts, between two countries A and B. The gap in H can be decomposed into two parts:

1. Group differences in the levels of Xs;
2. Group differences in the levels of βs.

For instance, country B may have lower levels of food distribution systems (first part), and also less accessible roads to procure the said distribution (second part). The differences are shown in the following graph:

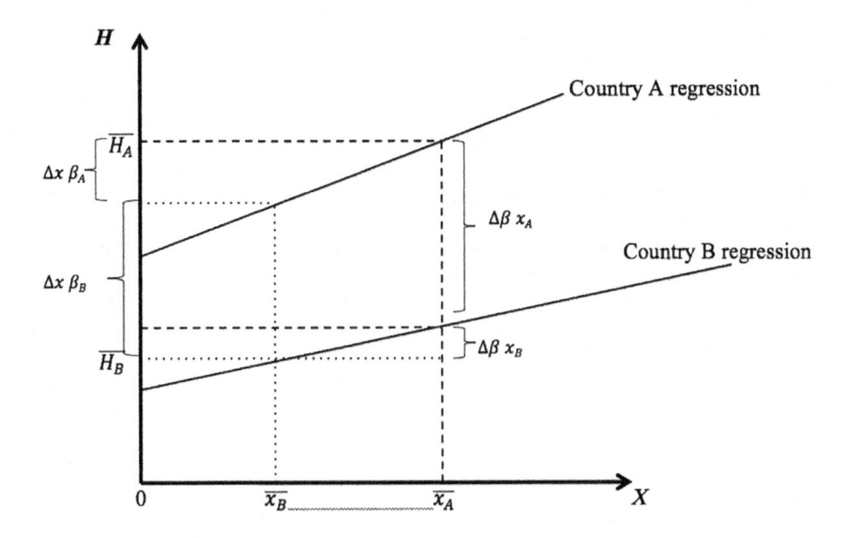

The graph pertains to regressions from the two countries A and B:

$$H = \begin{cases} \beta_A X_A + \epsilon_A, & if\ country = \text{A} \\ \beta_B X_B + \epsilon_B, & if\ country = \text{B} \end{cases}$$

For each value of X, country A has a higher score for H. Let $\overline{X_A}$ and $\overline{X_B}$ be the respective averages of Xs, and correspondingly, let $\overline{H_A}$ and $\overline{H_B}$ be the mean outcomes at these averages, such that the difference in mean outcomes satisfies:

$$\overline{H_A} - \overline{H_B} = \overline{X_A}\beta_A - \overline{X_B}\beta_B \tag{10.1}$$

The Blinder–Oaxaca decomposition regroups the difference in mean outcomes in Eq. (10.1) in two possible weighting schemes:

$$\overline{H_A} - \overline{H_B} = (\overline{X_A} - \overline{X_B})\beta_A + (\beta_A - \beta_B)\overline{X_B} + \{(\overline{X_A} - \overline{X_B})(\beta_A - \beta_B)\} \quad (10.2)$$

or as:

$$\overline{H_A} - \overline{H_B} = (\overline{X_A} - \overline{X_B})\beta_B + (\beta_A - \beta_B)\overline{X_A} - \{(\overline{X_A} - \overline{X_B})(\beta_A - \beta_B)\} \quad (10.3)$$

Eqs. (10.2) and (10.3) are known as threefold decomposition. Jaan (2008) points out that the decomposition indicates the following three parts:

1. $E = (\overline{X_A} - \overline{X_B})\beta_A$ is the first part which is the "endowments" (E) effect, due to the group differences in the levels of X;
2. $C = (\beta_A - \beta_B)\overline{X_B}$ is the impact of the "coefficients" (C); and
3. $I = (\overline{X_A} - \overline{X_B})(\beta_A - \beta_B)$ is the interaction between E and C.

The threefold decomposition is calculated based on the $\hat{\beta}$ estimate from two group regressions. Researchers also report a twofold decomposition, rewriting Eqs. (10.2) or (10.3) as:

$$\overline{H_A} - \overline{H_B} = \left(\overline{X_A} - \overline{X_B}\right)\beta^* + (\beta_A - \beta^*)\overline{X_B} + (\beta^* - \beta_B)\overline{X_B}$$

where β^* is known as a nondiscriminatory parameter, and the twofold decomposition is useful because it indicates the following two parts:

1. $Q = \left(\overline{X_A} - \overline{X_B}\right)\beta^*$ is the "quantity" (Q) effect, which is based on the group differences in the levels of X; and
2. $U = (\beta_A - \beta^*)\overline{X_B} + (\beta^* - \beta_B)\overline{X_B}$ is the unexplained component.

There are many alternative procedures to compute β^*. Sometimes, β^* is the average of the Ordinary Least Square (OLS) coefficients:

$$\beta^* = 0.5\hat{\beta}_A + 0.5\hat{\beta}_B$$

There are many weighting schemes discussed by Jaan (2008) and Spears (2013). β^* is also sometimes derived from an OLS estimation of the pooled sample.

A detailed decomposition that indicates the fraction of the gap in $\overline{H_A} - \overline{H_B}$ due to differences in individual exogenous variables can also be found by expanding $(\overline{X_A} - \overline{X_B})\hat{\beta}$ as:

$$(\overline{X_{1A}} - \overline{X_{1B}})\beta_{1A} + (\overline{X_{2A}} - \overline{X_{2B}})\beta_{2A} + \ldots$$

which corresponds to the explained portion of individual decomposition, while the unexplained portion of each variable is derived by expanding $(\beta_A - \beta_B)\overline{X_B}$ as:

$$\left(\hat{\beta}_{1A} - \hat{\beta}_{1B}\right)\overline{X_{1B}} + \left(\hat{\beta}_{2A} - \hat{\beta}_{2B}\right)\overline{X_{2B}} + \ldots$$

IMPLEMENTATION IN STATA

The following table has 40 observations from two countries on the following variables:

Z:	Height-for-age *z*-score from 20 individuals in each country.
S:	Percentage of open sanitation in the respondent's district in country A or B.
Educ:	Years of education obtained by the head-of-the-household.
Age:	Age of the head-of-the-household.
hh:	Total number of family members residing in the household.
Female:	Whether the mother in the household has a job outside the home.
Country:	A binary variable taking 0 (if country = A) and 1 (if country = B).

obs	Z	S	Country	educ	age	hh	Female
1	−0.294	0.810	1	5	57	4	1
2	−0.277	0.800	1	9	31	3	1
3	−0.254	0.810	0	5	31	3	0
4	−0.251	0.790	0	5	32	4	0
5	−0.245	0.770	1	9	32	3	1
6	−0.238	0.780	1	10.5	39	4	1
7	−0.230	0.750	1	9	33	3	1
8	−0.222	0.740	1	9	33	3	1
9	−0.204	0.720	0	9	33	4	0
10	−0.202	0.690	1	12.5	34	5	1
11	−0.190	0.680	0	17.5	60	4	0
12	−0.188	0.670	0	9	44	5	0
13	−0.175	0.660	0	9	41	4	0
14	−0.172	0.650	0	9	33	4	0
15	−0.172	0.640	0	10.5	33	4	0
16	−0.172	0.630	1	9	33	3	1
17	−0.153	0.620	1	10.5	38	3	1
18	−0.151	0.610	0	10.5	46	3	0
19	−0.134	0.600	1	11.5	35	4	1
20	−0.117	0.590	1	10.5	60	4	1
21	−0.114	0.580	1	5	35	3	1
22	−0.109	0.570	0	9	57	4	0
23	−0.104	0.560	1	10.5	48	4	1
24	−0.099	0.550	0	9	49	3	0
25	−0.096	0.540	1	10.5	58	3	1
26	−0.096	0.530	1	10.5	36	4	1
27	−0.096	0.520	1	9	46	3	1
28	−0.091	0.510	1	9	37	5	1
29	−0.073	0.500	0	9	38	3	0

(*Continued*)

Continued

obs	Z	S	Country	educ	age	hh	Female
30	−0.072	0.490	0	10.5	48	4	0
31	−0.070	0.480	1	10.5	42	4	1
32	−0.068	0.470	0	9	32	3	0
33	−0.063	0.460	0	12.5	58	4	1
34	−0.049	0.450	0	10.5	54	3	1
35	−0.049	0.440	0	5	37	5	1
36	−0.047	0.430	0	5	47	4	0
37	−0.043	0.420	0	9	46	3	0
38	−0.038	0.410	0	10.5	54	3	1
39	−0.020	0.400	0	9	36	4	1
40	−0.014	0.390	0	5	37	4	1

Jaan (2008) has developed and implemented a STATA program to estimate the Blinder − Oaxaca decomposition method for linear models. Suppose we wish to test the hypothesis posed by Spears (2013): are the differences in Z between two countries determined by access to sanitation? In other words, if the outcome variable is Z, then the slope coefficient of S must be negative in the regression:

$$Z = \alpha + \beta_1 S + \beta_2 \text{Educ} + \beta_3 \text{Age} + \beta_4 \text{hh} + \beta_5 \text{Female} + \text{Disturbance}$$

In order to divide the z-gap between countries A and B to a portion explained by the X-variables, namely, the determinants of Z, and to a portion that is left unexplained, Jaan (2008) provides the following `oaxaca` command:

```
oaxaca z s educ age hh female, by(country) noisily
```

The output from STATA is provided below. The `oaxaca` procedure first estimates two regressions, one for each country, and then performs the threefold decomposition. The option `noisily` produces all the results. From the regression results, we see that S is statistically significant in both countries, with the correctly hypothesized sign. In addition, the variables *age* and *female* are significant in the regression for country B.

The decomposition portion of the output is produced below the regressions. The mean of Z is −0.113 for country A, and −0.163 for country B, which is a z-gap of 0.05. This z-gap is divided into E, C, and CE fractions in the bottom panel of the results. The first part indicates that the mean B's z-score would increase, if it had A's features. This increase of 0.013 in our data is due to the differences in the X variables (educ, age, hh, female), which account for more than 25% of the z-gap. The second term (0.002) is the increase in A's z-score, if it had B's characteristics. The third part is the interaction term's contribution to the differences in the z-gap due to differences in endowments and coefficients. The total explained percent is roughly 0.04, which accounts for about 80% of all the differences in the first part (0.04/0.05).

Model for group 1

Source	SS	df	MS
Model	.119730861	5	.023946172
Residual	.00080991	16	.000050619
Total	.12054077	21	.005740037

Number of obs =	22
F(5, 16) =	473.06
Prob > F =	0.0000
R-squared =	0.9933
Adj R-squared =	0.9912
Root MSE =	.00711

z	Coef.	Std. Err.	t	P>|t|	[95% Conf.	Interval]
s	-.5683704	.0163639	-34.73	0.000	-.6030604	-.5336804
educ	-.0008976	.0007102	-1.26	0.224	-.002403	.0006079
age	.0001817	.0002268	0.80	0.435	-.0002991	.0006626
hh	-.002587	.002571	-1.01	0.329	-.0080373	.0028633
female	.0006676	.0044808	0.15	0.883	-.0088312	.0101665
_cons	.2114682	.0153683	13.76	0.000	.1788888	.2440475

Model for group 2

Source	SS	df	MS
Model	.087490873	5	.017498175
Residual	.000810077	12	.000067506
Total	.088300949	17	.005194173

Number of obs =	18
F(5, 12) =	259.21
Prob > F =	0.0000
R-squared =	0.9908
Adj R-squared =	0.9870
Root MSE =	.00822

z	Coef.	Std. Err.	t	P>|t|	[95% Conf.	Interval]
s	-.5824881	.0248694	-23.42	0.000	-.6366738	-.5283024
educ	-.0022181	.0016374	-1.35	0.200	-.0057857	.0013495
age	.0008094	.0002934	2.76	0.017	.00017	.0014487
hh	.00371	.0037233	1.00	0.339	-.0044023	.0118224
female	.0584388	.0167894	3.48	0.005	.0218579	.0950198
_cons	.1279324	.0381901	3.35	0.006	.0447234	.2111414

Blinder-Oaxaca decomposition Number of obs = 40

1: country = 0
2: country = 1

z	Coef.	Std. Err.	z	P>|z|	[95% Conf.	Interval]
Differential						
Prediction_1	-.1136818	.0161697	-7.03	0.000	-.1453738	-.0819898
Prediction_2	-.1639444	.0170196	-9.63	0.000	-.1973023	-.1305866
Difference	.0502626	.0234761	2.14	0.032	.0042503	.0962749
Decomposition						
Endowments	.013252	.0291634	0.45	0.650	-.0439073	.0704113
Coefficients	.0024129	.005565	0.43	0.665	-.0084943	.01332
Interaction	.0345977	.013986	2.47	0.013	.0071856	.0620098

Jaan (2008) also has the command for the twofold decomposition, which creates the weight of the nondiscriminatory coefficient (β^*) from a pooled regression. The input and the output screen shots from STATA for our example dataset are given below:

```
oaxaca z s educ age female, by(country) pooled
```

```
Blinder-Oaxaca decomposition                          Number of obs    =         40

      1: country = 0
      2: country = 1
```

z	Coef.	Robust Std. Err.	z	P>|z|	[95% Conf.	Interval]
Differential						
Prediction_1	-.1136818	.0161516	-7.04	0.000	-.1453384	-.0820253
Prediction_2	-.1639444	.0169848	-9.65	0.000	-.197234	-.1306549
Difference	.0502626	.0234383	2.14	0.032	.0043243	.0962009
Decomposition						
Explained	.04834	.023365	2.07	0.039	.0025455	.0941345
Unexplained	.0019226	.0048616	0.40	0.692	-.0076059	.0114511

The second part also confirms that the explained variation (0.04) is about 80% of the difference in the z-gap (0.05). Jaan (2008) also has a `detail` option that yields the individual contributions of each variable. The input and the output screen shots are given below:

```
. oaxaca z s educ age hh female, by(country) pooled detail

Blinder-Oaxaca decomposition                     Number of obs    =         40

      1: country = 0
      2: country = 1
```

z	Coef.	Robust Std. Err.	z	P>|z|	[95% Conf.	Interval]
Differential						
Prediction_1	-.1136818	.0161516	-7.04	0.000	-.1453384	-.0820253
Prediction_2	-.1639444	.0169848	-9.65	0.000	-.197234	-.1306549
Difference	.0502626	.0234383	2.14	0.032	.0043243	.0962009
Explained						
s	.0499761	.0228329	2.19	0.029	.0052244	.0947278
educ	-.0000545	.0004382	-0.12	0.901	-.0009134	.0008044
age	.0000892	.0007486	0.12	0.905	-.001378	.0015564
hh	-.0001059	.0003015	-0.35	0.725	-.0006969	.000485
female	-.0015648	.0033676	-0.46	0.642	-.0081652	.0050355
Total	.04834	.023365	2.07	0.039	.0025455	.0941345
Unexplained						
s	.0067235	.0146416	0.46	0.646	-.0219735	.0354205
educ	.0130068	.0166982	0.78	0.436	-.0197211	.0457347
age	-.0249642	.012683	-1.97	0.049	-.0498225	-.000106
hh	-.0229339	.0136479	-1.68	0.093	-.0496833	.0038156
female	-.0534453	.0123649	-4.32	0.000	-.0776801	-.0292106
_cons	.0835358	.0298693	2.80	0.005	.024993	.1420785
Total	.0019226	.0048616	0.40	0.692	-.0076059	.0114511

In this case we can confirm that the variable S has the highest impact on the observed differences, particularly with its contribution to the explained portion of the difference in the z-gap. Thus, the hypothesis proposed by Spears is validated for our example data set. From a policy angle, this example illustrates what researchers in recent years from the World Bank, NBER, and other health-based organizations are trying very hard to emphasize: water and sanitation should become a part of basic human rights, to generate better health outcomes and benefit from those gains. Rightly so, Sustainable Development Goal six is aimed at achieving this feat.

CONCLUSIONS

Rights to clean water and sanitation, along with affordable child care, may seem obvious. However, the real world experience is far from satisfying these basic rights. Policy makers have recently begun to discuss these issues, with specific reference to child nutrition indicators such as stunting and wasting. This chapter presented a host of policy proposals and programs that attempt to shape this policy agenda.

The influence of sanitation on health outcomes continues to receive a lot of attention among development economists, in particular. The STATA example follows one of the tools used, namely the analytical Blinder–Oaxaca decomposition, and illustrates the importance of socioeconomic variables, particularly access to sanitation. The STATA illustration shows that whether sanitation is the most important driver of well-being is ultimately an empirical question. We provide an example data set in the end-of-chapter exercises, and ask if the student can reject the null hypothesis for this data.

The overall purpose of this chapter is to integrate major themes connecting child nutrition, child care, and human rights pertaining to access to clean water and sanitation in the evidence generation and policy context. Research in this field, in recent years, produced by economists, health care specialists and public health officials, in conjunction with many global agencies, has generated a cohesive set of ideas that can form this chapter's take-away messages:

- Safe water and sanitation are key drivers of children's health outcomes.
- Stunting is a major problem, which is due to open defecation and lack of clean water.
- Improving overall health outcomes, including reduction in stunting, can come through a multifaceted approach, which includes education, promotional campaigns, and providing access to these basic human requirements.
- Successful programs are those that have buy-in from all the stakeholders in the system. Programs must integrate Ecohealth perspectives, be polycentric, and must be continuously monitored.

EXERCISES

1. Fig. 10.1 presents a set of linkages and pathways to obtain nutritional outcomes through better understanding of health, sanitation, water, and health care. Construct a conceptual framework that is more specific to your country of analysis or the region which you are studying as part of your research.
2. The data below captures the height-for-age (Z) scores for two countries (A = 1, B = 0), and other relevant independent variables as described in the text. Use STATA and decompose the

differences in Z between the countries, and test whether Spear's (2013) hypothesis holds for this data. What can you conclude about the role of S in health outcomes?

obs	Z	S	Country	educ	age	hh	Female
1	−0.294	0.660	1	5	57	3	1
2	−0.277	0.650	1	9	31	4	1
3	−0.254	0.660	0	5	31	4	0
4	−0.251	0.640	0	5	32	4	0
5	−0.245	0.620	1	9	32	3	1
6	−0.238	0.630	1	10.5	39	4	1
7	−0.230	0.600	1	9	33	3	1
8	−0.222	0.590	1	9	33	3	1
9	−0.204	0.570	0	9	33	2	0
10	−0.202	0.510	1	12.5	34	5	1
11	−0.190	0.500	0	17.5	60	4	0
12	−0.188	0.490	0	9	44	5	0
13	−0.175	0.480	0	9	41	3	0
14	−0.172	0.470	0	9	33	4	0
15	−0.172	0.460	0	10.5	33	4	0
16	−0.172	0.450	1	9	33	3	1
17	−0.153	0.440	1	10.5	38	3	1
18	−0.151	0.430	0	10.5	46	3	0
19	−0.134	0.420	1	11.5	35	4	1
20	−0.117	0.410	1	10.5	60	4	1
21	−0.114	0.400	1	5	35	3	1
22	−0.109	0.410	0	9	57	4	0
23	−0.104	0.400	1	10.5	48	4	1
24	−0.099	0.390	0	9	49	3	0
25	−0.096	0.380	1	10.5	58	3	1
26	−0.096	0.370	1	10.5	36	4	1
27	−0.096	0.360	1	9	46	3	1
28	−0.091	0.350	1	9	37	5	1
29	−0.073	0.340	0	9	38	3	0
30	−0.072	0.330	0	10.5	48	4	0
31	−0.070	0.320	1	10.5	42	4	1
32	−0.068	0.310	0	9	32	3	0
33	−0.063	0.300	0	12.5	58	4	1
34	−0.049	0.290	0	10.5	54	3	1
35	−0.049	0.340	0	5	37	5	1
36	−0.047	0.330	0	5	47	4	0
37	−0.043	0.320	0	9	46	3	0
38	−0.038	0.310	0	10.5	54	3	1
39	−0.020	0.300	0	9	36	4	1
40	−0.014	0.290	0	5	37	4	1

PART

PROGRAM EVALUATION AND ANALYSIS OF NUTRITION POLICIES

E

CHAPTER

11 METHODS OF PROGRAM EVALUATION: AN ANALYTICAL REVIEW AND IMPLEMENTATION STRATEGIES

RCTs may be appropriate for clinical drug trials. But for a remarkably broad array of policy areas, the RCT movement has had an impact equivalent to putting auditors in charge of the R&D department. That is the wrong way to design things that work. Only by creating organizations that learn how to learn, as so called Lean manufacturing has done for industry, can we accelerate progress.

—Ricardo Hausmann, 2016

INTRODUCTION

Suppose we want to answer questions such as: "What is the effect of providing free lunch at schools on attendance and grades?," "What is the effect of providing access to modern toilets and safe water, on stunting in children?," or "What is the effect of a cash-transfer or a pension program for the elderly on child nutrition in the household?" In order to correctly answer these questions, we should be able to find out how students who had the free lunch would have performed in the absence of the program, or the height of children in the absence of modern toilets, or similarly, we should first know, what would have happened to child nutrition in the absence of cash transfers. Questions such as: "What would have happened to these students in the absence of the free lunch program?" are known as counterfactual questions. Any of these questions is difficult to answer. At any point of time, a child is either exposed to this program or not. Similarly, a group is either exposed to the treatment (toilets or cash-transfer) or not. Answering these questions requires an understanding of program evaluation methods and their implementation strategies. In this chapter we review the most popular methods of program evaluation that are currently in use, and provide analytical examples. In doing so, we follow closely Babu et al. (2014).

Program evaluation estimates the impact of an intervention or a treatment. Evaluating the impact of food and nutrition programs on the targeted group is a key activity that can save resources and help in redefining the program to meet the necessary welfare goals. A sad truth in development practice is that the pilot programs are implemented poorly, and some of them are scaled-up even before they are evaluated for their impact on the targeted population. This is partly because the capacity for program evaluation is weak in many developing countries. In this chapter we review the methods of program evaluation, and present examples of the methods used to help practitioners take on such challenges on their own.

Nutrition Economics. DOI: http://dx.doi.org/10.1016/B978-0-12-800878-2.00011-6

Trying to answer the counterfactual question in practice and through data collection can be challenging. In order to answer the counterfactual question, say on free school lunches, we need information on a comparison group, or a control group which is not exposed to the program, but whose performance in attendance and exam grades is very similar to those of the students who availed themselves of the free lunch. While we may be able to find a group that did not participate in the free lunch program, it is not reasonable to compare this group's performance with those who received the treatment. Such a comparison is not reasonable, because the final performance on attendance and exam grades may be either due to the program, or due to pre-existing differences between the two groups.

In other words, programs and treatments such as free lunches, toilets, cash-transfers, medication, and information brochures are targeted to specific populations on locations, such as poorer or richer school children or neighborhoods. Usually, participants are screened for treatment based on specific measures such as height, weight, gender, or related metrics. Further, individuals are often asked to participate in these programs on a voluntary basis, which creates a problem of "self-selection." Families choose to send their children to school based on location, or on the school's reputation. For many of these and other reasons, the groups that do not participate in a given treatment are not a good comparison for those that are exposed to the said treatment. Consequently, any difference in the performance between these groups could be due to the program itself, or due to pre-existing differences, which creates the problem known as "selection bias."

In recent years, development economists have designed many new procedures to tackle the problem of selection "bias." They achieve this by decomposing the overall difference into a "treatment effect" and a bias term. The newer methods and research designs typically compare the results of a project between those who are exposed to the treatment and a control group, both before and after the said treatment. The recent approaches in the economics of impact evaluation are:

1. Randomization
2. Instrumental variables
3. Difference-in-difference
4. Regression discontinuity design.

Each of these methods seek to find the true impact of a policy or treatment by not only observing the participants' status before and after the treatment, but also by answering the counterfactual question "What would have been the participants' outcomes in the absence of the project?," both before and after the treatment.

Finally, which method should be used for evaluating a particular program or a policy depends upon a combination of factors: the program's design, implementation, the cost and difficulty of data collection, and the feasibility of implementation of the study. We will provide a short description of each of these methods in the next section. We also illustrate the implementation of some of these methods through simple examples using STATA.

ANALYTICAL METHODS

In this section we demonstrate various methods of program evaluation using STATA.

RANDOMIZATION

Recent research in development economics by Duflo, Kremer, and Robinson (2011), and Banerjee and Duflo (2007, 2008, 2009) indicates the importance of randomized trials and natural experiments. Duflo, Glennerster, and Kermer (2007) provide an excellent introduction to randomization and to all the recent econometric issues involved in impact evaluation.

Randomization is a technique that randomly assigns a pre-determined percentage of the eligible beneficiaries to the treatment, forming the treatment group. The remaining participants make up the control group. The difference in outcomes between the treatment and control group is the impact of the treatment.

We can use the example in Duflo et al. (2007) as a good starting point to illustrate the workings of randomization, and all the other methods. Suppose we decide to provide free textbooks to children in schools. What effect will this treatment have on average grades? That is, suppose we want to measure the impact of textbooks on learning. We can explore this issue by thinking about the average test scores of students in a given school i in two ways:

Y_i^T = the average test score if school i has free text books; and
Y_i^C = the average test score of the *same* school i if it has no free textbooks.

We are interested in the outcome:

$$Y_i^T - Y_i^C$$

While theoretically every school has these two *potential* outcomes, only one is observed. Now suppose in the population of schools, there are several with textbooks, while others are without textbooks. One easy method to evaluate the impact is to compute the averages from both groups, and then calculate the difference D, i.e.:

$$\begin{aligned} D &= E[Y_i^T|\text{school has textbooks}] - E[Y_i^C|\text{schools have no textbooks}] \\ &= E[Y_i^T|T] - E[Y_i^C|C] \end{aligned}$$

Now let us think of $E[Y_i^C|T]$ as the expected outcome of a student in the treatment group who was not treated, which is a theoretical possibility. Suppose we add and subtract this to D, then we can decompose the effects as follows:

$$= E[Y_i^T\,|T] - E[Y_i^C\,|T] - E[Y_i^C\,|C] + E[Y_i^C\,|T]$$

$$= E[Y_i^T - Y_i^C\,|T] + E[Y_i^C - Y_i^C\,|T]$$

"the treatment effect" + "the selection bias"

The first term is the *treatment effect*, which indicates the effect of treatment on the treated subject. It captures the average effect of textbooks in the schools where textbooks were offered. The second term is the *selection bias*. Selection bias, in this context, refers to the hidden score potential of some of the untreated schools, and refers to this hidden difference between treatment and control groups. In other words, it is possible that treated schools may have had different test scores on average, *even if they had not been treated.*

Suppose parents consider education as a high priority, push the children to do their homework, and undertake private tuition. In this case, $E[Y_i^C|T]$, would exceed $E[Y_i^C|C]$. In this case, the actual impact of D would be biased upward. The impact could also be biased downward if textbooks were given to schools in disadvantaged areas or to the control group.

As mentioned before, the entity $E[Y_i^C|T]$ is not observed, and hence, it is not possible for researchers to account for selection bias. Duflo et al. (2007) provide a good summary of the current research that is underway to correct this empirical problem. According to Duflo et al. (2007), randomization solves the selection bias issue.

Randomization, or assigning the treatment randomly, is a reasonable method that exposes the selected participants to a treatment. This is particularly useful if the finances required to support the project to the entire participant pool are insufficient. By definition, randomization places the same probability to each beneficiary being treated. The regression counterpart to D is:

$$Y_i = a + bT + e_i \tag{11.1}$$

where T is a dummy variable indicating assignment to the treatment group. The estimated coefficient $\hat{b}$ via OLS captures the impact of the policy. Duflo et al. (2007) point out how this estimation procedure solves the problem of selection bias, and further point out various strategies to correctly implement randomization. We end this section with a simple example to implement Eq. (11.1) using STATA.

IMPLEMENTATION IN STATA: TEST SCORES AND FREE LUNCH

Assume we have data on the test scores of students, some of whom have been treated to a free lunch program in their schools. In particular, suppose 1043 students were randomly treated to the free lunch program and 1346 were not, for a total of 2389 students in the sample. Then a regression counterpart of Eq. (11.1) in this context is:

$$Score_i = a + bD + e_i \tag{11.2}$$

where $Score_i$ is the score of the ith student in the sample, and D is a dummy variable which takes a value equal to 1, if the student in the observation was treated. The regression implemented in STATA produces the following output:

```
. regress score d

      Source |       SS       df       MS              Number of obs =    4778
-------------+------------------------------           F(  1,  4776) =    6.79
       Model |  45277.3565     1  45277.3565           Prob > F      =  0.0092
    Residual |  31855914.9  4776  6669.99893           R-squared     =  0.0014
-------------+------------------------------           Adj R-squared =  0.0012
       Total |  31901192.2  4777  6678.08085           Root MSE      =   81.67

------------------------------------------------------------------------------
       score |      Coef.   Std. Err.      t    P>|t|     [95% Conf. Interval]
-------------+----------------------------------------------------------------
           d |   6.206817   2.382272     2.61   0.009     1.536467    10.87717
       _cons |   1220.483   1.574075   775.37   0.000     1217.397    1223.568
------------------------------------------------------------------------------
```

The first line (`.regress score d`) is the command line in STATA that produces the OLS estimates of Eq. (11.2). We observe each student for two time periods, and hence, we have a total of 4778 observations. The important portion of the output is the regression which results in the last two lines. Based on the output we can write the estimated regression as:

$$Score_i = 1220.5 + 6.2D + e_i$$

From this simple example we see that participation improves the score by 6.2 points, hence indicating that the free lunch programs have some impact on test scores. The overall R-squared is 0.001, meaning that the model itself does not account for all variations in scores. The goal of this example is not to generate the best explanatory model, but to illustrate in the simplest terms the implementation of randomization with simplistic mock data on free lunches and scores.

INSTRUMENTAL VARIABLES ESTIMATOR

Instrumental variables (IV) is an econometric method that tracks those independent variables that determine program participation. Importantly, these variables at the same time have no influence on project outcomes. The IV method is used in statistical analysis to control for selection bias that arises due to the absence of variables that capture an individual's participation decision.

The IV method identifies the exogenous variation in outcomes attributable to the program, taking into account the possibility that the individual's decision to participate in the program may not have been random, but rather has an underlying motivation. The chosen instruments are first used to predict program participation. In the second step, we examine how the outcome indicator varies with the predicted values. That is, in the first step, the chosen instruments help us to predict those who would have been in the treatment group, and those who would have been in the control group, under the assumption that the participation decision was based on the chosen instruments.

The difference in outcomes between these predicted treatment and control groups is then the impact of the treatment. Often, one uses variation based on program location, availability, and characteristics as instruments, especially when endogenous program placement, or self-motivated participation decisions, seem to be a source of bias.

Chowhan and Stewart (2014) provide a methodical approach in their application of the IV estimator, to find out whether children shirk, while parents work. The researchers use data from the Canadian National Longitudinal Survey of Children and Youth. The observations are for adolescents between the ages of 12 and 17 in the years from 1996 to 2001. The researchers employ the IV method to test for the existence of endogeneity. Their results show that maternal employment is negatively related to TV viewing, physical and creative activities, eating breakfast, and allowances. These results show that the issue is not whether mothers work, but what children do when their mothers work.

You et al. (2016) is a recent study that investigates the paradoxical situation in China, where despite economic growth, there is a decline in mean nutrient intake. The econometric relation between household income and nutrient intake has two issues: household heterogeneity, and endogeneity. The authors tackle these issues using a quantile IV fixed-effects estimator, which controls for levels of income and nutrient intake in the population, and hence, addresses heterogeneity issues.

The results indicate that as income rises, there is a decline in fat intake for households with a moderate macronutrient intake. However, this is not true if a household has any member who is either obese or under-nourished. Finally, decline in overall nutrient intake in the face of economic growth is still possible, if price fluctuations are large, and if demand for nutrients is highly price elastic.

The efficiency of the estimator in Eq. (11.1) can be improved if we add some independent variables, say x_1 and x_2, such that we have:

$$Y_i = a + bT + \alpha_1 x_{1i} + \alpha_2 x_{2i} + e_i \tag{11.3}$$

Suppose we want to measure the weight of children who are in families that receive special food stamps that are valid only for a healthy diet, such as vegetables and fruits, and are not valid for unhealthy nutrition such as saturated fat in carbonated drinks, and excess sugar as in fruit loops.

Suppose our variable of interest in this outcome is Y_i for a given child in the observation. As usual, T reflects the time period of the treatment, and let us say that x_1 tracks the participation status ("received food stamps"). Unfortunately, in the real world, people do not follow the rules and find ways to hoodwink the system. In this selective food program assistance for instance, families may receive the food stamps, but may simply exchange them with their neighbors or friends, or with others for other goods such as cigarettes or even cash. Further, to complicate the process, the neighbors or friends receiving the said food stamps, who might end up using them, might unfortunately have been placed in our control group. This problem of *partial compliance* makes x_1 correlate with e_i, leading to a bias in the OLS estimator.

The IV estimation is useful if we have data on the treatment received (x_1) and also on another control variable, say z_1, which can be used as an instrument for the treatment actually received (x_1). In order for z_1 to qualify as a valid instrument for x_1, two conditions must be satisfied. First, it must be relevant, and this measure of relevance is usually given by a positive correlation between z_1 and x_1, or Corr $(z_1, x_1) \neq 0$. Second, the instrument must not be correlated with the random term, or the instrument must satisfy the exogeneity condition Corr $(z_1, e_i) = 0$.

Thus, an instrument that is relevant and exogenous can capture movements in x_1 that are exogenous. If the initial assignment of the treatment is random, then z_1 is distributed independent of e_i, and the problem of partial compliance can be addressed successfully. In practice, the IV estimator is implemented via a 2SLS procedure, where in the first stage the originally designed independent variable x_1 is regressed on all the exogenous variables and z_1. That is, in the first stage regression, we have:

$$x_{1i} = \gamma + \delta T + \beta_1 x_{2i} + \beta_2 z_{1i} + \epsilon_i \tag{11.4}$$

In the second stage, we go back and estimate Eq. (11.3), after replacing x_{1i} with $\hat{x_{1i}}$, the predicted value of x_{1i} obtained after the first-stage estimation in Eq. (11.4). That is, in the second-stage we estimate:

$$Y_i = a + bT + \alpha_1 \hat{x_{1i}} + \alpha_2 x_{2i} + e_i \tag{11.5}$$

We end this section with a simple illustration in STATA that implements the issues discussed in Eqs. (11.3–11.5).

IMPLEMENTATION IN STATA: IV ESTIMATION, TEST SCORES, AND FREE LUNCH

Recall that we have data on the test scores of students, some of whom have been treated to a free lunch program in their schools. We extend this example by adding observations on the same students for two time periods before treatment. Then a regression counterpart of Eq. (11.3) in this context is:

$$Score_{it} = \alpha + \beta_1 D_i + \beta_2 Treat + \delta D_i * Treat + \varepsilon_{it} \tag{11.6}$$

In Eq. (11.6) the variable D is a dummy variable that tracks whether the given student i in the sample "received the treatment" (or $D = 1$), and *Treat* is another dummy variable, that takes a value equal to 1 in the post-treatment period.

As discussed above, suppose we wish to use the IV method, and attempt to use an instrument for D, and let us say we have information on the family income level of each student. Consequently, we can regress D on all the variables in the system, including income, as indicated in Eq. (11.4), and we can obtain the first-stage estimates. We can then derive the second-stage results using the steps outlined in Eq. (11.5). We reproduce this portion of the STATA output below.

The STATA command in the first line (`ivregress 2sls score treat td (d = income), vce (robust) first`) is entered at the command line, which runs the IV regression using the 2SLS procedure.

Additionally, we specify that *income* is an instrument for D. The option `vce(robust)` is used to produce efficient standard errors to control for heteroscedasticity, and finally, the option `first`, produces the first stage estimates.

The first part of the output presents the first-stage regression of the exogenous variable D_i on all the exogenous variables, including *income*. The first-stage regression produces a decent fit of the model, and the parameter of interest, namely income, is significantly negative.

The second-stage regression results are in the second portion of the output. These results are key to the example, where $Score_i$ is regressed on all the exogenous variables. Note that the variable D is negative, but insignificant. Consequently, controlling for income levels, program participation does not necessarily have a significant effect on scores. The *Treat* effect is still positive and significant.

```
. ivregress 2sls score treat td (d = income), vce(robust) first

First-stage regressions
-----------------------

                                                Number of obs     =       7074
                                                F(   3,    7070)  =  953307.04
                                                Prob > F          =     0.0000
                                                R-squared         =     0.3342
                                                Adj R-squared     =     0.3339
                                                Root MSE          =     0.4047

------------------------------------------------------------------------------
             |               Robust
           d |      Coef.   Std. Err.      t    P>|t|     [95% Conf. Interval]
-------------+----------------------------------------------------------------
       treat |  -.4349979    .007227   -60.19   0.000    -.4491649   -.4208308
          td |   .9987585   .0007235  1380.54   0.000     .9973403    1.000177
      income |  -.0008559   .0002877    -2.98   0.003    -.0014198    -.000292
       _cons |   .4763401   .0155712    30.59   0.000     .4458159    .5068643
------------------------------------------------------------------------------


Instrumental variables (2SLS) regression          Number of obs =      7074
                                                  Wald chi2(3)  =   8030.66
                                                  Prob > chi2   =    0.0000
                                                  R-squared     =    0.2040
                                                  Root MSE      =    131.32

------------------------------------------------------------------------------
             |               Robust
       score |      Coef.   Std. Err.      z    P>|z|     [95% Conf. Interval]
-------------+----------------------------------------------------------------
           d |   -213.407   109.9931    -1.94   0.052    -428.9896    2.175474
       treat |   143.5663   47.93992     2.99   0.003     49.60583    237.5269
          td |   218.8511   110.0504     1.99   0.047     3.156234    434.5459
       _cons |   1107.694    47.9302    23.11   0.000     1013.752    1201.635
------------------------------------------------------------------------------
Instrumented:  d
Instruments:   treat td income
```

THE DIFFERENCE-IN-DIFFERENCE ESTIMATOR

To examine whether a particular intervention has an impact on our target population or on a specific target outcome, we use an econometric approach known as the difference-in-difference procedure. The difference-in-difference analysis helps us to answer the counterfactual question: what would have happened to the outcome, if the said intervention had not taken place? If the counterfactual question can be answered, then one can compare this answer to the factual situation, where the intervention or the treatment was initiated. The true impact of the treatment would then be the difference between the factual values and the answer to the counterfactual question.

The difference-in-difference strategy is often employed in policy analysis in the context of natural experiments. A natural experiment arises when some participants are exposed to a policy intervention, while others are not. The group that is exposed to the intervention is called the treatment group, and the group that is not exposed to the treatment is known as the control group. The difference-in-difference (DD) is a good econometric methodology to estimate the true impact of the intervention. Since it is not obvious a priori that an intervention is expected to have some outcomes, the DD method exposes the intervention to the treatment group, and leaves the control group out of the intervention. After the intervention, one can examine the differences in the outcomes from both the groups. The intervention is considered to have the necessary impact if differences in outcomes between the treatment and control group are significant.

The DD is a quasi-experimental technique that measures the causal effect of some nonrandom intervention. It is widely used in many branches of economics, to test the effectiveness of policy interventions. Angrist and Krueger (1991, 2001), Wooldridge (2002), and Stock and Watson (2011) provide several examples of such applications.

In order to apply the difference-in-difference method, one needs to ensure that some version of a natural experiment can be undertaken. We must have a clear idea about the intervention that has to be put in place. We must also identify the treatment group and the control group. We also need information about the timeline, that tells us when the intervention begins and ends. The timeline helps us to identify the characteristics of both the groups, before the intervention, and also after the intervention. Consequently, at the end of the intervention, we have four pieces of information:

1. The features of the control group before the intervention.
2. The features of the treatment group before the intervention.
3. The features of the control group after the treatment.
4. The features of the treatment group after the treatment.

From these four pieces of information, we can identify the changes that occur within each group, between the pre-treatment and post-treatment periods.

The difference-in-difference method captures the significant differences in outcomes across the treatment and control groups, which occur between pre-treatment and post-treatment periods. In the simplest quasi-experiment, an outcome variable is observed for one group before and after it is exposed to a treatment. The same outcome is observed for a second group (control group) that is not exposed to the treatment. The change in the outcome variable in the treatment group compared to the change in the outcome in the control group gives a measure for the treatment effect.

The difference in a given outcome of the project, between the treated members and the control group, can be computed *before* the treatment is implemented. This difference is usually referred to as the "first difference."

The difference in outcomes between the treatment and control groups can also be estimated sometime after the project has been in place, and this difference is referred to as the "second difference."

Under the difference-in-difference method, the impact of the treatment is the second difference minus the first difference. Basically, the impact of a treatment is the difference in outcomes between the treatment and control groups, after the project is implemented, taking into account all the already-existing differences in outcomes between the treatment and control groups.

It is usual to define a treatment as a form of policy intervention. The outcome variable can be defined as a child's height or weight, or a related index to denote nutritional status, which is observed for pre- and post-intervention time periods. We also have a control group, which received no intervention during the entire time period.

EXAMPLE: TEST SCORES AND FREE LUNCH

Let y be an outcome variable such as the BMI of a child. Then the effect δ_d of a policy intervention on BMI is given by:

$$\delta_d = (\bar{y}^d_{at} - \bar{y}^d_{bt}) - (\bar{y}^n_{at} - \bar{y}^n_{bt})$$

where the superscript d represents the observation where the child gets the treatment; the superscript n represents the observation when a child does not receive any treatment, i.e., the control group; the subscript bt denotes the time period, when there was no treatment; while subscripts at are observations captured post-treatment.

The expression $\bar{y}^d_{at}$ denotes the average BMI in the treatment group after the treatment, and the expression $\bar{y}^d_{bt}$ is the average BMI in the treatment group before the treatment; the expressions $\bar{y}^n_{at}$ and $\bar{y}^n_{bt}$ are the corresponding averages of the control group. The effect δ_d is the difference of the two differences between the treatment and control groups.

The first difference $(\bar{y}^d_{at} - \bar{y}^d_{bt})$ captures the difference in mean BMI before and after the implementation of the policy in the *treatment* group, while the second difference $(\bar{y}^n_{at} - \bar{y}^n_{bt})$ captures the difference in mean BMI between the two periods in the *control* group. The DD method cancels out the common trends in the control and treatment groups, and hence, the resulting difference between the two differences accounts for the effect of the promotion.

In practice, the DD estimator is implemented as a regression equation, where the level of the outcome variable, s, is used to estimate the model (see Stock and Watson, 2011, p. 493):

$$s_{it} = \alpha + \beta_1 X_i + \beta_2 T + \delta X_i * T + \varepsilon_{it} \tag{11.7}$$

where the level of the outcome variable, s_{it}, is the BMI of, say a given child i at time t, and X_i is a dummy variable that takes a value equal to 1, if the observed child belongs to the treatment group, and takes value equal to 0, if the child is in the control group. T is another dummy variable, which takes a value equal to 1 in the post-promotion period, and 0 otherwise. The DD estimator is δ, the coefficient of the interaction between X_i and T. Note this interaction term is a dummy variable, that takes a value equal to 1, only for the treatment group in the post-promotion period. ε_{it} is the error term assumed i.i.d Normal.

The other terms in the equation are the coefficients (α, β_1, β_2 & δ), of the regression model. The above regression is useful in deriving the "difference-in-difference" effects, namely, whether the treatment made a difference to s_{it} in the treatment group after the intervention. The regression coefficients help in deriving the four effects that were listed before for the control and treatment groups:

- "pre" effect for the control group: α
- "post" effect for the control group: $\alpha + \beta_2 T$
  - difference: β_2
- "pre" effect for the treatment group: $\alpha + \beta_1 X_i$
- "post" effect for the control group: $\alpha + \beta_1 X_i + \beta_2 T + \delta X_i * T$
  - difference: $\beta_2 + \delta$
    - difference-in-difference effect: δ

The important coefficient which provides the DD effect is δ, or the coefficient of the interaction between X_i and T. Note that this interaction term is a dummy variable that takes a value equal to 1, only if the observation belongs to the treatment group, during the post-treatment period.

Within this framework, additional treatment effects can easily be added, which is very useful for policy analysis, which we propose to add to our data. It is possible to extend the above equation accordingly, and estimate many general models to illustrate the importance of the DD method. We implement the DD method using STATA below.

IMPLEMENTATION IN STATA

Suppose we have data on the test scores of students, some of whom have been treated to a free lunch program in their schools, in which 1043 students were randomly treated to the free lunch program, while 1346 were not. We extend this example by adding observations on the same students for two time periods before treatment, for a total of 9556 students in the sample. Then a regression counterpart of Eq. (11.7) in this context is:

$$Score_{it} = \alpha + \beta_1 D_i + \beta_2\ Treat + \delta D_i * Treat + \varepsilon_{it} \tag{11.8}$$

We use the `regress score d treat td` command in STATA, to produce this portion of the output below. The estimated regression from the output is:

$$Score_{it} = 1009.2 + 12.14 D_i + 211.2\ Treat - 5.9 D_i * Treat + \varepsilon_{it}$$

The results tell us that being in the treatment group ($D = 1$), or in the treatment period ($Treat = 1$), increases the score of the ith student in the sample within the said groups.

However, our coefficient of interest ($\delta = -5.9$) is negative and insignificant. Note that the variable $D_i * Treat$ (variable td in the STATA output) is 1 only in the post-treatment period for the participants in the sample.

```
. regress score d treat td

      Source |       SS       df       MS              Number of obs =    9556
-------------+------------------------------           F(  3,  9552) = 3959.32
       Model |   104221169     3  34740389.5           Prob > F      =  0.0000
    Residual |  83812457.1  9552  8774.33596           R-squared     =  0.5543
-------------+------------------------------           Adj R-squared =  0.5541
       Total |   188033626  9555  19679.0817           Root MSE      =  93.671

------------------------------------------------------------------------------
       score |      Coef.   Std. Err.      t    P>|t|     [95% Conf. Interval]
-------------+----------------------------------------------------------------
           d |    12.1477   2.732345     4.45   0.000     6.791721    17.50367
       treat |   211.2415   2.553199    82.74   0.000     206.2366    216.2463
          td |  -5.940881   3.864119    -1.54   0.124    -13.51538    1.633613
       _cons |   1009.241   1.805384   559.02   0.000     1005.702     1012.78
------------------------------------------------------------------------------
```

REGRESSION DISCONTINUITY DESIGN

Regression discontinuity (RD) is an econometric method that can be employed when we have a cut-off point for a continuous variable, say a student's grade, or a female infanticide index. The cut-off point can be a deciding factor on who receives a given treatment. The outcome of the treatment is computed by comparing outcomes of the treated members who qualify for the treatment based on the cut-off, with outcomes for individuals who fail to meet the cut-off and, hence, are not exposed to the treatment.

Stock and Watson (2011) provide a good example to illustrate the application of the regression discontinuity estimators. Suppose students are mandated to enroll in a summer school for intensive study, if their grades from the previous year fall below a certain threshold. Suppose all the students whose grades were below the threshold attend the summer school. As a consequence, their grades in the next year should improve. Thus, we can define an outcome variable, which is basically the next-year's grade of all students, including all nonparticipants. If the minimum grade threshold was the only requirement to attend the summer school program, then it is reasonable to conclude that any jump in the next-year's grade can be attributed to summer school attendance. Since the cut-off occurs at some arbitrary threshold, the regression discontinuity can be estimated as:

$$GPA_i = a_0 + b_1 Program_i + b_2\, Grade_i + e_i \tag{11.9}$$

where GPA_i is the next-year's GPA of the ith student in the sample. $Grade_i$ is last-year's grade of the ith student. $Program_i$ is a dummy variable that takes a value equal to 1, if the student attended the summer program, and this depends on the mandated threshold ($Program_i = 1$, if $Grade_i < g_0$). Many alternative specifications and interactions could be included in Eq. (11.9) to capture the effect of the treatment. Finally, it is possible that many students who were supposed to attend the program did not, and further, many students who were above the cut-off attended the program anyway. Stock and Watson (2011, p. 495) discuss ways of dealing with these types of fuzzy situations. We provide a detailed example of Regression Discontinuity and an implementation using STATA in Chapter 13, when we discuss the economics of school nutrition.

PROPENSITY SCORE MATCHING AND PIPELINE COMPARISONS

Besides these methods, there are other methods that have also been used, such as Propensity Score Matching (PSM) and Pipeline Comparison. PSM is an econometric tool for identifying an appropriate group to compare with the recipients of the treatment. Specifically, the PSM method finds a comparison group made up of members who are not exposed to the treatment but, however, given their observable characteristics, had the same probability of receiving the project as individuals who were treated. The treatment's impact is then the difference in outcomes between these two groups.

Several interesting examples are available in the literature which demonstrate the application of PSM in the field of health and nutrition economics. Averett and Smith (2014) use the PSM method to investigate whether debt leads to obesity. This is an interesting research question, because the authors note that the average credit card debt in the United States was $5000, as of 2010. Using data from the National Longitudinal Survey of Adolescent Health, the authors examine "credit card debt" and "trouble paying bills," as two treatments to examine obesity, overweight, and BMI effects. Their results show that women's weight is driven primarily by the variable, "trouble paying bills."

Similarly, the Pipeline Comparison method evaluates outcomes of members who have already received the treatment with those members who have not yet received the said treatment, but are about to. This method assumes that the recipients, who are already exposed to the treatment, are similar to those who are about to receive the said treatment.

An interesting study conducted for Zambia by Chase and Sherburne-Benz (2001) from the World Bank that adopts the Pipeline Comparison method, asks the question: "What is the effect of the social fund on education and health outcomes?" The study concludes that the social fund impacts attendance positively and significantly; in the treated communities, 78% of children attended school, whereas only 71% of the children attended school in pipeline communities. The effect was significant only in urban areas, where 86% of children in treatment communities attended school, compared to 82% in PSM and 78% in pipeline communities.

Second, the study also finds that the social fund increases the percentage of children from appropriate age groups in school; 37% of children in social fund communities attend school at an appropriate age, compared to 25% in pipeline communities. Third, social funds also generate a higher share of education expenses in the total household budget; in social fund communities, households spend an average of 4.6% of earnings, whereas PSM communities spend 3.9% of their earnings on education. Randomization and these techniques have become very attractive ways to model program evaluation, and are currently used widely by development economists.[1]

Fan and Jin (2015) use an interesting combination of PSM and DD estimators to check if the SNAP program influences childhood obesity in the United States. The authors are not able

[1]Duflo, Kremer, and Jonathan (2011) discuss recent developments and applications. Also Duflo, Glennerster, and Kremer (2007) present the theoretical and methodological underpinnings, and also the limitations of these techniques.

to find any significant SNAP effects among 12-to-20-year-old participants. Their research highlights the importance of PSM techniques, and addresses selection bias and also alternative specifications.

Deininger and Liu (2009) use PSM, DD, and pipeline comparisons to determine the economic impact of women's self-help groups (SHGs) in India. This is an interesting study because subsamples of the population were exposed to the program at different points in time. That is, the program of interest started in 2001 and treated some sample households, while not treating others. The program was reintroduced in 2003 and introduced the unexposed households from the first round, while the previously treated households had already been in the program for three years. Finally, in 2006, when the impact was studied, the sample had households covered either for three or six years, depending upon the start date. This data feature allows for a pipeline comparison because the households who did not enter the program in 2001 could be used as a control, while those exposed to the program for six years are the "treated."

Using this method, Deininger and Liu (2009) note that longer exposure to the program benefits the participants. The benefits of the program also exceed the cost, where the poorest households also benefit from the treatment, a conclusion that is drawn due to the importance given to heterogeneity considerations.

Swain and Verghese (2014) also conducted a similar study using PSM and pipeline comparison methods to check for the impact of training provided by the SHGs in India. As in the previous paper by Deininger and Liu (2009), this particular research also confronts the selection bias issue, where participants in the SHG training programs can self-select based on their requirements. Swain and Verghese (2014) are able to track members who have been active in the program for a period time, and also new members who have not yet received any financial services. By using a pipeline method, the authors correct for self-selection, and resolve the endogeneity issue via PSM. They show that business training has a greater impact than general training.

Novak (2014) has also used the PSM technique to critically examine the public policy issue of clean water access in Senegal, and its effects on child health. In terms of informing policy, this paper shows that what is traditionally considered "clean" water is really no different from unsafe water, and in some instances is even worse in reducing child diarrhea. For example, public sources of water, say from taps and protected wells, are not qualitatively better than water from unimproved sources. The main reason for this is because of the outdated infrastructure of public sources, such as faulty pipes. Indeed, water pumped into the yard is better at reducing child diarrhea than water from other sources.

Torero et al. (2006) the "Accelerated Electricity Access Expansion" for Ethiopia conducted by the World Bank (year/reference) illustrates a good example of the PSM method. This intervention analysis asks several interesting questions concerning the impact of specific treatments: How does an expansion of electricity access help local development and improve consumer welfare? How does electricity usage change when energy efficient appliances are introduced? How do electricity connection rates change when financing mechanisms are altered? What is the efficient subsidy amount that maximizes usage rate per dollar invested in compact fluorescent lamps (CFLs)? Do alternative methods of delivering CFLs, including private purchase, vouchers, and public distribution channels, have an effect on household consumption?

The study selects towns with and without electricity, and applies the same selection criteria such as access to public phones, types of roads, types and the numbers of secondary schools, and population. Those households that have access to electricity are matched to those that do not have electricity, and these households are taken from towns that have comparable towns, to those without electricity. The PSM estimates are derived based on: size of the household, household's family composition, head of the household's ethnicity, head of the household's age, head of the household's education level, gender, access to water, and related variables. Additionally, 10 − 50% of eligible survey respondents are randomly assigned vouchers for a discount charge to cover the connection charges. Further, between 10% and 50% of eligible survey respondents are randomly assigned vouchers for a discount charge to cover the cost of the CFLs. Finally, the study also uses a subsample of towns randomly selected for the study where no vouchers will be distributed. The World Bank website also presents other interesting applications of this method.

Rosenbaum and Rubin (1983), Imbens (2000), and Dehejia and Wahba (2002) all provide detailed descriptions and methods associated with the development and application of PSM. The PSM method calibrates the treated observations with the control group, and produces a score that matches the groups as closely as possible, so as to reduce any bias in the way the treatment was initially administered.

Most of the time, the treatment may not be randomly administered, and it may be biased because of unobservables. PSM is a method that can reduce this bias. Let $D = 0, 1$ be an index of exposure to treatment, and X, a matrix of pre-treatment characteristics. Rosenbaum and Rubin (1983) define a propensity score $p(X)$ as the conditional probability of receiving a treatment, given pre-treatment characteristics:

$$p(X) = pr(D = 1X) = E(D|X)$$

Then, using Rosenbaum and Rubin (1983), Becker and Ichino (2002) express the Average effect of Treatment on the Treated (ATT) as:

$$EY_{1i} - Y_{0i}|D_i = 1, p(X_i) = E[E\{Y_{1i}D_i = 1, p(X_i)\} - E\{Y_{0i}D_i = 0, p(X_i)\}|D_i = 1]$$

The PSM method produces a balancing score, such that those observations with the same propensity score have the same distribution of characteristics, regardless of treatment status. Once we have the propensity score, we can assume that the treatment was randomized, and hence, all treated and control units are equivalent.

IMPLEMENTATION IN STATA

Becker and Ichino (2002) provide detailed procedures and routines to implement PSM in STATA. After calibrating the propensity score, the data is sorted in a way that places propensity scores in matching intervals, in order to produce ATT. The four usual methods of sorting are *Radius Matching*, *Kernel Matching*, *Nearest-Neighbor Matching*, and *Stratification Matching*.

With *Stratification Matching*, the variation in the propensity score is split into different intervals such that the treated and control units within each interval have the same propensity score on average. *Nearest-Neighbor* matching improves on the *Stratification* method, such that all treated units find a match, given that some blocks may not have enough matching pairs under the latter method. With *Radius Matching*, each treated unit is matched with the control units whose propensity score falls into a predefined neighborhood of the propensity score of the treated unit. With *Kernel Matching*, all treated are matched with a weighted average of all controls with weights that are inversely proportional to the distance between the propensity scores of treated and controls.

The STATA algorithm developed by Becker and Ichino (2002) implements the following steps:

1. Fits a probit or a logit model specified by the researcher to evaluate the likelihood of participation or treatment;
2. Splits the sample into k equally spaced intervals of the propensity score, where the researcher determines k, or where the program's default is 5;
3. Tests whether within each interval the average propensity score of treated and control units does not differ;
4. Tests the Balancing Hypothesis: whether within each interval, the means of each characteristic are different between treated and control groups.

EXAMPLE

Suppose we have the following observations on 32 students based on our example on Textbooks and Score. In table below, $D = 1$ indicates whether the student received the treatment, and the variable $Treat = 1$ indicates that the treatment was administered. Income and Score represent family income and the test scores. Our goal is to examine the ATT of providing free textbooks on exam scores.

obs	*i*	*D*	*Treat*	Income	Score
1	1	0	0	3276.009	45
2	1	0	1	5992.015	45.5
3	2	0	0	7304.947	55
4	2	0	1	9454.144	55.8
5	3	0	0	7188.219	63
6	3	0	1	2268.913	60.2
7	4	0	0	9344.465	55.5
8	4	0	1	2927.189	65.3
9	5	0	0	6691.779	45.6
10	5	0	1	3209.162	48.7
11	6	0	0	4107.006	54.7
12	6	0	1	7766.274	55.8
13	7	0	0	1546.991	60.2
14	7	0	1	7514.323	63.2

Continued

obs	i	D	*Treat*	Income	Score
15	8	0	0	1737.605	55.7
16	8	0	1	4001.105	65.3
17	9	1	0	4410.61	55.6
18	9	1	1	8482.669	58.7
19	10	1	0	8353.924	64.7
20	10	1	1	3675.96	65.8
21	11	1	0	2179.76	70.2
22	11	1	1	2280.997	73.2
23	12	1	0	10067.33	65.7
24	12	1	1	2783.191	75.3
25	13	1	0	3060.225	65.6
26	13	1	1	3036.27	68.7
27	14	1	0	2168.629	74.7
28	14	1	1	2626.472	75.8
29	15	1	0	2570.849	80.2
30	15	1	1	2837.423	83.2
31	16	1	0	7736.791	75.7
32	16	1	1	9037.849	85.3

To generate the propensity score, we begin with the `pscore` command:

```
pscore d treat income, pscore(ps1) blockid(blockf1) comsup level(0.001)
```

The `pscore` command estimates the propensity score and tests whether the balancing property holds. The propensity score is the probability of getting a treatment for each student, and the balancing property tests the assumption that the observations with the same propensity score have the same distribution of characteristics (income), independent of whether the student receives a textbook. The option `pscore(ps1)`specifies that the variable name for the estimated propensity score is *ps1*, while `comsup` specifies that the analysis be restricted to the region of common support, within the distributions of the treated and control groups. The output from STATA below shows that the regions of common support are [0.406, 0.506], and the probit estimation results are also reproduced in the first step:

```
****************************************************
Algorithm to estimate the propensity score
****************************************************


The treatment is d

          D |      Freq.     Percent        Cum.
------------+-----------------------------------
          0 |         16       50.00       50.00
          1 |         16       50.00      100.00
------------+-----------------------------------
      Total |         32      100.00


Estimation of the propensity score

Iteration 0:   log likelihood =  -22.18071
Iteration 1:   log likelihood = -22.008047
Iteration 2:   log likelihood = -22.008047

Probit regression                                 Number of obs   =         32
                                                  LR chi2(2)      =       0.35
                                                  Prob > chi2     =     0.8414
Log likelihood = -22.008047                       Pseudo R2       =     0.0078

------------------------------------------------------------------------------
           d |      Coef.   Std. Err.      z    P>|z|     [95% Conf. Interval]
-------------+----------------------------------------------------------------
       treat |  -.0117745   .4449567    -0.03   0.979    -.8838737    .8603246
      income |   -.000048   .0000816    -0.59   0.557     -.000208     .000112
       _cons |   .2453721    .522652     0.47   0.639     -.779007    1.269751
------------------------------------------------------------------------------


Note: the common support option has been selected
The region of common support is [.40610678, .55619935]
```

```
Note: the common support option has been selected
The region of common support is [.40610678, .55619935]


Description of the estimated propensity score
in region of common support

                 Estimated propensity score
-------------------------------------------------------------
      Percentiles      Smallest
 1%    .4061068        .4061068
 5%    .4129626        .4129626
10%    .4201823        .4196087       Obs                  30
25%    .4495198        .4207558       Sum of Wgt.          30

50%    .5179488                       Mean           .4957299
                        Largest       Std. Dev.      .0512459
75%    .5392598        .5494117
90%    .5495264        .5496412       Variance       .0026261
95%    .5559884        .5559884       Skewness      -.3949527
99%    .5561993        .5561993       Kurtosis       1.562483


*******************************************************
Step 1: Identification of the optimal number of blocks
Use option detail if you want more detailed output
*******************************************************


The final number of blocks is 3

This number of blocks ensures that the mean propensity score
is not different for treated and controls in each blocks
```

```
The balancing property is satisfied


This table shows the inferior bound, the number of treated
and the number of controls for each block

  Inferior |
  of block |         D
 of pscore |         0          1 |     Total
-----------+----------------------+----------
        .4 |        14         16 |        30
-----------+----------------------+----------
     Total |        14         16 |        30

Note: the common support option has been selected


*******************************************
End of the algorithm to estimate the pscore
*******************************************
```

The balancing property is satisfied, and the algorithm has identified three blocks to match the scores. It is of interest to see if the textbooks affect the scores between the matched treatment and control groups. The command `attnd score d treat dt income, pscore(ps1)comsup` estimates the ATT using the nearest-neighbor matching method, and the results are:

```
ATT estimation with Nearest Neighbor Matching method
(random draw version)
Analytical standard errors

---------------------------------------------------------
n. treat.   n. contr.        ATT    Std. Err.           t
---------------------------------------------------------

       16           9     11.600        3.343       3.470

---------------------------------------------------------
Note: the numbers of treated and controls refer to actual
nearest neighbour matches
```

The ATT estimate indicates that the average treatment for the treated is 11.6 points for the participants. The command `atts score d treat dt income, pscore(ps1) blockid(blockf1) comsup` generates the ATT using the stratification method. The result below indicates that the ATT in this case is 15.5 points:

```
ATT estimation with the Stratification method
Analytical standard errors

---------------------------------------------------------
n. treat.   n. contr.        ATT    Std. Err.           t
---------------------------------------------------------

       16          14     15.536        2.844       5.462

---------------------------------------------------------
```

For radius matching, the command `attr score d treat dt income, pscore(ps1) radius (0.001) comsup` generates the following output, with a positive ATT of 16.4 points on the treated:

```
ATT estimation with the Radius Matching method
Analytical standard errors

---------------------------------------------------------
n. treat.   n. contr.         ATT   Std. Err.           t
---------------------------------------------------------

        3           3      16.400       5.994       2.736

---------------------------------------------------------
Note: the numbers of treated and controls refer to actual
matches within radius
```

Finally, the command `attk score d treat dt income, pscore(ps1) comsup bootstrap reps (50)` is for the kernel matching method, and STATA generates the following output:

```
Bootstrapping of standard errors

command:      attk score d treat dt income , pscore(ps1) comsup bwidth(.06)
statistic:    attk       = r(attk)

Bootstrap statistics                         Number of obs    =         32
                                             Replications     =         50

------------------------------------------------------------------------------
Variable     |  Reps  Observed       Bias  Std. Err.  [95% Conf. Interval]
-------------+----------------------------------------------------------------
        attk |    50  15.60095 -.5274665  3.295831   8.977731   22.22417   (N)
             |                                       8.797622   21.67441   (P)
             |                                       10.83668   24.23825  (BC)
------------------------------------------------------------------------------
Note:  N   = normal
       P   = percentile
       BC  = bias-corrected
```

```
ATT estimation with the Kernel Matching method
Bootstrapped standard errors

---------------------------------------------------------
n. treat.   n. contr.         ATT   Std. Err.           t
---------------------------------------------------------

       16          14      15.601       3.096       5.039

---------------------------------------------------------
```

The kernel matching method indicates an ATT of 15.6 points for the treated group. The results from all matching methods are reasonably close, and are consistent with each other in producing a positive ATT, with significant t-values. STATA also has an nnmatch command which is used to test the robustness of the estimates. The nnmatch command applies the nearest-neighbor matching method directly to data without resorting to the pscore command in the first stage. This will enable a comparison of the ATT estimate to check for its reasonableness. The command and the output from STATA are given below:

```
nnmatch score d treat dt income, tc(att) m(1)
```

```
Matching estimator:  Average Treatment Effect for the Treated

Weighting matrix: inverse variance          Number of obs          =       32
                                            Number of matches  (m) =        1
```

score	Coef.	Std. Err.	z	P>|z|	[95% Conf. Interval]	
SATT	13.625	2.997914	4.54	0.000	7.749195	19.5008

```
Matching variables:  treat dt income
```

Once again, the ATT coefficient of 13.6 indicates that the results from the PSM method using the example data are consistent at a 5% level of significance.

THE DEATON–IMBENS DEBATE

Just as these methods have developed, they have also generated doubt and criticisms that expose some of the limitations in program evaluation. Heckman-Urzua (2009) and Deaton (2010) have taken exception to randomized control trials (RCTs) or experimental intervention methods, and question the theoretical foundations of the approach. Deaton (2010), in particular, poses some serious questions about research using RCTs with regard to the local average treatment effect, or LATE, developed by Imbens and Angrist (1994).

Deaton (2010) criticizes the instrumental variable estimation methods, because according to him, it is not clear whether the chosen instruments in this line of research are orthogonal to the disturbance terms of the original equation.[2] Further, in many instances, Deaton (2010) notes that randomization may be confounded by design. Deaton (2010) points to the deworming studies conducted in Kenya as an example of this aspect.

Deaton (2010) also mentions the problem of heterogeneity, where the parameter of interest can vary across the sample. Deaton (2010) notes that the local average treatment effect (LATE) may not be accurate if underlying heterogeneity is not considered in the analysis.

[2]The debate is also mired with alternative interpretations of technical terms, such as "structural" and "exogeneity" (see Imbens, 2010).

Imbens (2010) responds to these criticisms in detail, by noting the value of RCTs in satisfying internal and external validity concerns. Internal validity refers to the estimator's bias within a subsample of the population chosen for the experiment. External validity refers to the estimator's capacity to be replicated in other population groups (Box 11.1).

BOX 11.1 ARE RCTs THE GOLD STANDARD?

Pinstrup-Andersen (2013) has voiced concerns about the value of RCTs in helping food systems deliver appropriate nutritional value to the needy. These concerns are motivated by the following stylized facts:

- There are strong political and other interest groups, whose interests conflict with those of the malnourished poorer sections of the population.
- Policy interventions have lacked the necessary political momentum to make food systems nutrition-sensitive, particularly with the agro-sector supplying inputs to the food processing industry, which in turn, creates a profitable "ultra-processed" product in the value chain.
- Policies to promote research to aid output growth (of fruits, vegetables, and nutrition content) are very limited.
- Policies to reduce consumption of calorie-dense foods and sweeteners, alongside regulation of jazzy advertisements, "cool" marketing, and promotion are scarce.

Based on the above observations, Pinstrup-Anderson points out that policy makers are fixated with the idea that the only evidence for action is through RCTs. However, there are many limitations with this approach:

- RCTs are generally impossible to scale-up to the level of the current food system.
- RCTs are effective only in small, usually unimportant, projects.
- It is not possible to randomize many treatments, mainly because of long lags, or the "pathway effect."
- Existing evidence obtained from RCTs is inconclusive, and the relevant interventions are not obvious.

Standard pathways through which food systems can affect nutrition are already well established. Existing evidence convincingly shows that the main drivers of nutrient allocation are prices, income, women's time allocation, dietary diversity, advertising and promotion, etc. Why then should one only accept as evidence that which is derived via RCTs?

Even if one is fixated with RCTs, the evidence from this method will be limited to small food systems, such as kitchen gardens. However, this will miss the big picture, where really important improvement in nutrition lies in changing individual and industrial behavior through productivity improvements in fruit and vegetables, and other workplace policies to change women's time allocation. Based on the above observations, Pinstrup-Anderson doubts whether RCTs can help nutrition-sensitive issues move from rhetoric to actual action on a large scale.

With regards to the internal validity of IV estimators, Imbens (2009) points out that IV estimators can identify the LATE of subpopulations considered within the frame of the research question at hand.

Imbens (2010) uses Angrist's (2009) study on the earnings of Vietnam veterans to illustrate the usefulness of the IV approach. Imbens (2010) also notes that the assumptions underlying the application of IV are important, and in this context: the instrument is exogenous by design (like the draft lottery number); the instrument (or in this case, the lottery number) has no direct effect on the outcome; and monotonicity (any man who would serve if not draft eligible, will serve if draft eligible). Basically, Imbens (2010) argues that IV applications that satisfy the three assumptions, as in Angrist's (2009) study, can produce sufficient statistics to satisfy LATE interpretations. Imbens (2010) also illustrates how experimental estimates can be compared with structural estimates to test for external validity.

Dehejia (2013) presents an excellent summary of the methods discussed in this chapter, alongside a balanced review that evaluates the merits and limitations of each empirical approach. The following figure from Dehejia (2013) summarizes the relative merits of experimental and observational studies:

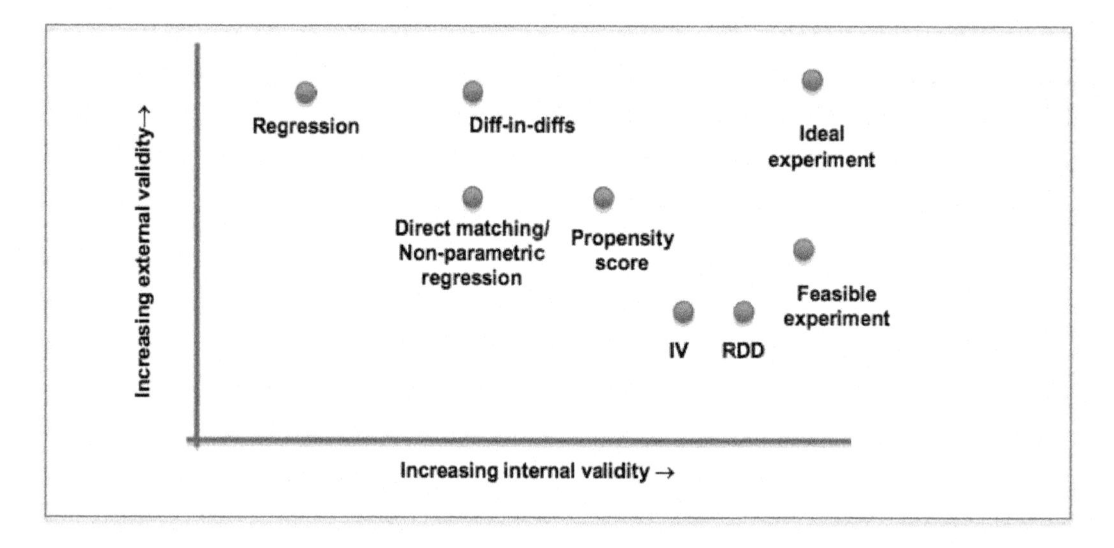

Based on Dehejia, R., 2013. The Porous Dialectic. WIDER Working Paper No. 2013/11, United Nations University., we present a brief summary of the empirical methods and note their relative merits in the following table:

No.	Methods	Pros	Cons
1.	Randomized Control Trials (RCTs)	Sample means are unbiased estimators of the their populations	Breakdown of the experimental protocol
2.	Instrumental Variables (IV)	Provides a consistent estimate of LATE on the compliers	Has a lower degree of external validity than an ideal experiment
3.	Regression-Discontinuity (RD)	Has better internal validity than IV, through internal tests	Difficult to test for external validity
4.	Direct Matching Methods (DMM)	Can be applied to any dataset, hence high external validity	If unobservables are important, then will not satisfy internal validity
5.	Propensity Score Matching (PSM)	This is better than PSM for internal validity	Potential for misspecification and small sample bias
6.	Linear Regression	Satisfies external validity	Sample-selection bias, and limited to linear specification
7.	Non-Parametric Regression	Not dependent on functional form, or self-selection. Has a high degree of internal validity	Not easy to implement if there are many covariates
8.	Difference-in-Difference (D-D)	Has a higher degree of internal validity than OLS	Linearity assumption is still questionable

As a way forward, for future empirical work, Dehejia (2013) points out a few methods that can combine experimental and nonexperimental techniques:

1. Regression and matching methods can be used to control for issues related to experimental implementation.
2. Random effects and other OLS estimators can be used to adjust for variation in treatment and in experimental implementation.
3. Nonexperimental methods may also set protocols and help design "experiment-like" observational studies. Pre-treatment units, treatment designs, etc. can be worked into the observational methods.

In this context, like Deaton (2010) and White (2014), the study from the International Initiative for Impact Evaluation (3ie), also notes the challenges that still need to be correctly addressed with regard to RCTs, because the results often fail to match expectations. However, like Dehejia (2013), White (2014) emphasizes the need for more studies. In order for RCTs to become more widely accepted, White (2014) notes the following challenges and guidelines:

- Most RCT studies are "nudges" that elicit responses from marginal changes in incentives, and hence, are not amenable to large-scale replication, particularly in other contexts.
- Most RCT studies are also not to be treated as substitutes for overall development proposals and projects.
- The theory of change still requires solid factual analysis with good assumptions and predictable causal chains.
- Most RCT studies still have difficulty in deriving the full costs of interventions, and the corresponding benefits, which makes political buy-ins very difficult.
- More RCT studies, which emphasize the need for internal validity and external validity are needed.
- Pooling of many studies into different categories for replication may be helpful.
- Controlling for self-selection and collection of time-invariant unobservable variables is necessary.

BOX 11.2 PRACTICAL TIPS FOR RCTs

In a sequence of papers, White (2014) has indicated some practical "how to" tips for researchers engaged in intervention studies, and evaluation methods.[a]

How to design a randomized evaluation:

- Define the treatment adequately, including eligibility criteria.
- Determine levels of assignment, treatment, and analysis.
- Identify an eligible population.
- Decide on the type of trial, and the randomized control mechanism.
- Fully write out the plan of analysis.
- Draw a sample for analysis.

How to manage an RCT:

- Construct an *ex ante* design, with better planning, timeline, and an early start.
- Avoid researcher capture.

(*Continued*)

(CONTINUED)

- Ensure all important stakeholders are invested in the project.
- Plan and anticipate all possible difficulties and have alternatives in place.
- Expect some opposition to the treatment and randomization, and make suitable plans.
- Maintain research ethics, provide adequate explanations, and make the process transparent.
- Monitor constantly, maintain the integrity of the design, and address changes if needed.

[a]See the details in http://www.3ieimpact.org/media/filer_public/2015/09/18/how_to_design_a_randomised_evaluation-hw.pdf, http://www.3ieimpact.org/media/filer_public/2016/02/17/how_to_manage_a_randomised_evaluation-hw.pdf, and also the corresponding youtube videos. Also see Khandker (2010) for detailed discussions of these topics and their implementation.

CONCLUSIONS

In recent years the development community has seriously begun to address the issue of the impact of development programs on targets. Using data from sample populations, they use different econometric techniques to control for self-selection and related problems with design.

In this chapter, the emphasis has been that the success of a public policy intervention program depends upon many factors, but is most crucially dependent upon the characteristics of the recipients, and nonrecipients. The chapter also describes the recent econometric advances that help evaluate the success of an implementation. For instance, the chapter discusses randomization, instrumental variables, difference-in-difference, propensity-score matching and regression discontinuity, and the value of policy implementation. In each case, the STATA illustration helps to position theory with results from data. The effect of the said treatment is captured in the estimation process, and policy makers are often interested in knowing whether the treatment has a significant and positive effect. The STATA implementation helps to capture this information in each section. We allow the students to explore the data and derive policy information in the exercises at the end of the chapter.

We also highlighted the pros and cons of various methods, and the debates that surround the choice and use of these methods under various data and program design situations. We invite the reader to explore these methods further.

EXERCISES

1. Use the data given in the section Propensity Score Matching and Pipeline Comparisons, and implement a program in STATA that satisfies an RCT, as in Eq. (11.1).
2. Use the **ivregress** command and estimate an IV regression as given in the implementation under the section Instrumental Variables Estimator.
3. How will you use the DD estimator to check for the importance of the treatment in this example? Is the DD estimate significant? In this exercise you are asked to combine the DD estimator with the PSM. This is a good method to control for observable heterogeneity in the data.

CHAPTER

NUTRITIONAL IMPLICATIONS OF SOCIAL PROTECTION: APPLICATION OF PANEL DATA METHOD

12

Strategy to alleviate poverty is as temporary as poverty itself.
—Anonymous

Economic growth is necessary, but not sufficient for poverty reduction or improving nutrition. Further poverty reduction is necessary, but not sufficient to reduce malnutrition. The economic development process may leave certain sections of the people behind, who may not have the opportunity to participate in the growth process. How do we protect these sections of the population? What policies and programs will help them to come out of poverty in the short and long run? What social protection and social safety net programs can help in improving the nutritional status of the poor and vulnerable groups when they do not benefit from economic growth?

In answering these questions, governments need evidence to design the programs and policies to protect their poor and vulnerable population. Policy makers are primarily interested in protecting the poor at the lowest cost possible and identifying the program design that will have longer term benefits for the poor, including human development. Further, in order for social protection programs with nutritional objectives to be seen as contributing to the national development objectives, in the short run to reduce acute poverty and in the long run to increase human capital development, they need to be evaluated and the documented evidence communicated effectively to the policy makers (Alderman and Mustafa, 2013). In a global context, how can social protection programs contribute to the Sustainable Development Goals? In this chapter we address these issues and provide the current status of research on social protection and its implications for nutritional outcomes.

A major concern for the policy makers is that social protection involves budgetary decisions made on an annual basis which have to be defended in parliament with the costs and benefits shown. The general policy debate at the national level often revolves around how much one could spend to protect the poor and to provide the safety net. Are there alternative investments in health and education that are more beneficial and cost-effective to support the poor and malnourished? How could one improve the cost-effectiveness of the program implemented in terms of reaching the real needy population with limited resources and to get the best results? Social protection program investments often compete for the resources needed for infrastructure investments and research in agriculture and technology development. Such investments can help rural communities to increase their productivity and market access, and in the long run can benefit them in terms of reduced poverty and increased food and nutrition security.

Nutrition Economics. DOI: http://dx.doi.org/10.1016/B978-0-12-800878-2.00012-8

Political motivations for implementing social protection programs are often challenged by opposition parties. Finally, social protection designed in one way in one country may not work in another country, or in another setting. Also, social protection may not be the best intervention in all countries, or in all the regions within a country. Understanding how social protection programs help achieve nutritional goals through different pathways, and how to study such pathways continues to be a key area of research. We take a closer look at these issues in this chapter.

PLACING SOCIAL PROTECTION PROGRAMS IN A LARGER DEVELOPMENT CONTEXT

In the large scheme of development interventions and in attempts to reduce hunger and poverty, social protection programs play a key role in both developed and developing countries. Timmer (2014) places social protection programs in the following framework. Ending hunger and poverty in any country requires consistent policies that help the poor to come out of poverty over a long period of time. In the short run, these policies and programs could be designed and implemented at the macro, meso, and micro levels.

At the macro level national emergency plans, disaster reduction strategies, and allocation of resources for food aid are used for emergency interventions. In the short run, e.g., managing food price volatility and the allocation of resources for food and nutrition interventions including social protection becomes important.

At the meso level, or at the market level, interventions in selected food markets in specific regions of the country affected by natural disaster or supply shortfalls, e.g., or in selected commodity markets, may be necessary.

At the micro level in the short run, social protection programs help in reducing the vulnerability to shocks, and enable households to cope better with short-run food and nutrition insecurity. Social protection programs could also help in increasing the resilience of households and individuals to shocks and emergencies.

In the long run, at the macro level, policies and programs that help in the inclusive economic growth and management of food markets, and provide stability to the food supply are the focus. At the meso and market levels, development of the food market infrastructure and liberalization of internal and external food markets remain key interventions. At the micro level in the long run, policies and programs focus on poverty reduction goals, with increased access to nutritious food and ensuring that food comes from sustainable systems (Timmer, 2014). Table 12.1 below maps these interventions at various levels.

Social protection programs are designed and implemented at the global, national, and decentralized levels. Global interventions by international agencies like the World Food Program largely address the emergency and disaster assistance needs of the affected populations in developing countries (WFP, 2015). Multilateral agencies like the World Bank have been involved in helping countries to design and implement social protection programs through development funding, and through technical assistance (Grosh and Del Ninno, 2005).

At the country level, major interventions such as Bosa Familia in Brazil; food stamp programs and the Women, Infant, and Children (WIC) program in the United States; the National Rural

Table 12.1 Mapping of Interventions and the Levels of Economy for Nutritional Outcomes

	Emergency Interventions	Short-Run Interventions	Long-Run Interventions
Macro level	• National emergency plans • Disaster intervention strategies • Allocation of resources for food emergencies	• Manage food price volatility • Budget allocations for food and nutrition interventions including social protection	• Inclusive economic growth • Management of food price stability
Meso/ market level	• Intervention in commodity markets • Release of food stocks for emergency purposes	• Food market stabilization • Market and price monitoring • Intervention in value chains	• Infrastructure and market development • Development of regional and export markets and connecting farmers to them • Food safety and value chain development
Micro/ household level	• Food aid distribution • Protect the children and vulnerable populations	• Safety nets • Prevention of vulnerability to shocks • Increasing mechanisms for resilience and development	• Household poverty reduction • Access to nutritious food • Sustainable food systems

Based on Timmer (2014).

Employment Guarantee program in India; and the Productive Social Safety Net program in Ethiopia are some examples of large-scale social protection interventions. In the next section we develop a conceptual framework to trace the nutritional benefits of social protection programs.

A CONCEPTUAL FRAMEWORK FOR EVALUATING SOCIAL PROTECTION PROGRAMS FOR NUTRITIONAL OUTCOMES

There is a large body of literature on social protection and its implication for welfare outcomes (Grosh et al., 2009; Alderman and Yemtsov, 2013; Gentilini, 2014; Gentilini, 2016). Yet the evidence on how social protection directly helps in nutritional outcomes is still limited. This is partly due to the varying objectives of social protection programs, and because most social protection programs do not have nutritional outcomes as their primary goal. As a result, there are both research and policy gaps in analyzing the contribution of social protection programs toward improved nutrition.

Social protection programs have wide ranging and sometime multiple objectives: poverty reduction, women's empowerment, resilience building, food security, and dietary quality, to name a few. Nutrition is addressed as the direct objective in only a few of the programs. Understanding how nutritional goals could be achieved through social protection interventions through both direct and indirect pathways can help policy makers re-think their intervention options and the approach to social protection. Some of the pathways to nutritional outcomes through social protection are depicted in Fig. 12.1.

As most of the poor and malnourished households live in rural areas and depend on agriculture for their livelihood, agricultural development strategies need to be inherently multi-sectoral in nature. They should cover a wide range of issues and approaches such as technology development and adoption, sustainable natural resource use, delivery of institutional services, and human capital development. All these interventions are needed for the holistic growth of the agricultural sector (WDR, 2008). Social protection programs can help in all these interventions.

In addition to the direct contribution through food and cash transfers, social protection has potential in the long run to achieve agricultural development goals which can further contribute to

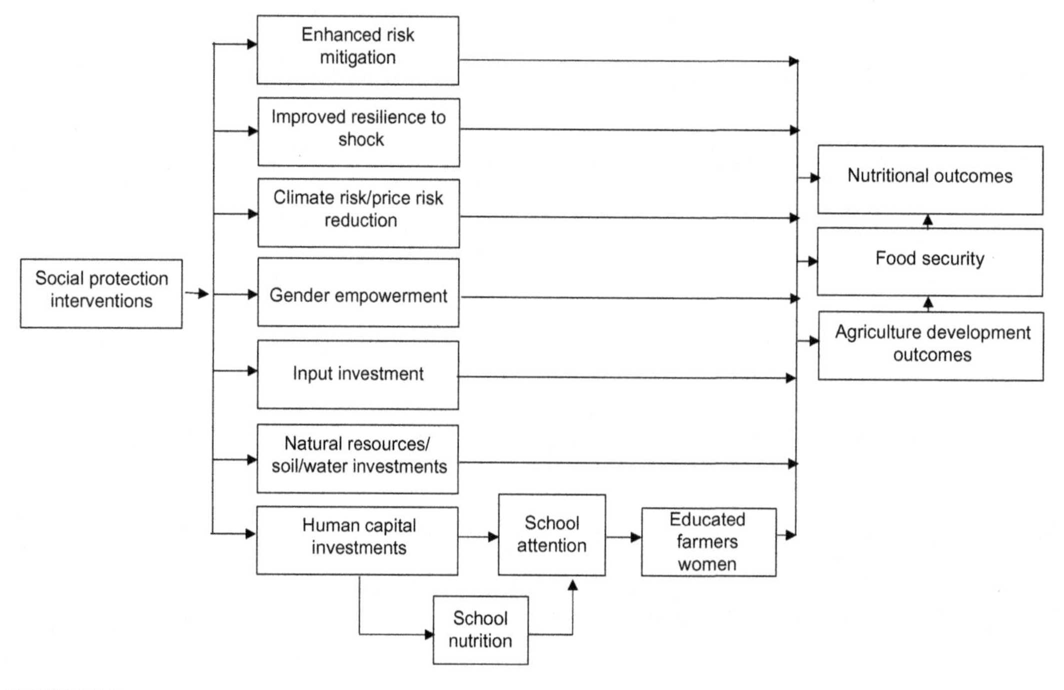

FIGURE 12.1

Conceptual pathways to link social protection and nutritional outcomes.

Based on Coady, P.D., 2004. Designing and evaluating social safety nets: theory, evidence and policy conclusions. Food Consumption and Nutrition Division Discussion Paper No. 172. International Food Policy Research Institute, Washington, DC; Harvey, P., 2007. Cash-based responses in emergencies. Humanitarian Policy Group report 24. Overseas Development Institute, London; Arnold, C., Conway, T., Greenslade, M., 2011. DFID Cash Transfers Literature Review; de Brauw, A., Hoddinott, J., 2011. Must conditional cash transfer programs be conditioned to be effective? The impact of conditioning transfers on school enrollment in Mexico. J. Dev. Econ. 96 (2), 359–370; and Ruel, M.T., Alderman, H., 2013. Nutrition-sensitive interventions and programmes: how can they help accelerate progress in improving maternal and child nutrition? Lancet 382 (9891), 536–551.

nutritional outcomes through the channels discussed in previous chapters. In some countries, social protection continues to focus on targeted poverty alleviation programs to reduce current poverty levels.

In addition, public interventions that aim at social protection also focus on increasing service delivery, enhancing labor productivity, improving livelihood strategies for the poor, and financial inclusion and sustainable insurance for the poor. The intermediate benefits of such interventions are reduced risk, improved resilience, gender empowerment, asset building and input investment in agriculture, sustainable natural resource management, and human capital investments, to mention a few. These intermediate benefits work through income increases and changes in consumption patterns to result in food security and nutritional outcomes.

As income increases, the demand for food increases, as we noted in Chapter 6, Consumer Theory and Estimation of Demand for Food. However, increased food intake does not necessarily mean good nutrition. How income transfers from social protection improve nutritional outcomes through diet changes has been of interest to policy makers for quite some time. Social protection programs can improve nutritional outcomes by empowering women, and by recognizing the role of women in the nutritional well-being of their children. Addressing intra-household dynamics, as seen in Chapter 9, Intrahousehold Allocation and Gender Bias in Nutrition: Application of Heckman Two-Step Procedure, can reduce the gender bias in nutrition (Babu et al., 1993).

The cost-effectiveness of social protection programs has been discussed in the literature (Gentilini, 2016). Programs and policies for poverty alleviation through social protection can improve their cost-effectiveness through improving their design. For example, transferring the funds to the beneficiaries through mobile technology and cell phones can reduce pilferage and reach the right target groups. The benefits of cash transfers could mean more improved nutrition if the program is also implemented with adult education and behavior changing activities toward good nutrition (Ruel and Alderman, 2013).

Skill development as part of the intervention could increase the chances of the target households moving away from dependency on social protection, as it has been shown that given opportunities households do find gainful benefits and do not continue to depend on social protection programs (FAO, 2015). Such complementary investments are necessary for the long-term sustainability of nutritional benefits.

In addition to the rural focus of social protection programs, urban focused interventions have been helpful in integrating the urban poor into the mainstream economy. Programs that help in providing rural credit and increased inclusion of the poor in employment guarantee schemes and other insurance programs help in increasing resilience and protecting the assets of the poor (FAO, 2015).

Policy makers often see social protection as a short-term solution to help the poor get out of the poverty trap. How long the program will have to be continued will depend on the opportunities that are created for households to become part of the mainstream economy. Thus, poor people's dependence on the public provision of social protection has been an issue in the development community. However, results of studies have shown that under certain institutional settings, the participants of the programs could be weaned off over time.

In general, most of the social protection interventions aim at increasing the income of the households in the long run. Thus, the design of the social protection program matters as a means of providing an enabling environment for the participants to move out of the program. This often comes in the form of human capital development. School feeding programs, such as "Food for

Education" in Bangladesh, aim at the provision of food under the condition that the children of the households will attend school for a certain number of days in a month (Ahmed and Babu, 2006).

Social protection programs that offer food in kind to the families have implications for local market development and the prices farmers get for the food they produce (Omamo et al., 2010). To avoid deleterious effects of such interventions, program designers sought to buy the food they provide to the household from local markets. Often when food is provided in kind, they tend to focus on reducing hunger. This has implications for dietary diversity and the nutrition intake of the program participants (Rabbani et al., 2006).

The provision of cash could help in diversifying their diets, and the targeted groups will have an increased ability to buy and consume high value commodities which are nutritionally rich. In addition, social protection programs could provide incentives for diversifying production systems, such as moving away from cereal crops to legumes, which can increase the protein content of the diet (Audsley et al., 2010).

The designers of social protection programs also face issues related to the combination of instruments such as food and education, food and health interventions, the level of rations to be given, and the combination of items in a ration. Timing of the intervention has implications for the benefits that accrue to the participants. Interventions during a lean season may have a better impact on nutrition (Babu et al., 1993). The nature of the intervention also differs based on the time intervals needed between the distribution of food or cash.

The target groups need to be specifically those who are most vulnerable to nutritional problems. In addition to food and other benefits for achieving best nutritional outcomes, programs need to have behavioral change and educational components. All these factors also affect how programs benefit the participants nutritionally. Finally, the economics of the interventions matter, in terms of how much the design costs, and how many people can be pulled out of hunger, poverty, and malnutrition challenges.

DESIGNING AND IMPLEMENTING SOCIAL PROTECTION: WHAT IS THE BEST APPROACH?

Most developing country governments design and implement targeted food rations or cash transfers to reduce food insecurity, and this has nutritional implications. International aid agencies such as WFP have also been using food aid as a social safety net to protect vulnerable populations. How to design and run cost-effective social protection programs? When does one switch between food to cash and vice versa, and under what conditions? Could a program be run with both cash and food side-by-side? Does it matter who receives the benefit—women, men, or children in a household? What does research show in terms of the choice of instruments and their nutritional implications? Does the program affect the local food market and create disincentives for the farmers to produce their own food? Is there a dependency created due to the program choices? Do the participants spend their benefits on nonfood and nonnutritional consumption items? What changes are observed in the food and nutritional consumption patterns of the recipients? These are some of the questions for which program managers and policy makers seek answers before designing interventions.

Research has shown that there is no one defined method of designing and implementing social protection programs in order to achieve food security and nutritional goals. The debate on "Food

Versus Cash" as a preferred instrument for social protection continues (Gentilini, 2016), although food distribution tends to be more expensive than cash delivery. In general, food consumption increases as more cash or food is made available to households through program interventions.

Cash is likely to have a larger impact if the market provides opportunities to access diverse food products, thereby increasing dietary diversity. The choice of instrument—cash or food or food vouchers—depends on the objectives and the context within which the social protection programs are designed. For example, when the food markets are thin, food rations are a better choice. And if the objective is to simply increase the calorific intake of food, direct food distribution could be a solution (Gentilini, 2016).

Research studies from the International Food Policy Research Institute (Hoddinott and Skoufias, 2004; Coady, 2004; Hoddinott et al., 2010; Hoddinott and Weisman, 2010) conducted over several years in many countries broadly show the following results. Cash transfers improve dietary diversity. Food transfers show increased consumption of food items supplied through the rations. Access to food markets and nonfood markets is a key factor in determining the comparative advantage of various designs in social protection programs (Harvey, 2007; Harvey and Bailey, 2011).

Food shares in total consumption do not change between cash and food recipients; food, cash, and vouchers all increase food share and generate a shift in the food Engel curves (see chapter: Microeconomic Nutrition Policy). There is no clear evidence that cash transfers are used for undesirable purposes, such as buying alcohol or other nonnutrition related consumption (FAO, 2015). However, who gets the cash transfers within the household makes a difference to the expenditure patterns and food consumption (Herrmann, 2009; FAO, 2015).

Cash is cheaper to operate and to deliver compared to food, as food deliveries involve high levels of logistics and personnel to operate. Cash transfers can also help in cognitive development by improving nutrition if the households spend their benefits on increased micronutrient consumption. There is generally a concern among policy makers that cash transfer programs can affect local market prices. There is no evidence for the market effects of program interventions. Price effects generally work through the demand for food, and hence the nature of the market decides these effects. If the market is thin then the design of programs may consider using food in place of cash. However, understanding the local market conditions and studying under what conditions cash transfers will have a price effect on the local food markets, and food production and consumption of the households is an important consideration in the design of social protection programs (Gentilini, 2016).

Specific insights from the IFPRI research team on social protection are worth noting (Alderman, Gilligan, and Leher, 2012). In Uganda, e.g., the markets are not thin for staple food crops. They argue the thinness of markets matters only in really thin markets. When they compare cash and food receiving households side-by-side with the control group, they find that the design of the social protection programs and their outcomes depend on what program managers care about as a final goal, and the context in which the program is implemented.

In terms of nutritional outcomes, if policy makers care about reducing hunger and food purchases are a problem for the communities, then the food supplied needs to be nutritious to meet the nutritional objectives. In such a situation, fortification of the foods distributed could be an option. However, in communities where sufficient calories are already obtained from the crops grown in the local areas or through staple food markets, distributing food may not necessarily help in improving nutrition. Cash transfers may work better in terms of both cost and dietary diversity as a conduit for better nutrition (Alderman and Mustafa, 2013).

WHAT DOES THE CURRENT LITERATURE SHOW?

In this chapter we look at various challenges related to social protection programs and their implications for nutritional outcomes. Developing new research areas involves seeing the potential of their programming to enhance the benefits of social protection (SP) for rural food and nutrition security. Making the case for a multi-sector approach to social protection is important to harmonize investment in the rural areas. Such complementary avenues of investment can help in rural areas. The evidence for such approaches is given in the recent FAO State of Food and Agriculture Report on Social Protection (FAO, 2015).

The FAO (2015) report highlights several issues that have nutritional implications. We discuss them briefly. Social protection programs continue in many countries in a pilot mode, and are funded and executed by donor-supported NGOs. Although there are several large-scale programs funded and implemented by the public sector that have proven successful, the lessons learned from these programs need to be communicated better.

Some highlights of the FAO report include the following. Social protection programs, in general, are handled by the ministry of social welfare, and yet involve a multi-sectoral approach bringing together agriculture, health, finance, and other ministries. A major challenge is how to organize them better to harmonize their programs and interventions on the ground.

The relationship between the social protection programs and the food and agricultural systems needs to be understood fully. For example, the conditionality related to program participation can help in technology adoption among rural households if they involve modern inputs and farm level advice. Incentives for producing food and the effect of such decisions on food and cash transfers need to be studied further in different contexts. In some instances, the macro level and sectoral policies need to be changed to help the rural population, particularly the vulnerable sections such as women, who can participate in and benefit from the social protection programs. Bosa Familia in Brazil is a good example of this. In Haiti, social protection interventions using food vouchers are able to increase nutritional access through increasing the demand for locally produced food. Improving access to food and nutrition through locally produced food can help further in better understanding the complementarity between nutrition and agricultural outcomes.

In the context of agricultural systems, seasonal migration can be reduced through social protection programs if they help in investment toward agricultural inputs and in natural resource management, and if they are connected to production activities, as in the Ethiopian productive social protection programs (Gilligan et al., 2009, 2008).

The long lasting impact of food aid in reducing stunting for children from rural Ethiopia has been investigated by Porter (2010). What makes Porter's (2010) study interesting is that she collects anthropometric information for children twice; first, when the children are below 5 years old in 1995, and second, when they are roughly 10 years older, in 2004. Two important conclusions emerge from Porter's (2010) study: malnutrition at early ages can have long-lasting effects, even a decade later, hence, early intervention is key; and food aid, as a safety net, substantially improves the nutritional outcomes of child growth, with positive spillover effects into the future.

In the context of Ethiopia, Broussard (2012) shows that adults also benefit from food aid. Adult subsistence farmers, and adults in poor households who are dependent on nonfarm incomes are all affected by adverse conditions. Broussard (2012) uses a very rich panel data set, covering 9000 adults from several survey rounds. The data contains information on consumption, assets, aid receipts, labor-saving arrangements, female empowerment, and livestock holdings. Broussard (2012) estimates several versions of panel data regressions to demonstrate that food aid benefits male adults. However, women in lower-income households are adversely affected by aid. This is similar to the observation that we found in Chapter 8, Socioeconomic Determinants of Nutrition: Application of Quantile Regression, where bargaining power can determine intra-household allocation. The findings of Broussard (2012) and Porter (2010) stress the importance of safety nets, alongside other public assistance programs for all groups in the population. We further explore this data under STATA implementation, and in the exercises at the end of this chapter.

SAFETY NET IN ETHIOPIA: POLICY LESSONS

In recent years economists and food policy analysts have started to examine the conditions under which social protection and safety nets allow the poor to overcome adverse conditions. In this context, recent research by Hoddinott et al. (2012) has contributed a massive amount of information to enable policy makers to design appropriate interventions. In an earlier study, Gilligan et al. (2009) assess the impact of Ethiopia's Productive Safety Net Programme (PSNP). PSNP is the largest safety net program, covering more than seven million people. PSNP provides food transfers to worst-hit areas and to food-insecure populations.

Alongside PSNP is another complementary activity in Ethiopia called Other Food Security Programme (OFSP). The OFSP provides a productivity transfer service, which may include any one of the transfers that enhance household productivity: access to credit, agricultural extension services, and advice on technology, crop rotation, irrigation, and water harvesting. The study by Gilligan et al. (2009) is interesting because their data shows that a portion of the households received both types of treatment—PSNP and OFSP. This feature in the data allows the researchers to tease out the effects of just the PSNP participants, relative to others.

The researchers find that those participants who received treatment from both PSNP and OFSP are more likely to be food secure. Relative to the non-PSNP participants, or the comparison group, the participants of PSNP and OFSP borrow for productive purposes, use improved technologies, and own nonfarm businesses. The participants also did not suffer from disincentive work effects.

The researchers also note that the impact of just PSNP alone was minimal. In a related study, Hoddinott et al. (2012) extend the Gilligan et al. (2009) study to include participants in a revamped OFSP called the Household Asset Building Programme (HABP). Once again, in the extended study Hoddinott et al. (2012) find that participation in both PSNP and OFSP/HABP enhanced investment in agriculture and produced higher yields, compared to those who received either PSNP or just OFSP/HABP.

In a more recent follow-up study, Berhane et al. (2014) provide evidence that just PSNP improves food security by 1.29 months. Five years of participation increases tropical livestock by 0.38 units. The joint impact of PSNP and OFSP/HABP is much larger than participation in any one program. Having both programs increases food security by 1.5 months and tropical livestock by almost 1 full unit.

The studies cited above indicate that the social safety net programs must combine income transfers and productivity enhancing investments. Finally, it is not easy to mimic any of these programs directly in another setting without taking institutional and social features into account. Even within Ethiopia, the PSNP proved to be vulnerable when it was unconditionally transplanted to pastoral areas (Sabates-Wheeler et al., 2013).

The vulnerability of social safety net programs arises due to two reasons. First, informal authorities seize hold of transfers and implement the targets arbitrarily, and manipulate the system. Second, there is a lot of exchange, and give-and-take among pastoral households, and this network-based affiliation leads to program-dilution. As the researchers conclude, one must be cognizant of the unique social, political, and livelihood characteristics of the setup that needs intervention, so as to develop and design appropriate programs.

BRAZIL'S BOLSA FAMILIA: CASH TRANSFERS ARE NOT ENOUGH

Conditional Cash Transfer (CCT) programs known as Bolsa Familia (Family Grant) are now standard social policies to fight poverty in Brazil. Hall (2008) evaluates this policy. The cash transfers help mitigate long-term poverty by mandating school attendance, participation in health care, strengthening human capital, and stimulating effective demand (see Hall, 2008, p. 800).

Hall (2008) notes that the Bolsa Familia program was popular and supported approximately two million families, with 73% of them in the lowest two quantiles. The nutritional and food security status among participants had improved, compared to nonparticipants. Participants also improved in terms of education and clothing, and experienced lower income inequality. Women participants have become stronger in civil rights and empowerment.

However, similar to Ethiopia's PSNP in the pastoral areas, Hall (2008) finds that in the pastoral areas, Brazil's Bolsa Familia is also fraught with political manipulation. Monitoring of participants and checking on their requirements has been very ineffective. Beyond the political manipulation of the organizers and participants, CCT programs are also prone to politicization for election purposes. This is also evident with Bolsa Familia. Finally, the program is also viewed as creating a culture of dependence.

Hall (2008) points out, similar to other research in this area, that CCT programs in the absence of adequate income-generating schemes may backfire; namely, the costs of Bolsa Familia may outstrip the benefits. Such programs require an adequate social infrastructure that promotes access to employment, health care, and sanitation facilities, which are all in keeping with the other themes in this book.

NUTRIA-COOKIES, CHILD LABOR, AND SAFETY NET PROGRAMS IN INDIA

In this section, we review some interesting findings regarding the safety net programs in India. As we have mentioned in previous chapters, the incidence of malnutrition is severe in India. Shah (2011) presents an interesting case study from the urban slums in Mumbai, where about 50 children below 6 years of age were placed in a program that provided nutria-cookies for three months. Even though the scope of Shah's (2011) study is limited, the results are of interest. Consumption of nutria-cookies was positively associated with height and weight gains among the participants.

Higher gains were observed for the most malnourished participants. At the very least, these results call for an enlargement of the program at other locations, with adequate monitoring and reporting.

We have also presented different evaluations of India's Integrated Child Development Services (ICDS) in the previous chapters. As mentioned before, ICDS is a very large program that targets long-term nutrition and development in children. Kandpal (2011) presents excellent econometric techniques to examine newer data concerning participants in ICDS. Kandpal's (2011) estimates account for two important features in the data. First, Kandpal (2011) controls for the endogeneity of program placement, and second, addresses the negative skewness in data coverage.

What do these additional econometric novelties tell us about ICDS? Contrary to the findings mentioned in the previous chapters, Kandpal (2011) demonstrates that ICDS improves children's nutritional outcomes, and increases HAZ scores by about 6%. As Kandpal (2011) points out, once a village gets ICDS, chronic malnutrition is lower. But which village gets the program? If placement is based on "average" rather than negative skewness in the original distribution, then placement may not uniformly target poor areas. Indeed, Kandpal (2011) shows that sex ratios and landholdings do not play a significant role in placement. The program also targets areas which have more educated mothers, rather than the opposite case, where intervention might be particularly needed. Finally, political affiliations rather than child under-nutrition rates influence placement.

The targeting of other food distribution systems in India, besides ICDS, has also sparked interesting debates. Suryanarayana and Silva (2007) point out that a food distribution system that targets the poor is prone to suffer from type 1 and type 2 errors. That is, the system will cover the poor who are food-secure, and will fail to cover the nonpoor who are food insecure. The problem arises because the set of food insecure is larger than the set of poor in India. The authors note that in recent years, the richer sections have shifted away from cereals, and consume more noncereal and nonfood items. At the same time, cereal consumption among the poor has increased. Consequently, monetary measures of poverty may not adequately capture nutritional deficiency. In their revised estimates, the researchers find that consumer choices are dictated by variety, rather than nutrition. Hence, more concentrated efforts should be toward education, rather than direct income or indirect food transfers.

BANGLADESH SOCIAL PROTECTION WITH A BEHAVIORAL CHANGE COMPONENT

A recent study in Bangladesh has shown human capital investment is possible with social protection programs that provide cash transfers. This is a very interesting study (Sunny Kim and Phuong Nguyen, 2016) that tracks the relative impact of five different social safety net programs for Bangladesh: (1) only cash, (2) only food, (3) food and cash, (4) nutritional behavior change communications (BCC) and cash, and (5) BCC and food. The program was implemented as a pilot for 50 villages covering a total of 500 households, for each treatment design.

The results, based on randomized control design and a difference-in-difference estimator, indicate that treatment (5), which is BCC and food, has the highest impact of the treatments. This treatment had a significant impact on calories and reduction in stunting in the village clusters positioned in the north of the sample area. Providing only food, or treatment (2), had a significantly better influence on diet quality than providing only cash. The authors note that significant improvement is noted whenever a treatment is combined with BCC.

The study concludes that one can make social protection more nutritionally impactful by giving food and cash along with behavior change communication. This is partly because cash allows the recipients to consume diversified foods due to its flexibility. The study also shows that while social protection has much to offer to improve food and nutrition security, it need not always be based on food. The results also show that ineffective programs should be removed and the cost savings must be transferred to those programs that combine BCC alongside other treatments.

SAFETY NETS AND NUTRITION IN INDIA

India's National Rural Employment Guarantee Scheme (NREGS) is a major undertaking which guarantees 100 days of paid work. Ravi and Engler (2015) provide a historical background and rationale for NREGS; Amaral (2015) analyzes the impact of NREGS on gender-based violence, kidnapping, and dowries. NREGS has become a well-researched workfare program as a means of providing direct transfers to the poor. It was launched in 2006, and currently covers all of India. The program provides employment within 15 days of application and within 5 km of where the household resides. Further, the program also guarantees a minimum wage, and aims to reach poor rural households to provide immediate relief. Several research papers have examined the economic effects of NREGS, and nutrition economists are also interested in seeing whether the program has any beneficial effects on food security and calorific intake. Ravi and Engler (2015), for instance, examine a huge panel of data from over 190 villages in the state of Andhra Pradesh, and show that NREGS increases per capita monthly spending on food items, thereby promoting food security. The study shows that NREGS has a significant positive impact on the intake of energy and protein, and on asset accumulation.

Shah and Steinberg (2015), on the other hand, show that NREGS has the unintended consequence of reducing human capital formation, because the exposure to the program causes school enrollment to drop by 2%, as well as reducing math scores by 2% among 13 − 16 year old children. Most of the adolescents substitute into market work or domestic work, which leads to a lowering of potential human capital.

Jha et al. (2011) examine NREGS and also the Public Distribution System (PDS), which are two important safety net programs that address poverty and malnutrition. Jha et al. (2011) hypothesize the existence of a vicious cycle of a poverty–nutrition trap, and test both the programs with data from rural areas in three Indian states: Rajasthan, Maharashtra, and Andhra Pradesh. In particular, the researchers examine whether the programs have an impact on a household's nutritional status. Their focus is on two macronutrients (calories and protein) and other micronutrients.

The NREGS guarantees 100 days of employment a year to at least one member of any rural household who is willing to perform unskilled work for a minimum wage. India's Public Distribution System (PDS) refers to the distribution of some essential commodities (e.g., wheat, rice, kerosene) by the government at subsidized rates. There is a general hope that these transfer programs help reduce calorific under-nutrition among the recipients. In order to test whether program participation improves nutritional status, the authors estimate the following regression for each nutrient intake:

$$n_i = \beta_1\, PDS\ Participation_i + \beta_2\, NREGWage_i + \beta_3\, NonwageIncome_i + \mathbf{X}_i\gamma$$

where n_i is intake of nutrient n for household i, and $\mathbf{X}_i$ is a vector of household characteristics. The regression equation also includes the role of participation in PDS, wages earned through NREG, and non-NREG earned wages. However, since participation and wages are endogenous, the researchers adopt the Instrumental variable (IV) estimator. The regression is estimated for 13 nutrients (protein, fat, minerals, carbohydrates, fiber, calories (energy), phosphorus, iron, carotene, thiamine, riboflavin, niacin, and vitamin C) for each of the three states.

The results indicate that both NREG Wage and PDS Participation significantly increase the intake of protein, carbohydrates, calories, phosphorus, iron, thiamine, and niacin in all three states. The impact of NREG Wage and PDS Participation also influences the intake of minerals and calcium, in two out of the three states.

Thus, the two policy interventions have varied impacts on the intake of various nutrients in the three states studied here. This is a reflection, among other factors, of the ways in which the income from NREGS and the income transfer implicit in PDS are spent, and the dietary preferences of households in the three states. Overall, the results indicate that the two safety net programs in Indian have positive effects on nutrient intakes. The authors also note that the impact of these programs varies across states, and by nutrient intakes. The viability of both programs depends crucially on the pre-existing proportion of under-nourished in each location.

In related studies, Jha et al. (2009, 2013, 2015) examine the effect of the social safety net programs on BMI. They include the role of participation and duration of employment on BMI. They also control and test for the endogeneity between BMI and the participation decision in their tobit estimation. Interestingly, the higher the predicted BMI, the higher was the probability of participating in NREGS. The probability of being overweight reduced the likelihood of program participation. Based on the results, Jha et al. (2013) conclude that along with NREGS, a broader supplementary subsidy program targeted to children, lactating women, and the elderly will help the under-nourished break out of the vicious circle of poverty (Box 12.1).

BOX 12.1 SHOULD IMMIGRANTS RECEIVE SUPPLEMENTAL NUTRITION ASSISTANCE?

Absorbing immigrants and providing for them has become a major international challenge, as is seen by the recent events in Europe. How do local policies affect the nutritional status of immigrants? In the context of the United States, Skinner (2012) observes that as of 2009, more than one child in four below 18 years was either born abroad or lived with a foreign born parent, and that by 2020, this number will be one in three children. The fastest growing population segment in the United States is children living with immigrant parents. What makes this an incredible statistic is the fact that about 24% of these children live below the poverty line, and 51% below double the poverty line.

What is perplexing is that immigrant families that are eligible for the Supplemental Nutrition Assistance Program (SNAP) do not access it, particularly when compared to their native counterparts. Skinner (2012) notes that only 44% of the immigrants participate in SNAP, compared to 65% of all eligible families. Skinner (2012) points to the range of state laws that discourage participation among immigrants. For instance, just from 2005 to 09, there was a more than fourfold increase in the number of immigration-related bills introduced and enacted by different state legislatures.

Skinner (2012) points out that these bills have discouraged participation in SNAP, and recommends substantial outreach services to increase accessibility to the working poor. While the federal government spent almost $18 million in matching funds to support outreach services, many states with a huge immigrant population spent less than the national average, while 16 states spent no funds at all on outreach.

Skinner (2012) makes the case that a modest outlay to help SNAP participation has large multiplier effects that mitigate much of the hardship among nonparticipating poor immigrants.

THE GREAT RECESSION AND THE SOCIAL SAFETY NET IN THE UNITED STATES

The Supplemental Nutrition Assistance Program (SNAP) and the Unemployment Insurance Program (UI) are two important safety net programs in the United States. SNAP is designed to provide food assistance for the poor, and as of 2013 had about 48 million recipients. UI is designed to provide assistance to middle-income households suffering from unemployment. Heflin and Mueser (2013) show that both SNAP and UI swelled in numbers following the economic slowdown that started in 2008. They evaluate the relative importance of both SNAP and UI for Florida residents. The data set they calibrate is unique, since it identifies individuals entering both programs. They show that the number of people receiving SNAP grew dramatically, following the Great Recession.

Further, the researchers show that many of the SNAP recipients were also receiving UI, with the latter source being of primary importance. The study notes that for the recipients who received both UI and SNAP, UI was of primary importance for about a third of them, prior to the recession. However, after the recession, UI was of primary importance for two-thirds of the recipients.

The authors note that the increase in the number of SNAP recipients who were already receiving UI benefits is attributed purely to the economic slowdown. However, for those receiving both UI and SNAP, the primary importance of UI was driven by legislation that extended UI benefits. A third of the recipients would have had UI benefits run out, had it not been for the legislation.

The findings of Heflin and Mueser (2013) show the importance of safety nets in the US. For instance, the number of SNAP-UI claims increased by 57% after the recession, which amounts to roughly three-quarters of a million, or about one out of every 15 Florida residents. Given that the growth of SNAP alone was over three times more than the growth in joint SNAP-UI, the authors conclude that there are serious limits to the cushion that UI provides during hard times.

In a related study for Michigan, O'Leary and Kline (2014) find that SNAP usage increased every year from 2006 to 10. UI applications also increased dramatically in 2008 by 23.6%. An average of 20% of UI applicants had received SNAP in the previous year. Among UI applicants, prior SNAP recipients were the highest. Among those who had not received SNAP in the year before applying for UI, 13% received SNAP one year after applying for UI. SNAP receipt after UI application was highest among those who were discharged from work, exhausted their UI benefit entitlements, were between 25 and 44 years of age, were less educated, and were not employed in the retail trade, hospitality, or health care services. The Michigan data also shows that during the Great Recession, UI applicants entered SNAP faster than before the official start date of the economic decline in December 2007.

The above studies selected and reviewed for this chapter set the stage for policy analysis related to the nutritional implications of social protection programs. Studies have also addressed other related issues including asset building, resilience building, women's empowerment, combining programs, and improving extension systems to work with nutritional field workers.

The chapter also notes some tradeoffs in different safety net programs. Social protection programs can make people dependent on the benefits, or make them independent. Lessons from Ethiopia, India, and Bangladesh indicate that income-boosting programs may work best, when they take into account the availability of basic needs and support systems, including nutritional education. In the next section we show how these policy issues could be addressed using field data and undertaking statistical analysis. In particular, we demonstrate the usefulness of panel data analysis and provide an example from field data in Broussard (2012).

ANALYTICAL METHODS

In this section, we focus on panel data estimation to analyze social protection programs. Panel data estimation methods have become very important and are applied extensively in development economics, dealing with issues related to nutrition and health outcomes. Following Cameron and Trivedi (2010), Woolridge (2009), and Greene (2012), we represent a simple linear equation as follows:

$$y_{it} = \propto + x_{it}^{'}\beta + u_{it} \tag{12.1}$$

where y_{it} is the dependent variable in the data set observed for the ith individual at time period t. x_{it} represents regressors, and u_{it} is the disturbance term with standard OLS assumptions. The usual tests of autocorrelation, and between-correlation over individuals have to be conducted, and there are a variety of approaches to perform these diagnostics, including FGLS estimation procedures.

An important assumption in Eq. (12.1) associated with the standard OLS estimator is that the x_{it} terms are not correlated with u_{it}. However, when we have many observations over time, spanning for each individual i in the data set, some OLS assumptions, including $E(x_{it}, u_{it}) = 0$, may not be true. To account for such deviations, researchers have modeled several possible error structures for different characteristics within the data. In this context, the fixed-effects (FE) and the random-effects (RE) models are most commonly discussed within panel data estimation.

Assume that u_{it} in Eq. (12.1) can be further split into an individual-specific component, and a random error term, such that we now have:

$$y_{it} = \alpha_i + x_{it}^{'}\beta + \epsilon_{it} \tag{12.2}$$

where α_i is an individual-specific effect which varies across individuals in the data, but stays fixed over time. In a sample covering adults, the variable height-for-age might be an individual specific effect, that may qualify as α_i. In this case the α_i term will be correlated with x_{it}, and the term $u_{it} = \alpha_i + \epsilon_{it}$, provides an extension of Eq. (12.1). This variation of Eq. (12.1) given in Eq. (12.2) is the FE model. While x_{it} is correlated with α_i, Eq. (12.2) still assumes that $E(x_{it}, \epsilon_{it}) = 0$.

Cameron and Trivedi (2010, p. 237) provide an example where in an earnings regression, independent variables (say experience, age, education) are correlated with the unobserved ability of the worker, only inasmuch as the time-invariant portion of this unobservable entity, α_i.

The panel data estimation methods exploit the variation that arises across different i units in the sample, and also across different t for each individual i. Holding i constant, and checking its variation over t, is called within-variation. Variation across i is called between-variation. If N is the number of individuals with each individual observed for T_i periods, with individual mean $\overline{x_i} = 1/T\sum_T x_{it}$, and the grand mean $\overline{x} = 1/NT\sum_i\sum_t x_{it}$. These variances are calculated as follows (Cameron and Trivedi, 2010, p. 244):

$$\text{Within Variance: } s_W^2 = \frac{1}{\sum_{i=1}^{N} T_i - 1}\sum_i\sum_t (x_{it} - \overline{x}_i + \overline{x})^2$$

$$\text{Between Variance: } s_B^2 = \frac{1}{N-1}\sum_i (\bar{x}_i - \bar{x})^2$$

$$\text{Overall Variance: } s_W^2 = \frac{1}{\sum_{i=1}^{N} T_i - 1}\sum_i \sum_t (x_{it} - \bar{x}_i)^2$$

It is standard practice to check and produce these values to help in data exploration and inspection of time-invariant variables. The FE estimator takes into account the within-variation and between-variation in the data, and the estimation proceeds by using Eq. (12.2) to produce:

$$\overline{y_i} = \overline{x_i'}\beta + \overline{\epsilon_i} \tag{12.3}$$

Subtracting Eq. (12.2) from Eq. (12.3) yields what is known as the mean-difference model:

$$(y_i - \overline{y_i}) = (x_{it} - \overline{x_i'})\beta + (\epsilon_{it} - \overline{\epsilon_i}) \tag{12.4}$$

This version in Eq. (12.4) is called the within-estimator of β. Cameron and Trivedi (2010) provide a full discussion of the advantages of this estimator, and indicate that the within-estimator of β is also known as the Least-Squares Dummy Variable (LSDV) estimator. The LSDV estimator is tantamount to introducing a dummy variable for each i in the sample. Cameron and Trivedi (2010, p. 259) provide additional details.

Similarly, the between-estimator (BE) of the FE model exploits the cross-sectional variation in the data, and produces the OLS estimator of the model:

$$\overline{y_i} = \alpha + \overline{x_i'}\beta + (\alpha_i - \alpha + \overline{\epsilon_i}) \tag{12.5}$$

It is also possible that α_i is completely random, and is not correlated with x_{it}. This version of Eq. (12.2) is called the RE model, and is usually written as:

$$y_{it} = x_{it}'\beta + (\alpha_i + \epsilon_{it}) \tag{12.6}$$

In Eq. (12.6), we have $\alpha_i \sim N(0, \sigma_\epsilon^2)$, and $u_{it} \sim N(0, \sigma_\alpha^2 + \sigma_\epsilon^2)$, with $Cov(u_{it}, u_{it+s}) = \sigma_\alpha^2$. Researchers are also interested in the calculated value of the serial correlation term:

$$\rho_u = Cor(u_{it}, u_{it+s}) = \frac{\sigma_\epsilon^2}{\sigma_\alpha^2 + \sigma_\epsilon^2}$$

The β in Eq. (12.6) is estimated using FGLS and the RE estimator is:

$$\left(y_i - \hat{\theta}_i\overline{y_i}\right) = \left(1 - \hat{\theta}_i\right)\alpha + \left(x_{it} - \hat{\theta}_i\overline{x_i}\right)'\beta + \left\{\left(1 - \hat{\theta}_i\right)\alpha_i + \left(\epsilon_{it} - \hat{\theta}_i\overline{\epsilon_i}\right)\right\} \tag{12.7}$$

where $\hat{\theta}_i$ is the evaluation of $\theta_i = 1 - \sqrt{\frac{\sigma_\epsilon^2}{T_i\sigma_\alpha^2 + \sigma_\epsilon^2}}$

Finally, the Hausman test compares the FE and RE estimators. Greene (2012, p. 419) provides the details of this test. Under the null hypothesis, the FE and RE estimators do not diverge, the individual effects are random, and OLS estimates are inefficient. Under the alternative, these estimators diverge. Let $V(\hat{\beta}_{FE})$ and $V(\hat{\beta}_{FE})$ represent the variance of the estimates under FE and RE. The Hausman tests computes the difference $V(\hat{\beta}_{FE}) - V(\hat{\beta}_{FE})$, which is distributed as a chi-square with k degrees of freedom.

There have been several extensions and developments in panel data methods in recent years. An interesting extension arises in a dynamic model, where y_{it} is influenced by its past values, say y_{it-k}. For instance, in a simple version Eq. (12.2) can be rewritten as:

$$y_{it} = \alpha_i + x_{it}^{'}\beta + \gamma y_{it-1} + \epsilon_{it} \tag{12.8}$$

Even in a simple dynamic model, as in Eq. (12.8), we note that the FE estimators from the standard procedures are automatically inefficient, due to the presence of the lagged dependent variable on the RHS of Eq. (12.8). A version of the IV estimation is applied to derive the estimators in this context. The procedure that is useful in this context is the Arellano–Bond estimator, which is often applied in this framework.

EMPIRICAL EXAMPLE IN STATA

The `xtreg` command in STATA is used for implementing linear panel data estimation procedures.

Broussard (2012) has an excellent example from rural Ethiopia, which we can implement using the `xtreg` procedure.The data and the related STATA .do files for Broussard (2012) can be accessed from http://onlinelibrary.wiley.com/doi/10.1111/j.1574-0862.2011.00564.x/suppinfo

We implement this example in STATA by typing the following command:

```
xtreg bmi aidmpc aidfpc fdid lnliv hhsize lncons frac_female frac_male lost_work
days_labor pa2_1 pa3_1 pa8_1 pa10_1 pa14_1 pa2_2 pa3_2 pa8_2 pa10_2 pa14_2 pa5_1
pa5_2 if sampl==1 & male==1, fe cluster(hhid)
```

In the above context, we have a dependent variable (*bmi*) which is the Quetelet index of a health outcome, which is dependent on several factors capturing the health endowment status in the *i*th household in village v in the survey round t, which is the variable identifying each sample observation (*hhid*).

Whether free distribution of food aid to adults in Ethiopia has an impact on their nutritional outcomes is an empirical question, and the above regression can be used to verify the implications of this policy. The variables *aidmpc* and *aidfpc* are the per capita amounts of aid received by the males and females.

Whether or not the individual in the sample was an aid recipient is tracked by the dummy variable (*fdid*), as this role itself may influence the observed BMI, and also how much others get. The model also captures information on the livestock value (*lnliv*), household size (*hhsice*), per capita consumption (*lncons*), fraction of the household that is male and famale (*frac_male, frac_female*), the number of days not working (*lost-work*), and the number of days in labor sharing activities (*days_labor*).

The model also includes additional dummy variables to capture time-varying trends that track whether a village v was surveyed in round 1 or 2. The `fe` model is estimated for the males in the data (`male ==1`), and the `fe cluster(hhid)` produces the following output for the fixed-effects model with robust standard errors:

```
Fixed-effects (within) regression               Number of obs      =      1012
Group variable: pid                             Number of groups   =       363

R-sq:  within  = 0.1508                         Obs per group: min =         2
       between = 0.0225                                        avg =       2.8
       overall = 0.0012                                        max =         3

                                                F(22,291)          =      7.16
corr(u_i, Xb)  = -0.2854                        Prob > F           =    0.0000

                                  (Std. Err. adjusted for 292 clusters in hhid)
```

bmi	Coef.	Robust Std. Err.	t	P>|t|	[95% Conf.	Interval]
aidmpc	.0287061	.0124959	2.30	0.022	.0041122	.0532999
aidfpc	.0295936	.0152562	1.94	0.053	-.0004329	.0596201
fdid	.1265905	.1839174	0.69	0.492	-.2353865	.4885675
lnliv	-.0349666	.064115	-0.55	0.586	-.1611545	.0912213
hhsize	.1131478	.0694733	1.63	0.104	-.0235862	.2498817
lncons	-.0337695	.1006085	-0.34	0.737	-.231782	.164243
frac_female	-1.623966	1.092592	-1.49	0.138	-3.774351	.5264186
frac_male	.4887894	.787952	0.62	0.536	-1.062018	2.039597
lost_work	-.0214815	.0099889	-2.15	0.032	-.0411412	-.0018218
days_labor	-.1114249	.0536665	-2.08	0.039	-.2170487	-.0058012
pa2_1	-.0780871	.3149886	-0.25	0.804	-.6980318	.5418576
pa3_1	.1594669	.3913142	0.41	0.684	-.6106979	.9296318
pa8_1	-.4180449	.1379792	-3.03	0.003	-.6896087	-.1464812
pa10_1	-.4734151	.3957527	-1.20	0.233	-1.252316	.3054854
pa14_1	.0949303	.244069	0.39	0.698	-.385434	.5752945
pa2_2	-.5410832	.2520476	-2.15	0.033	-1.037151	-.0450159
pa3_2	.2823577	.3829243	0.74	0.461	-.4712945	1.03601
pa8_2	-.0706846	.2300331	-0.31	0.759	-.5234241	.3820549
pa10_2	.8229282	.2553687	3.22	0.001	.3203245	1.325532
pa14_2	1.150351	.2652855	4.34	0.000	.6282291	1.672472
pa5_1	-.4853533	.2800189	-1.73	0.084	-1.036472	.0657657
pa5_2	-.8408908	.2480925	-3.39	0.001	-1.329174	-.3526076
_cons	19.36141	.8264075	23.43	0.000	17.73491	20.9879
sigma_u	1.8836504					
sigma_e	1.2197858					
rho	.70455272	(fraction of variance due to u_i)				

The F-value indicates that the overall model is significant. In the males' sample, the aid received by both males and females is significant at 95% and 90%. Days spent in labor sharing and lost-days have the expected signs, and are also significant. We can also test whether the slope coefficients *aidmpc* and *aidfpc* are equal. The input command and the output portions of STATA are reproduced below:

```
. test aidmpc=aidfpc

 ( 1)  aidmpc - aidfpc = 0

       F(  1,   291) =    0.00
            Prob > F =    0.9534
```

The F-test in this case indicates that the null hypothesis of equality can be rejected. Broussard (2012) estimates a similar model for the females in the sample, with an additional dummy variable to capture the status of pregnancy or breast feeding (*lact_bre*), and the STATA implementation is:

```
. xtreg bmi aidmpc aidfpc fdid lnliv lact_bre hhsize lncons frac_female
frac_male lost_work days_labor pa2_1 pa3_1 pa8_1 pa10_1 pa14_1 pa2_2 pa3_2 pa8_2
pa10_2 pa14_2 pa5_1 pa5_2 if sampl==1 & male==0, fe cluster(hhid)
```

```
Fixed-effects (within) regression               Number of obs      =       970
Group variable: pid                             Number of groups   =       346

R-sq:  within  = 0.1233                         Obs per group: min =         2
       between = 0.0091                                        avg =       2.8
       overall = 0.0323                                        max =         3

                                                F(23,291)          =      4.17
corr(u_i, Xb)  = -0.1094                        Prob > F           =    0.0000

                              (Std. Err. adjusted for 292 clusters in hhid)
------------------------------------------------------------------------------
             |               Robust
         bmi |      Coef.   Std. Err.      t    P>|t|     [95% Conf. Interval]
-------------+----------------------------------------------------------------
      aidmpc |  -.0033293   .0069473    -0.48   0.632    -.0170026    .0103441
      aidfpc |    .020969   .0247004     0.85   0.397     -.027645    .0695831
        fdid |  -.2259588   .1873993    -1.21   0.229    -.5947887    .1428711
       lnliv |   .0058628   .0674142     0.09   0.931    -.1268184     .138544
    lact_bre |   .4853335   .1584704     3.06   0.002     .1734401    .7972268
      hhsize |  -.1299795   .0847423    -1.53   0.126     -.296765     .036806
      lncons |  -.0246799   .1112774    -0.22   0.825    -.2436905    .1943306
 frac_female |  -.7588874   1.151707    -0.66   0.510     -3.02562    1.507845
   frac_male |   1.584986   .9693805     1.64   0.103    -.3228995    3.492872
   lost_work |   -.024671   .0120529    -2.05   0.042    -.0483928   -.0009491
  days_labor |   .0879014   .0545205     1.61   0.108    -.0194031     .195206
       pa2_1 |   .9804731   .3631399     2.70   0.007     .2657594    1.695187
       pa3_1 |  -.4126158   .4646117    -0.89   0.375    -1.327041    .5018096
       pa8_1 |  -.3767912   .2174849    -1.73   0.084    -.8048339    .0512516
      pa10_1 |  -1.216006   .6683769    -1.82   0.070    -2.531471    .0994601
      pa14_1 |   .2235198   .2733709     0.82   0.414    -.3145149    .7615546
       pa2_2 |   .3828406   .2633033     1.45   0.147    -.1353796    .9010608
       pa3_2 |  -.2474584   .3980453    -0.62   0.535    -1.030871    .5359543
       pa8_2 |   .3643752   .1870547     1.95   0.052    -.0037765    .7325269
      pa10_2 |   .5824278    .510084     1.14   0.254    -.4214937    1.586349
      pa14_2 |    1.22686   .2574907     4.76   0.000     .7200803    1.733641
       pa5_1 |  -.2759374   .2392636    -1.15   0.250    -.7468439    .1949691
       pa5_2 |  -.3820759   .2204443    -1.73   0.084    -.8159432    .0517914
       _cons |   20.77194   .9009838    23.05   0.000     18.99867    22.54521
-------------+----------------------------------------------------------------
     sigma_u |  1.8423913
     sigma_e |  1.3020726
         rho |  .66690373   (fraction of variance due to u_i)
------------------------------------------------------------------------------
```

The FE estimates for the females' regression model indicate that the aid receipts for either gender are not significant. This is also verified by the `test` command:

```
. test aidmpc=aidfpc

 ( 1)  aidmpc - aidfpc = 0

       F(  1,   291) =    1.00
            Prob > F =    0.3186
```

In the next section, we illustrate how to generate the BE and the RE estimator. We first define `xlist` as a `global` command to list all the exogenous variables:

```
global xlist aidmpc aidfpc fdid lnliv hhsize lncons frac_female frac_male lost_work
days_labor pa2_1 pa3_1 pa8_1 pa10_1 pa14_1 pa2_2 pa3_2 pa8_2 pa10_2 pa14_2 pa5_1
pa5_2
```

We use `quietly` to suppress the output from different models, and generate a table with all the results. We restrict attention to the sample with males, as in our first FE estimation. We begin with the `regress` command for the OLS estimates with robust standard errors using the `vce (cluster hhid)` option, and then generate the BE, the FE, and finally the RE with robust standard errors:

```
. quietly regress bmi $xlist, vce (cluster hhid)
. estimates store OLS_rob
. quietly xtreg bmi $xlist, be
. estimates store BE
. quietly xtreg bmi $xlist, fe
. estimates store FE
. quietly xtreg bmi $xlist, fe vce(cluster hhid)
. estimate store FE_rob
. quietly xtreg bmi $xlist, re
. estimates store RE
. quietly xtreg bmi $xlist, re vce(robust)
. estimates store RE_rob
```

Finally, we produce a table that captures the key features of all the models, using the following command in STATA:

```
estimates table OLS_rob BE FE FE_rob RE RE_rob, b se stats(N r2 r2_o r2_b r2_w sig-
ma_u sigma_e rho) b(%7.4f)
```

The output from STATA is produced below. For simplicity, we suppress the estimates of the dummy variables. For most of the variables, the estimates from alternative models are not different.

Once we have the estimates from the FE and RE estimators, we can perform the Hausman test using the `hausman FE RE, sigmamore command`. We have also reproduced the output from the Hausman test below. The result $\chi^2(22)$ with $p = 0.00$ leads to a rejection of the null hypothesis that RE provides consistent estimates.

Variable	OLS_rob	BE	FE	FE_rob	RE	RE_rob
aidmpc	0.0161	0.0338	0.0134	0.0134	0.0153	0.0153
	0.0079	0.0140	0.0053	0.0059	0.0050	0.0059
aidfpc	0.0024	-0.0129	0.0240	0.0240	0.0197	0.0197
	0.0175	0.0206	0.0096	0.0082	0.0089	0.0086
fdid	0.3716	0.4500	0.0748	0.0748	0.1804	0.1804
	0.1256	0.2344	0.0965	0.0888	0.0905	0.0866
lnliv	-0.0073	-0.0243	-0.0039	-0.0039	0.0207	0.0207
	0.0275	0.0305	0.0327	0.0419	0.0216	0.0229
hhsize	-0.0024	0.0247	-0.0037	-0.0037	-0.0510	-0.0510
	0.0198	0.0232	0.0444	0.0479	0.0191	0.0189
lncons	0.2909	0.3822	-0.0227	-0.0227	0.1393	0.1393
	0.0677	0.1116	0.0539	0.0590	0.0481	0.0483
frac_female	-0.4005	-0.1890	-0.8543	-0.8543	-0.8828	-0.8828
	0.4612	0.4684	0.5826	0.6495	0.3628	0.3906
frac_male	-0.9564	-1.1657	0.2334	0.2334	-0.9130	-0.9130
	0.3845	0.4173	0.5173	0.5719	0.3176	0.3178
lost_work	-0.0184	-0.0039	-0.0200	-0.0200	-0.0184	-0.0184
	0.0096	0.0151	0.0063	0.0062	0.0059	0.0064
days_labor	-0.1094	-0.1006	-0.0307	-0.0307	-0.0652	-0.0652
	0.0263	0.0496	0.0275	0.0298	0.0247	0.0239
_cons	18.8464	17.8850	19.9169	19.9169	19.7620	19.7620
	0.3622	0.5088	0.4874	0.5464	0.2940	0.2942
N	3171	3171	3171	3171	3171	3171
r2	0.1349	0.1758	0.1449	0.1449		
r2_o		0.0943	0.0229	0.0229	0.1066	0.1066
r2_b		0.1758	0.0014	0.0014	0.1061	0.1061
r2_w		0.0139	0.1449	0.1449	0.1232	0.1232
sigma_u			2.0972	2.0972	1.7247	1.7247
sigma_e			1.2121	1.2121	1.2121	1.2121
rho			0.7496	0.7496	0.6694	0.6694

legend: b/se

```
. hausman FE RE, sigmamore

                 ---- Coefficients ----
             |      (b)          (B)            (b-B)     sqrt(diag(V_b-V_B))
             |       FE           RE         Difference          S.E.
-------------+----------------------------------------------------------------
      aidmpc |    .0133696     .0153151       -.0019456         .002022
      aidfpc |    .0240369     .0196987        .0043383        .0041613
        fdid |    .0748209     .1803588       -.1055378        .0393408
       lnliv |   -.0039301     .0206945       -.0246247        .0254375
      hhsize |   -.0036735    -.0509686        .0472951        .0411876
      lncons |   -.0226774     .1393109       -.1619884        .0269151
 frac_female |   -.8543287    -.8827688        .0284401        .4718921
   frac_male |    .2334175    -.9130282        1.146446         .422545
   lost_work |    -.019994    -.0183506       -.0016434        .0024661
  days_labor |   -.0306621    -.0651791         .034517        .0133684
------------------------------------------------------------------------------
                         b = consistent under Ho and Ha; obtained from xtreg
          B = inconsistent under Ha, efficient under Ho; obtained from xtreg

    Test:  Ho:  difference in coefficients not systematic

                  chi2(22) = (b-B)'[(V_b-V_B)^(-1)](b-B)
                           =      172.09
                Prob>chi2 =      0.0000
```

Broussard (2012) also estimates a dynamic panel data model, under the assumption that past health status influences aid receipts and, hence, the current health outcome. To verify this feature, the following dynamic model is set up:

```
xtabond bmi aidmpc aidfpc fdid lnliv hhsize lncons frac_female frac_male lost_work
days_labor pa2_1 pa3_1 pa8_1 pa10_1 pa14_1 pa2_2 pa3_2 pa8_2 pa10_2 pa14_2 pa5_1
pa5_2 if sampl==1 & male==1, lags(1) artests(1) nocons twostep vce(robust)
```

The `xtabond` produces the Arellano–Bond estimates of the dynamic panel data model, where a one-period lag for *BMI* is set up with `lags(1)`. The output from STATA in this case is:

```
Arellano-Bond dynamic panel-data estimation  Number of obs      =       290
Group variable: pid                          Number of groups   =       290
Time variable: rnd
                                             Obs per group:   min =       1
                                                              avg =       1
                                                              max =       1

Number of instruments =      0               Wald chi2(17)      =     60.90
                                             Prob > chi2        =    0.0000
Two-step results
                                  (Std. Err. adjusted for clustering on pid)
------------------------------------------------------------------------------
             |              WC-Robust
         bmi |      Coef.   Std. Err.      z    P>|z|     [95% Conf. Interval]
-------------+----------------------------------------------------------------
         bmi |
         L1. |   .2330271   .1657915     1.41   0.160    -.0919182    .5579724
             |
      aidmpc |   .0406183   .0245185     1.66   0.098    -.0074371    .0886737
      aidfpc |   .0330678    .020505     1.61   0.107    -.0071212    .0732569
        fdid |   .1757765   .2146965     0.82   0.413    -.2450209    .5965739
       lnliv |   -.068738   .1455712    -0.47   0.637    -.3540523    .2165763
      hhsize |   .2049372   .1412398     1.45   0.147    -.0718877    .4817621
      lncons |  -.1559416   .2245927    -0.69   0.487    -.5961351    .2842519
 frac_female |  -4.846888   1.959568    -2.47   0.013     -8.68757   -1.006206
   frac_male |   1.586644   1.223098     1.30   0.195    -.8105841    3.983873
   lost_work |  -.0251307   .0174743    -1.44   0.150    -.0593798    .0091184
  days_labor |   .0007396   .0845565     0.01   0.993    -.1649881    .1664673
      pa2_2  |  -.6090621    .313544    -1.94   0.052    -1.223597    .0054728
      pa3_2  |   .3155677   .4031347     0.78   0.434    -.4745617    1.105697
      pa8_2  |  -.2936994   .3218059    -0.91   0.361    -.9244273    .3370285
     pa10_2  |   1.460776   .5702331     2.56   0.010     .3431393    2.578412
     pa14_2  |   1.500947   .4249343     3.53   0.000     .6680915    2.333803
      pa5_2  |  -1.028671   .3147991    -3.27   0.001    -1.645666   -.4116761
------------------------------------------------------------------------------
Instruments for differenced equation
        GMM-type: L(2/.).bmi
        Standard: D.aidmpc D.aidfpc D.fdid D.lnliv D.hhsize D.lncons
                  D.frac_female D.frac_male D.lost_work D.days_labor D.pa2_1
                  D.pa3_1 D.pa8_1 D.pa10_1 D.pa14_1 D.pa2_2 D.pa3_2 D.pa8_2
                  D.pa10_2 D.pa14_2 D.pa5_1 D.pa5_2
```

The model is estimated by using lag(2) of BMI as a suitable instrument for the first-differences in BMI. The estimates of the dynamic model are roughly the same as that of the FE model without lags.

How do policy makers know that income-boosting programs are the best, only under certain conditions? Why can a simple increase in income transfers not work to generate a higher nutritional status across all households? This chapter has identified several pieces of the puzzle, to show why initial conditions and assumptions are important, and why a single, unique driver of a said nutritional outcome is nonexistent. The analytical methods using panel data are very important in this context, particularly when information about several households is gathered over a period of time. It is important to control for variation across time and households, to capture the correct underlying cause of the dependent variable, say BMI.

The STATA example presented in this chapter illustrates the application of the econometric panel-data methods using real-world data. The STATA estimation helps us to find out which estimation method, whether OLS, Fixed Effect, or Random Effect gives the correct estimator. The STATA commands can also identify and test for the significance of individual drivers of BMI, such as household income or the fraction of females in the household. Policy makers can use this information to find out which households are more prone to economic shocks versus household-specific shocks. Such information is useful to determine whether food supplementation should come in cash or in food aid. Are women in low-income households particularly prone to lower BMI? How would one frame the regression and reject the null-hypothesis for this research question? We allow students to explore such angles in the end of chapter exercises.

CONCLUSIONS

Sustainable Development Goals recently established by the global development community delve into social protection as a key intervention for poverty reduction strategies in several areas. Researchers and practitioners have looked at social protection from several perspectives. From the rural poverty perspective, social protection interventions are considered as a quick solution to relieving people deeply rooted in poverty, and where no immediate productive opportunities exist. They also help vulnerable populations and the destitute, who are left behind by the normal economic growth process. From the food security perspective, social protection programs that increase income can increase food security. Direct food aid and food distribution systems, such as the one in India, can be effective in achieving basic food security for all the population. From the perspective of nutrition, while there is potential for better nutrition, complementary interventions may be needed to translate income and food transfers into nutritional outcomes. Cash made available with well-functioning markets for nonstaple foods can increase dietary diversity and the quality of food consumed.

In the context of livelihood protection, productive social safety nets link benefits to sustainable food systems. The rural poor depend on agriculture. The food security implications of rural poverty could be addressed through social protection. Close to 80% of the poor live in rural areas in many developing countries. In this context, social protection is considered also as an agricultural risk reduction strategy, and strengthened social networks have improved livelihoods. In periods of food crisis such as the one seen during the 2007 – 08 food crisis, social protection programs can respond to crises quickly due to the presence of implementation mechanisms on the ground. South Africa and Brazil are good examples (Babu, 2015) where such programs buffer the poor from falling below a certain level of poverty.

Social protection programs can enhance women's access to income, food, and nutrition, and thereby contribute to reduced gender bias in nutrition and to improved gender empowerment. They also can contribute to human capital development as a part of health related interventions, and keep children in school as a condition for receiving the program benefits. In some cases, reduced reliance on borrowing has also been reported. Resilience against climate shocks has implications for nutritional outcomes (Jones, 2015). In societies where there is outmigration for work, informal transfers and remittances from relatives complement the program benefits on the ground.

Large-scale social protection interventions will continue for some time. For example, in India public food distribution systems still continue as a major intervention for food and nutrition security. They are complemented by targeted interventions in the form of Integrated Child Development Services, in which child nutrition enhancement is a major goal (Jain, 2015). Such large-scale programs have shown that behavior change and transfer modalities matter in realizing better nutritional outcomes. Food rations are narrow in design and the income gains are not substitutable, so they can negatively affect dietary diversity.

If nutrition and diverse diets are important, both food and cash may be needed. Cash transfer, along with behavioral change with information, has been shown to result in a 7% reduction in child stunting, as observed in Bangladesh (Sunny Kim and Phuong Nguyen, 2016). Empowerment of rural populations is key for the long-run benefits of social protection. Finally, long-term economic growth is needed to uplift the poor and the malnourished. Social protection programs could at best be considered as short-term interventions, and cannot be expected to address the problems of the poor and hungry on a long-term basis.

Several research issues still need to be addressed in generating evidence for effective policy making. They include the effect of social protection on seasonal migration, asset accumulation, and investment strategies of the beneficiary households, building better resilience against shocks, diversification to nonfarm activities, the contribution of children and women's labor to agriculture, predictability of transfers, multiplier benefits to local economy, complementarity of interventions, modern input use in agriculture, and avoiding over-nutrition. We leave them for the readers and budding social protection researchers to handle.

EXERCISES

1. Use the data from Broussard (2012) and estimate a model separately for the females in the sample with low assets and high assets. The two panel data regressions can be estimated with these commands:

```
.xtreg bmi aidmpc aidfpc fdid lnliv lact_bre hhsize lncons frac_female frac_male
lost_work days_labor pa2_1 pa3_1 pa8_1 pa10_1 pa14_1 pa2_2 pa3_2 pa8_2 pa10_2
pa14_2 pa5_1 pa5_2 if sampl==1 & male==0 & low_asst==1, fe cluster(hhid)
.estimates store FE4
.xtreg bmi aidmpc aidfpc fdid lnliv lact_bre hhsize lncons frac_female frac_male
lost_work days_labor pa2_1 pa3_1 pa8_1 pa10_1 pa14_1 pa2_2 pa3_2 pa8_2 pa10_2
pa14_2 pa5_1 pa5_2 if sampl==1 & male==0 & low_asst==0, fe cluster(hhid)
.estimates store FE6
```

Note that the estimates are stored as FE4 and FE6. Produce a table to show these estimates. Broussard (2012) notes from these results from low-asset female households, that the household decisions about aid allocation matter a lot for *BMI*. Examine the coefficients of *aidmpc* and *aidfpc* between these regressions and see if you can come to a similar conclusion.

2. Based on the previous problem and its conclusion, Broussard (2012) wishes to test whether women in low-income households are adversely affected. In order to do this, Broussard (2012) defines a new variable `land*aidmpc` which receives a value of >0 if the wife gets half of the land holdings if the marriage were to end in a divorce, the variable receives a value = 0. Define this variable as follows using the `gen` command and run the FE model specified below:

```
gen land_val_div = land_div_wife*aidmpc*land94
replace land_val_div = land_div_wife*aidmpc*(land94/2) if landhf_div_wife==1
xtreg bmi land_val_div aidmpc aidfpc fdid lnliv lact_bre hhsize lncons
frac_female frac_male lost_work days_labor pa2_1 pa3_1 pa8_1 pa10_1 pa14_1 pa2_2
pa3_2 pa8_2 pa10_2 pa14_2 pa5_1 pa5_2 if sampl==1 & male==0 & low_asst==1,
fe cluster(hhid)
```

What is the sign of the *land_val_div* variable? Is it significant? What does this suggest about household bargaining power and nutritional outcomes?

CHAPTER

ECONOMICS OF SCHOOL NUTRITION: AN APPLICATION OF REGRESSION DISCONTINUITY 13

The Tamil Nadu Government is presently managing what has been described as the largest school and preschool feeding program in the world... To read this Noon Meal Scheme as the outcome of the policy advocacy of the Tamil Nadu Nutrition Study would be to err. Indeed, it is actually opposed by the World Bank nutritionists and many welfare economists who regard its commitment as excessive, its opportunity costs too high.

—Barbara Harriss (1991)

INTRODUCTION

How food provided to school children affects the household economy has been of interest to policy researchers for a long time (Babu and Hallam, 1989; Long, 1991; Ahmed and Babu, 2006; Alderman and Bundy, 2011; Gelli et al., 2015). Although school nutrition programs are considered as part of the social protection programs covered in Chapter 12, Nutritional Implications of Social Protection: Application of Panel Data Method, they deserve a chapter on their own for several reasons. The definition of food security includes key aspects of availability, accessibility, utilization, and stability. School feeding programs (SFPs) as a safety net strategy help in increasing access to food to school-going children and their families. Thus, in the context of a broad development intervention process, linking social protection and access to food is done through SFPs. Increasingly, safety net policies of international organizations such as the World Food Program combine human capital development with food security objectives (Gelli et al., 2015).

Nutrition-sensitive safety nets also help to link local food production to the supply of food in schools. Thus, SFPs have potential for enhancing and reorienting the local food production impact toward nutritional objectives in the community. Further, by bringing together education and social protection goals along with the local economy linkages of SFP, multiple sectors work together at community levels. For example, the quality of education and improvements in enrollment and attendance at school brings the education sector to work with the social welfare and agricultural sectors (Ahmed and Babu, 2006).

SFPs help with social intervention and provide several key socioeconomic benefits (Ahmed and Babu, 2006). Although aimed at increasing the food and nutritional security of school-going children, they help in transferring food to other household members when the children are allowed to take food rations home. At certain levels of school food contributions, SFPs help to release income of households for other nonfood expenditures. It can increase resilience and reduce the vulnerability of the poorest households. By bringing girls to school, SFPs help increase girls'

Nutrition Economics. DOI: http://dx.doi.org/10.1016/B978-0-12-800878-2.00013-X

education, and reduce gender inequality in the long run. Home-grown school feeding can help in stimulating local agricultural production for new crops, and building value chains for the crops. They also help in increasing dietary diversity, and invest in human capital building and the country's longer-term human development goals (Alderman et al., 1999; Afridi, 2010).

Innovations are also introduced within SFPs, such as cash-based transfers for accessing food in the market (Bhattacharya et al., 2006; Conner et al., 2012; Todd and Winters, 2011; Ishdorj et al., 2013). Procurement of food from farmers groups helps to connect subsistence farmers to the market chains. Behavioral changes are also introduced through the education of children who take knowledge home to their families. This involves provision of the right type of food in the school, and nutritional education through introducing good eating habits. The cost and benefits of SFPs indicate that $1 invested in school feeding returns $3 – 10 in health, education, and productivity benefits in the children's adulthood (Singh et al., 2012).

In this chapter we will look at the economic analysis of school feeding programs as a special case of social protection program, and show how they can be evaluated using selected analytical methods.

WHAT DO WE KNOW FROM EXISTING LITERATURE?

Countries have used SFPs to ensure minimum levels of food security for their populations. For example, the Supreme Court of India ruled in 2001 that all government schools should provide cooked meals. The ruling was enacted by most states in India by 2003, covering roughly 120 million school-going children. Singh et al. (2012) note that India's Midday Meals Scheme (MDMS) is the largest school lunch program in the world.[1]

In contrast to India's Supreme Court ruling, Alderman and Bundy (2011) point out that Food For Education (FFE) programs are best viewed as income transfer programs. FFE does not qualify as the best investment in nutrition. FFE funds may also not be the best method to improve educational outcomes. Indeed, Alderman and Bundy (2011) cite the report of the United States General Accounting Office, which suggests that school feeding programs may not be the most cost-effective intervention.

Given the two opposing viewpoints, it is important to find out just how well the school feeding programs work, and what mechanisms are still required to strengthen their positive effects. Singh et al. (2012) provide an extensive evaluation of the MDMS implemented in India. The researchers use an interesting data set from the state of Andhra Pradesh in India, involving children in two different cohorts, observed in two different rounds, 2002 and 2007, respectively. Further, they also incorporate anthropometric measures on outcome variables, to capture the health effects of MDMS.

An important feature in the data is that the period between the two rounds coincides with severe drought experienced by the households in the survey. Consequently, the data enables the researchers to evaluate the impact of school feeding on households' ability to deal with environmental shocks.

[1]A large set of research studies evaluates India's MDMS in recent years. See Afridi (2010, 2011), Khera (2006), and for further references see Singh, Dercon, and Smith (2012).

The authors find that children exposed to the drought suffered from severe height and weight loss. However, MDMS acts as a safety net in these cases, and compensates for the environmental shock. Indeed, the authors estimate that the benefits of MDMS in drought-affected areas more than exceeded the costs of drought. A WFP—World Bank combined report (2009) also mentions the SFPs capability to be scaled-up during social crises.

While the experience with MDMS in India's Andhra Pradesh has been noteworthy, the survey evidence provided by Alderman and Bundy (2011) regarding SFPs is not all that positive. Alderman and Bundy (2011) recommend that the SFPs be viewed as a form of income transfer. The authors' observation, which is also made by the WFP report, is that in some low-income countries, the feeding cost per child is the same as the cost of basic education. This means that SFPs merely crowd-out other investments in education. Hence, it is difficult to make the case that SFPs increase educational outcomes.[2]

Similarly, McEwan (2013), using a regression-discontinuity design for data from Chile, examines the effect of higher-calorie meals on school enrollment and attendance, students' first-grade enrollment age and grade repetition, and fourth-grade test score outcomes including national mathematics and language tests. Overall, McEwan (2013) finds no effects of the said intervention on any of the outcome variables.

Alderman and Bundy (2011) gather evidence from research conducted in Uganda, Burkina Faso, and Lao PDR. Both in Uganda and Burkina Faso, children enter schools at a younger age in places where SFPs were available, when compared to control groups which had no such programs. Based on a randomized control evaluation in Northern Uganda, Alderman et al. (2010) suggest that school meals induce hungry children to delay completing primary school.

Further, in Burkina Faso and in Western Kenya, researchers identified an increase in attendance. Vermeersch and Kremer (2005) note an increase in the scores in written and oral exams, albeit the schools had a better stock of teachers and class size. Further, SFPs in Kenya are often bundled along with other programs, such as malaria reduction. Consequently, the influence of SFPs on test scores may be confounded with other intervention effects. For example, Kazianga et al. (2008) use a randomized evaluation procedure to assess the impact of two school feeding schemes on educational and health outcomes of children from low-income households in northern rural Burkina Faso. While school enrollment improved for girls, and the siblings of beneficiaries showed weight gains, academic performances and attendance were lower for those children exposed to the treatment. The researchers conclude that alternative wage employment opportunities and family size are crucial drivers of SFP success.

Alderman and Bundy (2011) also acknowledge the importance of SFPs as a social safety net in a time of crisis. They cite evidence that links SFPs to positive externalities, where the health outcomes of siblings improve through household reallocation. But the authors caution that programs often fail to reach children during their vulnerable period, which is between conception and two years of age.

In a related study for Lao PDR, Buttenheim et al. (2011) find no effect of school feeding on either enrollment or nutritional status. Interestingly, some of the reasons for nonparticipation among

[2]Gelli et al. (2016) also do not find any substantial differences from school feeding in the outcomes variables from several interventions, although the authors note that the results are very sensitive to the design and the selection of the control group.

villagers, the authors find, are the distance to the food delivery point, and the difficulties in building storage silos. The authors correctly conclude that a basic level of social capital is needed if SFPs must reach vulnerable groups.

LESSONS FROM THE EVALUATIONS OF THE INTERNATIONAL AGENCIES

Given the divergence in findings and viewpoints from the above two streams of research, it is very important to get a comprehensive understanding of the evaluation of school feeding programs. With this objective in mind, the World Bank (WB) and the World Food Program (WFP), in collaboration with the Partnership for Child Development (PCD), have published many analytical reports in recent years.[3] The *State of School Feeding Worldwide* (2009) and *Rethinking School Feeding: Social Safety Nets, Child Development, and the Education Sector* by Bundy et al. (2009) are two excellent research reports that present a comprehensive evaluation of global practices in school feeding. The reports have gathered exhaustive information from several research findings and collaborators. The *State of School Feeding Worldwide* (2009) is very helpful in letting us know the advantages of SFPs, and lists all the challenges faced by practitioners in this field. Both research reports provide several insights regarding global school feeding programs.

HOW MANY KIDS ARE FED IN SCHOOLS?

According to the WFP report (2009), at least 368 million pre-primary, primary, and secondary-school children from 169 countries receive food in schools. This corresponds to roughly \$47–\$75 billion dollars as allocations from government budgets.

THE BOTTOM LINE

The WFP report (2009) performs a cost–benefit analysis, and the conclusion is noteworthy: the benefits from school feeding (improved children's health, education, and increased productivity) greatly exceed the costs of a program.

Among a sample of nine countries, the WFP researchers show that the cost–benefit ratio ranges from 1:3 to 1:8. Hence, every dollar the state spends on school feeding generates a minimum of \$3 in benefits.

[3]The analysis was undertaken to better understand the growing demand from countries for school feeding programs, particularly after the food, fuel, and financial crises of 2008.

THE IMPORTANCE OF SCHOOL FEEDING PROGRAMS

According to the WFP report (2009), there are two main reasons why countries choose SFP:

1. To address social needs and provide a social safety net during crises;
2. To support child development through improved learning and enhanced nutrition.

SFPs provide good support as a social safety net in the short-term. These programs indirectly transfer income to poor families. SFPs also secure education for the children, especially for girls. SFPs also promote human capital formation, and long-term cognitive and productive abilities, particularly for girls.

The WFP report (2009) indicates that, in spite of low budgets, eight low-income countries have SFPs, and globally, everyone wants to scale-up the programs, to improve nutritional outcomes, quality, reduce costs, and improve efficiency. Likewise, an important feature of SFPs is that these programs can be scaled-up in response to crises. The report points out that the SFPs have helped vulnerable populations during social crises due to armed conflicts, natural disasters, and during the financial crisis in 2008. The report presents evidence from 38 countries where SFPs have responded adequately to social shocks.

An important question that is often asked by researchers is whether SFPs enhance educational attainment. SFPs can help with educational attainment, but that success crucially depends upon the pre-existing educational setup—such as the quality of the teachers, textbooks, classroom size, curriculum, and the overall learning environment. In some countries, teachers and staff prepare and serve the food, which unduly taxes the system and creates additional opportunity costs. In this context, the best results are obtained when SFPs are cleverly combined with other safety net programs for poor families, as in Brazil and Mexico.

THE CHALLENGES WITH SCHOOL FEEDING PROGRAMS

School feeding is present in almost every country in the world, but is not always efficiently delivered. In some low-income countries the per-child cost of school feeding is more than the per-child cost of education. We have to look for opportunities to reduce these costs in these countries.

The WFP–WB report (2009) points out that countries with the lowest coverage are the ones with the greatest need. In other words, middle and low-income countries with the greatest need in terms of food cover only 18% of school children, compared to 49% in developed countries. Once again, this low coverage is due to cost-constraints and hence, we need a good understanding of the cost structures to attain allocative efficiency.

SFPs can add to educational attainment, only if the learning environment is also conducive. The WFP–WB report (2009) points to this prerequisite that has also been mentioned by other researchers, and caution that teachers and staff should not be used in food preparation and incidental activities.

Most importantly, the WFP–WB report (2009) mentions that the crucial period of investment occurs during the first 1000 days of life, and SFPs can only help in sustaining the growth that occurs during the early development phases. SFPs can enhance the health outcomes if the program includes fortified meals, deworming, and most importantly, safe water.

According to the WFP–WB report (2009), a formal partnering system among countries is also a big need. The collaboration will disseminate knowledge, coordinate action, and increase the support needed to make school feeding an important social support system and a safety net. This aspect becomes even more important when we note that the SFPs in developed countries are well established through regulatory frameworks. Low-income countries do not have such institutionalized settings, and hence have to rely on partners, which makes the SFPs in these countries unsustainable.

The WFP–WB report (2009) also notes that a major effort is needed to link the local suppliers, farmers, and the agricultural sector to the school feeding programs. This effort will improve the supply chain, and will ensure the sustainability of the program. Indeed, Brazil, Chile, and Scotland are good examples, where food is purchased from local farmers with small farm sizes. The WFP–WB report (2009) also points out that locally grown foods may not have all the necessary fortification required.

While SFPs can work and achieve a lot more, the WFP–WB report (2009) points out that programs must have clear objectives, proper design, and targeting. Monitoring and evaluation must also be a crucial part of every design. Bundy et al. (2009) provide an excellent guide for initial program setup, design tools, assessment techniques, evaluation guidelines, and tools for updating SFPs (Box 13.1).

BOX 13.1 FOOD FOR GRADES: LACK OF ACCOUNTABILITY AND MANIPULATION IN VIRGINIA AND MUMBAI

The United States passed the No Child Left Behind Act of 2001, a federal law, which required implementation of significant accountability systems for schools. The law enforced accountability based on the fraction of students that meet proficiency standards in state examinations. Most importantly, the law can also sanction schools that do not meet the goals. Sanctions may be in the form of reduced funding and tighter budget scrutiny. Hence, the new law and the accountability requirements place a considerable burden on students' performance. Can schools "game" the system?

Figlio and Winicki (2005) demonstrate in an interesting paper that the schools manipulate the test performance by artificially improving the nutritional content of school lunches around test dates. Figlio and Winicki (2005) use data from schools in Virginia and link the nutritional content of school lunches to information on test dates and scores. Further, the researchers also match this information with schools that are on the margin in terms of test performance.

Figlio and Winicki (2005) show that the nutritional content of school meals is altered on test dates, interestingly in districts with at least one "failing," or sanctioned, school. Does the menu manipulation work? The researchers show that by increasing the calorie content by 100 units, scores in math, English, and history/social studies increased by 7%, 4%, and 7% points.

The authors argue that since the National School Lunch Program (NSLP) is subsidized to poorer student groups, and since this is the group in the first place that does not meet the accountability standards, the schools find it natural to pursue the gaming strategy to avoid further sanctions. Finally, the USDA, which oversees the NSLP is disconnected from the Department of Education that requires educational attainment. Consequently, provisions introduced by the USDA within the NSLP, allow schools to game the accountability provisions required by No Child Left Behind.

Findings similar to those in Virginia schools are also observed by Linden and Shastry (2012) for schools in Mumbai, India. The conditional transfer in Mumbai is the Grain Distribution Program, which is a part of the Nutritional Support to Primary Education (NSPE), which was set up in 1995. Students who exceed 80% in attendance are provided with

(Continued)

BOX 13.1 (CONTINUED)

3 kg of grain at the end of each month. The researchers compare official daily attendance records with periodic daily attendance data and demonstrate misreports of true student attendance.

Using two years of school data, Linden and Shastry (2012) find that teachers inflate attendance records. The data also points to selective discretion: teachers inflate the records less frequently for boys, children with low test scores, Muslim children, and students belonging to high castes. Linden and Shastry (2012) note that teachers may have more information about the local conditions, which allows them to use the information more efficiently as they implement the program. Hence, teachers' discretionary powers may improve the nutritional outcomes of children. But, by the same token, this behavior can also influence school attendance. Further, the teachers may discriminate against specific students or groups. Hence, whether local agents can always use their discretionary powers correctly toward nutritional and educational outcomes still remains an open area for research.

SCHOOL FEEDING PROGRAMS IN THE UNITED STATES

Ishdorj et al. (2013) note that the National School Lunch Program (NSLP) and the School Breakfast Program (SBP) influence children's diets and food habit formation, even after the participants go home. That is, the researchers examine children's intake of fruits and vegetables when they are away from school and are at home. This is an important enquiry because it is possible that nutritional intakes at school may be substitutes for intakes at home. Ishdorj et al. (2013) use data from the 2004 – 05 School Nutrition Dietary Assessment Study-III (SNDA-III), on 2096 school-age children and data from 256 schools. The authors provide several insights regarding school–home nutritional intake substitutability. Their results show that increased consumption of vegetables at school is also associated with reduced consumption away from school. Overall, they find that school lunch programs encourage program participants to consume healthier foods, particularly fruit and vegetables.

Ishdorj et al. (2013) also note that school policies such as not offering French fries or dessert, not offering high-fat milk, not offering a la carte food and beverages, or offering fresh fruit and raw vegetables, all had no effect on children's decisions to participate in school meal programs. The authors conclude that these policies therefore do not discourage participation in NSLP. The results also show that students in middle and high school are less likely to participate in the school meal programs. Similarly, a "no store or snack bar" policy increases fruit intake in school, while "no high-fat milk" increases both fruit and vegetable consumption in school. "No French fries" discourages fruit consumption and does not influence vegetable consumption in school (Ishdorj et al. 2013; page 357). These observations indicate that we need to fully understand the drivers of nutrition, and the authors recommend support systems through information awareness among families and children who participate in these important programs.

As discussed before, the WFP–World Bank report *State of School Feeding Worldwide* (2009) indicates that the sustainability of the SFPs can be improved by linking school food procurement to local farmers. Conner et al. (2012) note that the USDAs initiative, "Know Your Farmer, Know Your Food" seeks to implement this goal. Conner et al. (2012) examine these efforts and have made several key observations from the USDAs program. To assess the efforts, the researchers use the experience of two public K–12 school districts: Saint Paul (Minnesota) Public Schools (SPPS), and Denver (Colorado) Public Schools (DPS). These schools obtained more healthful, regionally sourced, and sustainably grown foods for their school meals program.

The authors assess the extent to which value chain partnerships can contribute to the sustainability of SFPs. This study is important because the link between schools and farmers is not as direct as it would seem. The incentives of the distributors and farmers have to be aligned, and this is usually very difficult in a complicated value chain organization. An important part of who gets what contract depends upon the price, which is related to the opportunity cost of supplying to schools, and not to restaurants and retailers. Vendors and farmers are under tough competition, rather than locking a contract on a fixed price. The situation is also made more difficult by vertical integration, particularly if the first mover is a powerful corporation, as in the case of the poultry sector. The authors recommend that there is a need to improve the cooperation between farmers and vendors so as to share risks and rewards within the system.

DO SCHOOL BREAKFAST PROGRAMS WORK?

In the United States, the USDA, through its Food and Nutrition Service (FNS), has established the SBP to provide nutritionally balanced, low-cost meals to children in more than 89,000 schools. As of 2012, about 12.9 million children participated in the program every day, for an annual cost of about $3.3 billion. Breakfasts meet minimum daily dietary requirements through juice, fruit, cereal, and milk, covering vitamin C, folate, calcium, protein, and other nutrients. Children from families with incomes at or below 130% of the Federal poverty level are eligible for free meals. Those with incomes between 130% and 185% of the poverty level are eligible for the meal at a subsidized rate (see http://www.fns.usda.gov/sites/default/files/SBPfactsheet.pdf).

Bhattacharya et al. (2005) examine the effect of SBP not only on the children, but also on other household members. Similar to the observations made by Ishdorj et al. (2013), Bhattacharya et al. (2005) note that SBP may alter the budget constraints of the household, allowing reallocation of food expenditure toward other household members.

In order to examine the effect of SBP on nutritional outcomes, Bhattacharya et al. (2005) use the NHANES III, a national survey, which contains information about dietary intakes besides health and related socioeconomic variables. The authors estimate a difference-in-difference model for 4841 children, by setting children from schools without an SBP as the comparison group. Overall, Bhattacharya et al. (2005) find that SBP leads to better dietary habits. They find that SBP increases the scores on a healthy eating index, with reductions in fat consumption and vitamin deficiencies. SBP also increases the likelihood of high vitamin C, vitamin E, and folate levels. The authors are also able to find partial support for SBP increasing the health scores of other family members.

In an interesting study about SFPs in the United States, Long (1991) examines the critical issue of food supplementation. It is possible that SFPs supplement normal food consumption expenses at the household level. Supplementation arises when SFP benefits do not reduce household expenses on food. Using data from the NSLP and the SBP, Long (1991) finds that SFPs do supplement food expenditures incurred by the beneficiaries. NSLP benefits are offset by a smaller amount, while SBPs are supplemented by the full amount of benefits. Long's (1991) results indicate that households use less than one-half of each additional dollar from NSLP benefits to supplement their food expenditures. Further, all of the additional dollars of SPB are allocated to food expenditures. Long's (1991) estimation procedure includes control of selection bias through participation equations using probit analysis. In a recent study, Amin et al. (2015) observed that requiring children to take fruit and vegetables in their school trays will also not necessarily work, because most of the selection is wasted.

SCHOOL FEEDING PROGRAMS GLOBAL EVOLUTION

Skoufias (2005) evaluates Mexico's major poverty alleviation program called PROGRESA. In 1999 the PROGRESA program served about 2.6 million families. PROGRESA aims to improve human capital development and has schemes in education, health, and nutrition in an integrated effort to decrease poverty. For example, the program gives mothers authority toward allocation decisions in families.

An important scheme in PROGRESA is cash transfers and nutritional supplements to children, conditioned on regular school attendance and visits to health care centers. Skoufias (2005) uses a panel data of 24,000 households and shows that the program has significant impact on enrollment, particularly of girls at the secondary school level. Indeed, children have about 0.7 years of extra schooling because of PROGRESA, and as a consequence, will enjoy 8% higher lifetime earnings.

Overall, PROGRESA has beneficial consequences on diet and nutrition, and is a good example of how SFPs work well when combined with an integrated poverty alleviation program. In spite of the overall success of the program, Skoufias (2005) notes that there was no measurable impact on the participant's test scores. Hence, as under all SFPs, more attention needs to be directed to the quality of education provided in schools. Todd and Winters (2011) also examine Mexico's Oportunidades (the old PROGRESA program) program and note that early health nutrition interventions impact school enrollment positively, and also have a negative effect on absenteeism.

The experience with SFPs in Africa is also very similar to that in other countries. Overall for Mozambique, Burchi (2010) finds that schooling, and particularly the nutritional knowledge of the mothers, are large drivers of children's health status. Beesley and Ballard (2013) examine the role of SFPs initiated by the KwaZulu–Natal provincial government. Since its inception, small, medium, and micro-enterprises were responsible for the provision of ingredients. However, the researchers show that when the government replaced these enterprises with women's cooperatives, the SFPs began to struggle because of inflexible guidelines and institutional rigidities.

The experience in South Africa has been more encouraging. Oldewage-Theron and Napier (2011) show that when nutritional education is a part of education in primary schools, SPFs have better outcomes. The researchers provide the tools to develop and implement simple strategies that are very successful in improving children's health outcomes in the Vaal Region in South Africa.

Babu and Hallam (1989) quantify the impact of a SFP. Using regression analysis and linear programming, Babu and Hallam (1989) show a significant linear relation between school participation and its determinants—household income, adult literacy, and the presence of school nutrition. Their study also demonstrates a reduction in the Gini-coefficient of intake inequality, after the implementation of the SFP. In another comprehensive study for India, Afridi (2010) demonstrates that the SFPs increased the daily nutrient intake by 49% to 100% of the transfers. Indeed, Afridi (2010) estimates that for as low a cost as 3 cents per child per school day, the SFP reduced the daily protein deficiency of a primary school student by 100%, the calorie deficiency by almost 30%, and the daily iron deficiency by nearly 10%. These studies strongly support the nutritional benefits of SFPS.

Likewise, Takeuchi (2015) notes that substantial progress has been made globally on many fronts with respect to improvements in children's welfare. Takeuchi (2015) uses data from two periods for several indicators of welfare for 20 developing countries. The data is used to construct a total inequality index using the L-Theil measure. The measure is broken down to "between households" and "within-household" components.

Although there have been welfare improvements, Takeuchi (2015) also notes that this progress has not been equitable. Moreover, the patterns of inequality vary across dimensions of well-being. Significant differences are noted between girls and boys for school attendance and stunting. These inequalities appear large across households.

Particularly of interest are situations where intra-household inequality in birth registration and school attendance tends to be higher in countries where total inequality is lower. Takeuchi (2015) suggests that the gaps are going to be difficult to address, because they appear inside households. Within-household disparities in stunting and birth registration between boys and girls are small. However, households have a bias for girls with respect to school attendance, whereas there is a pro-boys bias when it comes to working hours. Takeuchi (2015) recommends specific policy targeting to address the gender gaps.

With the inequality and gender gap issues raised by Takeuchi (2015) in mind, it is important to consider the study by Alderman et al. (2001) for Pakistan. Alderman et al. (2001) show that improved nutrition increases enrollment more for girls, thus closing a portion of the gender gap. The researchers, as a part of the World Bank Poverty and Human Resources Division, construct a panel data of about 800 households residing in 45 villages, with each household observed for 5 years in 15 rounds of data collection. An important aspect of their estimation is to take into account the idea that a child's health and school performance are influenced by factors within private household decision-making, regarding human capital investments. In order to overcome this limitation, the authors use price shocks when children are of preschool age. This variable provides information about the child health stock. Since the price shocks observed at this time are uncorrelated with price shocks at subsequent ages when enrollment decisions are made, the estimation is able to separate child health influencers from enrollment factors.

Correcting for these unobserved influences, the authors find that children's health and nutrition is three times more important for enrollment than is suggested by OLS estimates that assume that children's health and nutrition is predetermined, rather than determined by household choices. This is an important study because it suggests that private behaviors and public policies that affect the health and nutrition of children have a much greater effect on school enrollment and on eventual productivity.

ANALYTICAL RESULTS: APPLICATION OF REGRESSION DISCONTINUITY

As discussed in Chapter 12, Nutritional Implications of Social Protection: Application of Panel Data Method, Regression Discontinuity (RD) treatment is a quasi-experimental design, where subjects are placed in the treatment group based on some rule set up by the researcher. Usually, subjects are placed in the treatment group if they are at or above a certain minimum threshold of a particular metric. Lee and Munk (2008) provide several examples, where intervention programs use a cut-off point to administer the treatment. For instance, remedial reading and math classes may be developed for those who fall below a certain SAT score. McEwan and Shapiro (2008) also use a similar setup for studying the impact of delaying school enrollment on student outcomes in Chile.

In the context of this chapter, we can think of a program design where free meals are provided to children whose family income is below a certain threshold. Suppose the relationship between the outcome variable, say exam scores (Y), and an independent variable, say X, is given by the following simple regression:

$$Y = a + bX + \varepsilon$$

After providing free meals, some of the treated individuals may end up with higher scores, which may shift the regression line for these individuals by a factor d, so as to affect the above regression equation in the following manner:

$$Y = a + + dT + bX + \varepsilon$$

where T is a dummy variable, with a value equal to 1, if the student gets free meals. The idea behind RD is to estimate the program's effect or arrive at d. Lee and Lemieux (2010), in their excellent survey of the method, use the following illustration to show the jump at the cut-off point given by C on the x-axis. In Fig. 13.1, the treatment effect is given by τ, and observations such as $(A'' \& B')$ close to the cut-off at C, would yield the value of τ, as C is also assumed to be close to $(C'' \& C')$.

Lee and Lemieux (2010) point out many limitations of this approach. Data limitations, linear specification, smoothness, and behavior away from the cut-off are all important considerations that have to be dealt with. In recent years, however, program evaluation and intervention studies have produced many refined methods to produce many different techniques for estimating τ. The starting point is to note that, for an individual i in the sample, there are two "potential" outcomes: $Y_i(1)$ if i is exposed to the treatment, and $Y_i(0)$ if not. The basic idea is to relate this to estimating τ, by computing $Y_i(1) - Y_i(0)$.

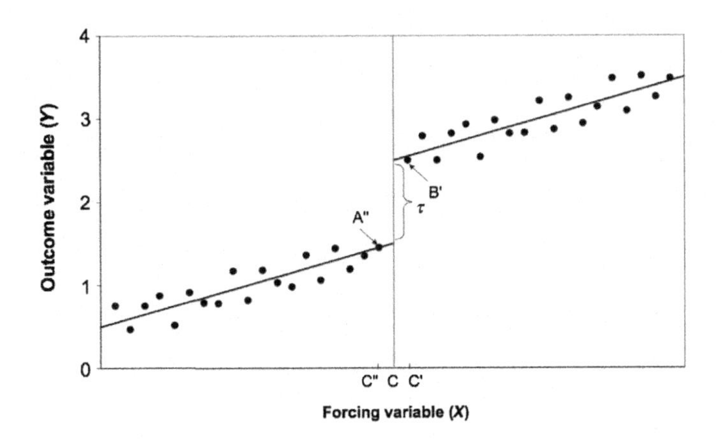

FIGURE 13.1

Treatment effect and the jump at the cut-off points.

From *Lee, D.S., Lemieux, T., 2010. Regression discontinuity designs in economics. J. Econ. Lit. 48, 281–355.*

However, it is not as straightforward as it seems, because either an individual is exposed to the treatment or not, and hence, we do not observe the pair ($Y_i(1)$, $Y_i(0)$). As Lee and Lemieux (2010) point out, only those individuals to the right of C are exposed to the treatment, and hence, we observe $E\ [Y_i\ (1)\ |X]$, on the right, and $E\ [Y_i\ (1)\ |X]$ to the left of C, which captures all untreated individuals. Hence, the RD procedure tries to generate the average effect of the treatment on the treated by examining the averages of the subgroups, by examining the limits from both sides of the following expression, where the disturbance approaches zero:

$$B - A = \lim_{\varepsilon \downarrow 0} E[Y_i | X_i = C + \varepsilon] - \lim_{\varepsilon \uparrow 0} E[Y_i | X_i = C + \varepsilon]$$

which by construction yields:

$$E[Y_i(1) - Y_i(0) | X_i = C]$$

Further, for our point of interest,

$$B - A = \lim_{\varepsilon \downarrow 0} E[Y_i | X_i = C + \varepsilon] - \lim_{\varepsilon \uparrow 0} E[Y_i | X_i = C + \varepsilon] = \tau$$

Researchers often graph the data very much similar to the figure presented above, and divide the assignment variable (or the variable that provides the cut-off value) into a number of bins. The average value of the outcome variable is computed for each bin and graphed against the mid-points of the bins (see Lee and Lemieux, 2010). In order to construct the bins, researchers also define the number of bins (usually written as K_0, K_1) to the right and left of the cut-off value, and also specify the corresponding bandwidth (usually written as h), such that:

$$b_k = C - (K_0 - k + 1).h$$

and the averages and the number of observations within each bin can be computed. The graph produced in this fashion provides a guideline for the regression functional form. The standard approach is to estimate polynomial regressions of different orders.

Lee and Lemieux (2010) point out that the behavior off the cut-off point can also be gleaned from the scatter plot generated by construction of different bins for the data. There are several alternative methods to construct the bins and the mean outcomes, via kernel estimation. Using the means within the bins is referred to as a rectangular kernel. Triangular, Epanechnikov, and other kernel methods are also adopted in practice, so as to yield comparable estimates.

There are many substantive tests that have been developed over the years to check for the size of the bandwidth, and the robustness of the specification. Recently, Calonico, Cattaneo, and Titiunik (2014), or CCT (2014), have developed robust methods to construct nonparametric estimators and confidence intervals (CIs), near the cut-off. CCT (2014) use kernel-based polynomials on both sides of the cut-off and arrive at $\hat{\tau}_p(h_n)$. They produce CIs using a common approach used in the literature, and also produce bias-corrected CIs, based on:

$$\hat{\tau}_p(h_n) \pm \textit{value of the Gaussian Distribution weighted by the estimator's variance}$$

Finally, CCT (2014) provide three data-driven RD treatment-effect point estimators, based on different constructions of bandwidth estimators:

$$\hat{\tau}_p(\hat{h}_{IK,n,p}),\ \hat{\tau}_p(\hat{h}_{CCT,n,p}) \text{ and } \hat{\tau}_p(\hat{h}_{CV,n,p})$$

The first estimator $\hat{\tau}_p\left(\hat{h}_{IK,n,p}\right)$ is from Imbens-Kalyanaraman (2012), the second is from CCT (2014), and the last is the cross-validation alternative produced by Ludwig and Miller (2007). CCT (2014) show that all three estimators are consistent and MSE-optimal.

In CCT (2016), these methods have been further refined to produce globally robust estimates with new bandwidth selection procedures and confidence bands in RD plots. We illustrate these methods below.

IMPLEMENTATION IN STATA

For the purposes of illustration, we use the following data on 66 children whose scores in the school exam (given by Score) are recorded. Some children are eligible for free meals, and this is based on an income range. If the income range is above the cut-off (given by cut-off when it is positive), then that child is not eligible for free meals. All children with positive cut-offs have access to free meals. Our goal is to see if there is a jump at the crucial point where cut-off = 0.

Obs	Score	Cut-off	Obs	Score	Cut-off	Obs	Score	Cut-off
1	48.44	−0.97	23	54.59	−0.03	45	56.62	0.67
2	29.48	−0.85	24	29.96	−0.02	46	53.14	0.68
3	33.36	−0.77	25	57.65	0.04	47	65.97	0.69
4	39.31	−0.73	26	56.85	0.09	48	52.75	0.70
5	47.24	−0.69	27	61.08	0.10	49	48.84	0.71
6	42.46	−0.59	28	62.49	0.16	50	54.72	0.72
7	49.34	−0.59	29	56.79	0.23	51	37.30	−0.37
8	51.84	−0.55	30	40.46	0.25	52	43.21	−0.19
9	41.45	−0.53	31	48.37	0.34	53	53.59	−0.12
10	43.55	−0.48	32	51.70	0.34	54	28.96	−0.08
11	41.27	−0.43	33	47.46	0.35	55	49.84	−0.07
12	44.84	−0.42	34	51.06	0.40	56	55.72	−0.06
13	47.81	−0.42	35	39.05	0.43	57	38.30	−0.05
14	29.95	−0.39	36	45.58	0.43	58	44.21	−0.04
15	37.30	−0.37	37	47.45	0.44	59	54.59	−0.03
16	43.21	−0.19	38	55.52	0.46	60	29.96	−0.02
17	53.59	−0.12	39	52.04	0.56	61	57.65	0.04
18	28.96	−0.08	40	64.87	0.56	62	56.85	0.09
19	48.81	−0.07	41	51.65	0.61	63	61.08	0.10
20	30.95	−0.06	42	47.74	0.64	64	62.49	0.16
21	38.30	−0.05	43	53.62	0.65	65	56.79	0.23
22	44.21	−0.04	44	48.55	0.66	66	40.46	0.25

We begin with the rdplot command developed by CCT (2014), which can graph the data into different groupings, based on the options. We can use this to describe our data and also identify the regression we wish to estimate.

```
. rdplot score cutoff, graph_options(title(RD Plot - Score-Meals data) ytitle(Score) xtitle(cutoff))

RD Plot with evenly spaced mimicking variance number of bins using spacings estimators.

         Cutoff c = 0 | Left of c  Right of c            Number of obs  =          66
----------------------+----------------------            Kernel         =     Uniform
        Number of obs |        34          32
   Eff. Number of obs |        34          32
  Order poly. fit (p) |         4           4
     BW poly. fit (h) |     0.970       0.720
 Number of bins scale |     1.000       1.000

Outcome: score. Running variable: cutoff.
---------------------------------------------
                        |   Left of c   Right of c
------------------------+--------------------------
 Selected number of bins |          5            8
              Bin length |      0.194        0.090
------------------------+--------------------------
       IMSE-optimal bins |          3            7
  Mimicking Variance bins |          5            8
------------------------+--------------------------
Relative to IMSE-optimal:
           Implied scale |      1.667        1.143
    WIMSE variance weight |      0.178        0.401
        WIMSE bias weight |      0.822        0.599
---------------------------------------------
```

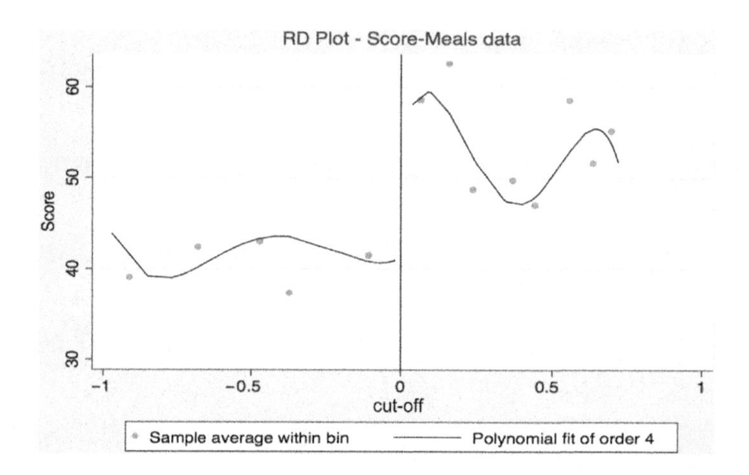

The above figure is the default, which has evenly spaced bins (−1, −0.5, 0, +0.5, +1). "Mimicking Variance bins," tells us that the number of optimal bins for control and treatment is $\hat{J}_{-,n} = 5$ and $\hat{J}_{+,n} = 8$, with bin lengths of 0.19% and 0.09%. The polynomial fit is set for the 4th degree [$p = 4$]. The additional information has to do with the algorithm's selection of bins and suitable weights.

Using the `binselect(es)` option, we can generate another evenly spaced plot that tracks the underlying regression function. The command and the output are given below:

```
. rdplot score cutoff, binselect(es) graph_options(title(RD Plot - Score-Meals data) ytitle(Score) xtitle(cutoff))

RD Plot with evenly spaced number of bins using spacings estimators.

         Cutoff c = 0 | Left of c  Right of c            Number of obs  =         66
----------------------+----------------------            Kernel         =    Uniform
        Number of obs |        34          32
   Eff. Number of obs |        34          32
  Order poly. fit (p) |         4           4
     BW poly. fit (h) |     0.970       0.720
 Number of bins scale |     1.000       1.000

Outcome: score. Running variable: cutoff.
-----------------------------------------------------
                          |   Left of c    Right of c
--------------------------+--------------------------
   Selected number of bins |           3             7
                Bin length |       0.323         0.103
--------------------------+--------------------------
         IMSE-optimal bins |           3             7
   Mimicking Variance bins |           5             8
--------------------------+--------------------------
Relative to IMSE-optimal: |
             Implied scale |       1.000         1.000
     WIMSE variance weight |       0.500         0.500
         WIMSE bias weight |       0.500         0.500
-----------------------------------------------------
```

The figure shows that the regression fits the data well, and also indicates a slight discontinuity at the cut-off.

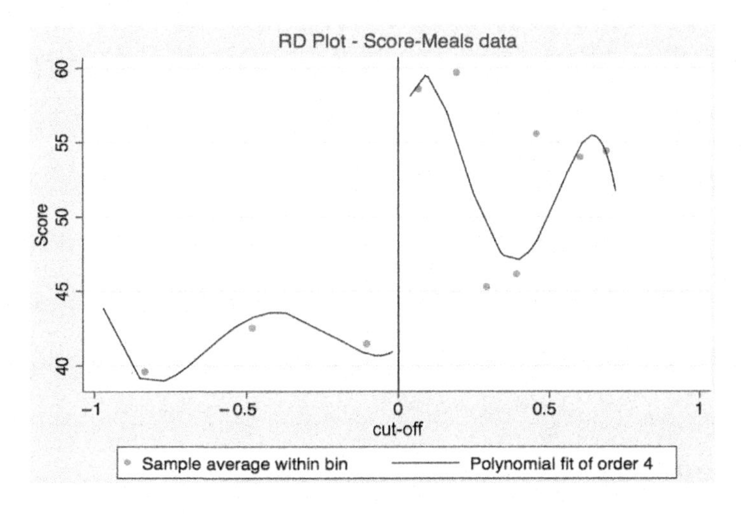

Both figures can be used to describe the data and capture the variability in the scores. In the `binselect(es)` option, the figure fits the data well, and takes into account the squared bias and variance in the data (see CCT, 2014, p. 937).

We can also use the `binselect(qsmv)` option to generate a quantile-spaced plot. Note here the optimal number of bins is $\hat{J}_{-,n} = 6$ and $\hat{J}_{+,n} = 8$.

```
. rdplot score cutoff, binselect(qsmv) graph_options(title(RD Plot - Score-Meals data) ytitle(Score) xtitle(cutoff))

RD Plot with quantile spaced mimicking variance quantile spaced using spacings estimators.

         Cutoff c = 0 | Left of c  Right of c            Number of obs  =         66
----------------------+----------------------            Kernel         =    Uniform
        Number of obs |        34          32
   Eff. Number of obs |        34          32
  Order poly. fit (p) |         4           4
     BW poly. fit (h) |     0.970       0.720
 Number of bins scale |     1.000       1.000

Outcome: score. Running variable: cutoff.
---------------------------------------------
                    |   Left of c  Right of c
--------------------+------------------------
  Selected number of bins |     6           8
             Bin length   | 0.162       0.090
--------------------+------------------------
      IMSE-optimal bins   |     4           7
  Mimicking Variance bins |     6           8
--------------------+------------------------
Relative to IMSE-optimal:
          Implied scale   | 1.500       1.143
  WIMSE variance weight   | 0.229       0.401
      WIMSE bias weight   | 0.771       0.599
---------------------------------------------
```

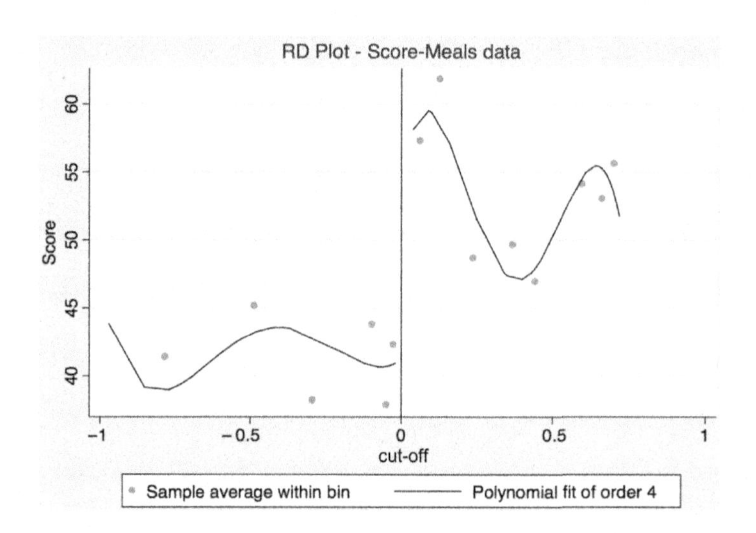

We can also use the following command from CCT (2016):

```
rdplot score cutoff, binselect(es) ci(95) graph_options(title("RD Plot: Scores
Meals") ytitle(scores) xtitle(cutoff) graphregion(color(white)))
```

which produces the following output for the same data:

```
RD Plot with evenly spaced number of bins using spacings estimators.

          Cutoff c = 0 | Left of c  Right of c          Number of obs  =        66
-----------------------+----------------------          Kernel         =   Uniform
         Number of obs |        34          32
    Eff. Number of obs |        34          32
   Order poly. fit (p) |         4           4
      BW poly. fit (h) |     0.970       0.720
 Number of bins scale  |     1.000       1.000

Outcome: score. Running variable: cutoff.
---------------------------------------------------------
                             |  Left of c    Right of c
-----------------------------+---------------------------
    Selected number of bins  |          3             7
                 Bin length  |      0.323         0.103
-----------------------------+---------------------------
          IMSE-optimal bins  |          3             7
    Mimicking Variance bins  |          5             8
-----------------------------+---------------------------
Relative to IMSE-optimal:    |
              Implied scale  |      1.000         1.000
     WIMSE variance weight   |      0.500         0.500
         WIMSE bias weight   |      0.500         0.500
```

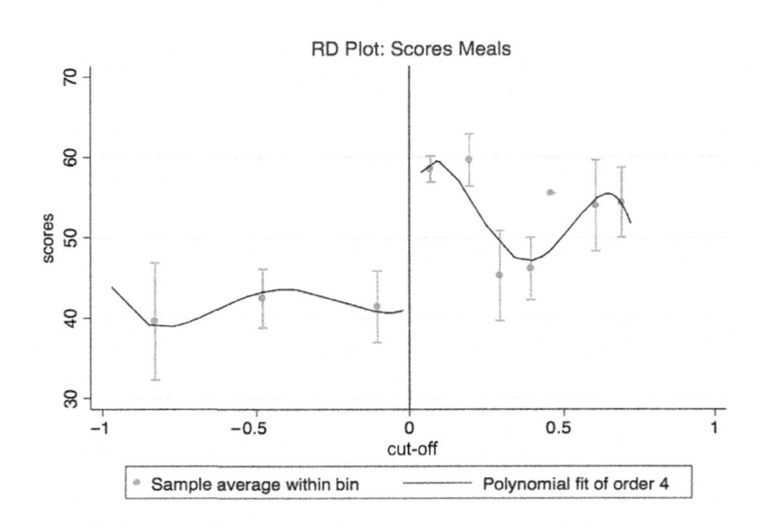

The command `rdrobust` implements the RD treatment effects estimation and inference procedure. The default command and output from STATA are given below:

```
. rdrobust score cutoff

Sharp RD estimates using local polynomial regression.

      Cutoff c = 0 | Left of c  Right of c            Number of obs =         66
-------------------+----------------------            BW type       =      mserd
     Number of obs |        34          32            Kernel        = Triangular
 Eff. Number of obs|        16           6            VCE method    =         NN
 Order loc. poly. (p)|       1           1
    Order bias (q) |         2           2
  BW loc. poly. (h)|     0.130       0.130
       BW bias (b) |     0.287       0.287
         rho (h/b) |     0.452       0.452

Outcome: score. Running variable: cutoff.
--------------------------------------------------------------------------------
            Method |   Coef.    Std. Err.    z     P>|z|    [95% Conf. Interval]
-------------------+------------------------------------------------------------
      Conventional |  15.947    8.8135    1.8094   0.070    -1.32693    33.2213
            Robust |     -         -      1.1506   0.250    -9.21835    35.4265
--------------------------------------------------------------------------------
```

The results indicate that there are 34 observations in the control group, and 32 in the treated group. The estimator is a local linear one with $p=1$ with a local quadratic bias-correction estimate, with a triangular kernel.

The bandwidth selection procedure as developed by CCT (2014) shows that $[\hat{h}_{CCT,n,p} = 0.13$ and $\hat{b}_{CCT,n,p+1,q} = 0.287]$, where $p=1$, and $q=2$. Importantly, the point estimator and CI:

$$\hat{\tau}_p\left(\hat{h}_{CCT,n,p}\right) = 15.947, \quad \hat{CI}^{rbc}_{1-\alpha}\left(\hat{h}_{CCT,n,1}, \hat{b}_{CCT,n,2,2}\right) = [-9.21, 35.42]$$

The command `rdrobust score cutoff, all` generates more detailed information regarding the variance and CIs. The command line and the output for that portion from STATA are reproduced below. The Conventional Method produces the coefficient $\hat{\tau}_p(h_n)$, while the Bias-corrected Method generates $\hat{\tau}^{bc}_{p,q}(h_n, b_n)$, and finally, the Robust Method produces $\hat{\tau}^{bc}_{p,q}(h_n, b_n)$.

```
. rdrobust score cutoff, all

Sharp RD estimates using local polynomial regression.

      Cutoff c = 0 | Left of c  Right of c            Number of obs =         66
-------------------+----------------------            BW type       =      mserd
     Number of obs |        34          32            Kernel        = Triangular
 Eff. Number of obs|        16           6            VCE method    =         NN
 Order loc. poly. (p)|       1           1
    Order bias (q) |         2           2
  BW loc. poly. (h)|     0.130       0.130
       BW bias (b) |     0.287       0.287
         rho (h/b) |     0.452       0.452

Outcome: score. Running variable: cutoff.
--------------------------------------------------------------------------------
            Method |   Coef.    Std. Err.    z     P>|z|    [95% Conf. Interval]
-------------------+------------------------------------------------------------
      Conventional |  15.947    8.8135    1.8094   0.070    -1.32693    33.2213
    Bias-corrected |  13.104    8.8135    1.4868   0.137    -4.17001    30.3782
            Robust |  13.104    11.389    1.1506   0.250    -9.21835    35.4265
--------------------------------------------------------------------------------
```

The bandwidth selection procedures can be generated using the command `rdbwselect score cutoff, all`. The command and the output from this portion are reproduced below. The procedure generates five MSE- and CER-optimal bandwidth selectors for the RD treatment estimator. These are based on whether one common MSE- or CER-optimal bandwidth or two different MSE-optimal bandwidth selectors are used.[4] MSE refers to mean squared error and CER refers to coverage error rate. For details see http://www-personal.umich.edu/~cattaneo/software/rdrobust/stata/rdbwselect.pdf (Calonico et al., 2016).

```
. rdbwselect score cutoff, all

Bandwidth estimators for sharp RD local polynomial regression.

         Cutoff c = 0 | Left of c  Right of c            Number of obs =          66
----------------------+----------------------            Kernel        = Triangular
        Number of obs |        34          32            VCE method    =          NN
        Min of cutoff |    -0.970       0.040
        Max of cutoff |    -0.020       0.720
  Order loc. poly. (p)|         1           1
       Order bias (q) |         2           2

Outcome: score. Running variable: cutoff.
--------------------------------------------------------------------------------------
                      |     BW loc. poly. (h)      |          BW bias (b)
               Method |  Left of c    Right of c   |   Left of c    Right of c
----------------------+----------------------------+-----------------------------------
                mserd |     0.130          0.130   |       0.287         0.287
               msetwo |     0.231          0.104   |       0.399         0.203
               msesum |     0.129          0.129   |       0.315         0.315
            msecomb1  |     0.129          0.129   |       0.287         0.287
            msecomb2  |     0.130          0.129   |       0.315         0.287
----------------------+----------------------------+-----------------------------------
                cerrd |     0.105          0.105   |       0.226         0.226
               certwo |     0.187          0.084   |       0.314         0.160
               cersum |     0.104          0.104   |       0.248         0.248
            cercomb1  |     0.104          0.104   |       0.226         0.226
            cercomb2  |     0.105          0.104   |       0.248         0.226
--------------------------------------------------------------------------------------
```

As we see, the bandwidth selectors produce estimates of h_n and b_n depending upon the selections and bandwidth options. As shown above, the RD method and the estimation procedures have been refined over the years. The application of RD goes beyond the school nutrition example given here; we encourage readers to explore this further in other program evaluation settings.

The major theme in this chapter has been set up to understand whether school feeding programs improve nutrition and related outcomes. Policy makers would be confident that the SFPs are good safety net investments, only if there is adequate evidence that the said intervention works well in

[4]MSE refers to mean squared error and CER refers to coverage error rate. For details see http://www-personal.umich.edu/~cattaneo/software/rdrobust/stata/rdbwselect.pdf (Calonico, Cattaneo, Farrell, and Titiunik, 2016).

terms of the intended objectives. The Regression Discontinuity technique is another econometric method that allows us to evaluate the success of a policy intervention, such as free school lunches.

The examples from STATA help to narrow the focus of this question by providing information about the relevant treatment effects. The example in STATA and the estimates suggest that free school meals have a positive treatment effect for the treated groups. Such information is useful to all the stakeholders in the system. The Regression Discontinuity methods have become highly refined in recent years and we encourage students to explore the implications of these refinements on policy in the exercises at the end of the chapter. This is a rich area for further research, as the school feeding programs in various forms are expanding as a key human development intervention.

In this context, the WFP (2009) report lists the main areas in which further research is needed:

- A database on school feeding programs in high-income countries is crucial. This database must be constructed in such a way that it replicates the information that is available for middle- and low-income countries. Data must be collected for all the programs, including technical details regarding size and coverage.
- Currently, a good analysis that evaluates the efficiency of SFP in low-income countries is missing. Are the programs reaching the poor? It is important to replicate the SFP evaluation in several contexts and countries.
- The cost structure of the SFPs is not available. An analysis of the cost drivers of SFPS will be a good way to ascertain the efficiency of the programs, and make cross-country comparisons easier. For example, in Zambia the cost of school feeding is about 50 % of annual per capita costs for primary education, while in Ireland it is only 10% (see Bundy et al., 2009).
- Countries must conduct impact evaluations for different programs, and develop robust monitoring and evaluation systems.
- We also need studies that assess the impact of purchasing food from smallholder farmers for school feeding operations. This will provide information on supply chain management and help in improving food system efficiency.
- Children from high-income countries pay for their meals and cover a portion of the costs, and subsidize those children from poorer families. Middle and low-income countries must also develop appropriate targeting to cover some of the costs, and there is a need to identify some of these cost-recovery mechanisms.
- There is a need to understand how countries tackle the issue of food quality standards and nutritional guidance for school feeding.

CONCLUSIONS

The chapter starts with the role of school feeding programs in increasing food accessibility at the household level, a major pillar for achieving food security for all. In addition to providing a potential source for nutritional security for school-going children, school feeding programs have the potential to increase the human capital benefits for the participating household. National and local governments are increasingly investing in school nutrition programs. Economists and nutrition specialists are interested in generating evidence on whether public funds expended toward school meals generate the necessary benefits for the society. The issues of feeding millions of school children do not follow a simple design or focus on specific objectives. The objectives and design vary depending on the local context and resources availability. Based on the experiences from several countries including Uganda, Laos, and Burkina Faso, the WFP (2009) notes that substantial work is still needed to fully understand the costs and benefits of SFPs.

This chapter also presents several policy proposals that are currently undertaken by the global community to address child malnutrition through effective school feeding programs. We have also demonstrated the usefulness of the analytical results involving the Regression Discontinuity procedure and its corresponding STATA implementation.

As Bundy et al. (2009) note, if we have to take full advantage of SFPs as nutrition improving interventions, then a more systematic and policy-driven approach by both governments and development partners is required.

EXERCISES

1. Prepare a review of the studies focusing on the school feeding programs in the country of your interest. What has been the history and how has the program design changed over the years, and why? What are the budgetary implications for this program, and how does this crowd out other social welfare programs?
2. Use the above data and attempt the following commands in STATA:
   1. `rdrobust score cutoff, kernel (uniform)`
   2. `rdrobust score cutoff, p(2) q(4)`
   3. `rdrobust score cutoff, vce(resid)`

   Refer to CCT (2014) and CCT (2016) to see if there are differences within the estimated CIs and note their differences.

PART

ECONOMICS OF TRIPLE BURDEN: UNDER-NUTRITION, OVER-NUTRITION, MICRONUTRIENT DEFICIENCIES

CHAPTER

ECONOMIC ANALYSIS OF OBESITY AND IMPACT ON QUALITY OF LIFE: APPLICATION OF NONPARAMETRIC METHODS

14

In the end, as First Lady, this isn't just a policy issue for me. This is a passion. This is my mission. I am determined to work with folks across this country to change the way a generation of kids thinks about food and nutrition.

—First Lady Michelle Obama in www.letsmove.gov/about

INTRODUCTION

Most of the chapters in this volume deal with malnutrition in the context of under-nutrition. In this chapter we deal with over-nutrition problems. In the context of developing countries economic growth, income increases, and urbanization jointly drive the process of transition into increases in overweight and obesity. The focus on food security through cereal crops and self-sufficiency in cereals limits crop diversity, and hence dietary diversity. Vegetable oil became an inexpensive source of fat. An increase in household income increases their consumption of sugar and eating away from home, and particularly increases consumption of foods that are rich in oil and sugar. Increased urbanization is also accompanied by jobs that are not physically demanding, motorized transportation, and less opportunity for physical activity. These factors jointly contribute to the dietary transition in developing countries seen in the last 20 years (Hawkes et al., 2007; Alston et al., 2010).

In the context of developed countries, in the United States, e.g., the National Center for Health Statistics reports that presently two-thirds of adult Americans are either overweight or obese. Obesity rates have grown by 35% since the 1960s in the United States. The statistical trends noted by the US Department of Health are truly a cause for concern. The department also notes that about 67% of whites are overweight or obese, and about 34.3% of them are obese. Among the black population, 77% are overweight or obese, about 50% are obese; among Hispanics, 79% are overweight or obese, and 40% are obese; among adults in the United States in all racial categories, 68% are overweight or obese, and about 36% are considered obese (USDH, 2016). The National Institute of Diabetes and Digestive and Kidney Diseases (NIDDK), which is a part of the US Department of Health, notes from the National Health and National Examination Survey that about one-third of children and adolescents aged 6 to 19 are considered to be overweight or obese, and more than 1 in 6 children and adolescents aged 6 to 19 are considered to be obese.[1]

[1]See niddk.nih.gov for recent updates on the status of the problem in the United States.

Nutrition Economics. DOI: http://dx.doi.org/10.1016/B978-0-12-800878-2.00014-1

The following two graphs indicate the nature of the obesity epidemic in the United States, by capturing the trends over the last 50 years (Duffy et al., 2012).

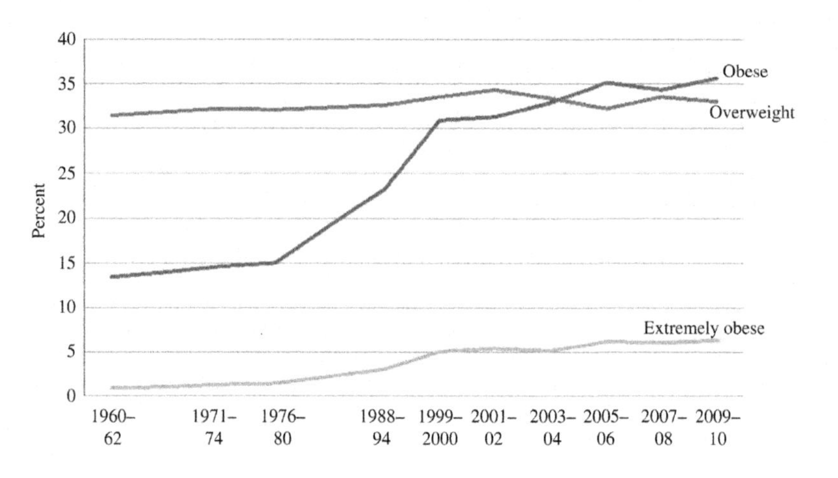

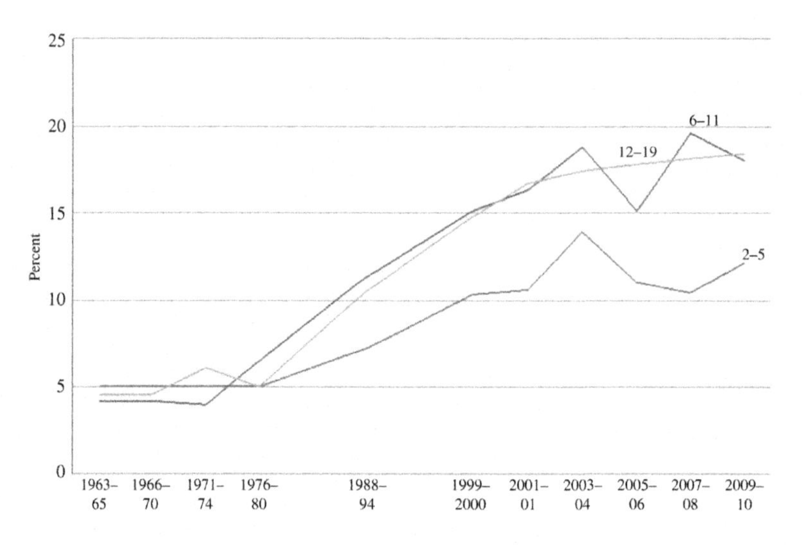

The first graph shows US adult overweight and obesity rates since the 1960s. The second graph shows the obesity rate for children and adolescents. Both graphs depict upward sloping lines, particularly after the 1970s. The above facts have caused much concern among public health officials, medical practitioners, health economists, nutrition experts, and many other concerned groups. Recent research has addressed several of these issues and, overall, for the United States, there is a general agreement that obesity in childhood is associated with obesity in adulthood. Himes (2011a,b) summarizes several of these findings. In the United States, women are more likely to be obese than men. Further, obesity is more common in the central and southern parts of the United States. Finally, obesity rates are determined via a complex interaction between economic, social, and cultural factors. In this chapter, we examine relationships between obesity and food prices, the influence of social networks and peer-effects, the availability of healthy, junk, and processed foods, and physical activity, and try to understand the causes and economic consequences of obesity.

THE ECONOMIC EFFECTS OF OBESITY ON HUMAN CAPITAL

The link between obesity and wages, and on long-term human capital formation is an important economic factor. Averett and Stifel (2013) point out that the long-lasting effect of obesity on human capital may be due to the earlier effects of childhood obesity on cognitive development. In order to test this hypothesis, Averett and Stifel (2013) examine the children of the National Longitudinal Survey of Youth 1979 cohort, in particular, with a given child's weight and cognitive ability.

The fixed-effects model shows that there are race and gender differences that are coupled with weight, that influence cognitive abilities, test scores, and academic performance, and hence, potential human capital. For instance, the researchers show that overweight white boys have math and reading scores that are one standard-deviation lower than the population mean.

Overweight white girls also have lower math scores. Finally, overweight black boys and girls also have lower reading scores. These results have implications for health – nutrition – education policies tied to potential life-time earnings.[2]

Although researchers have also started to examine whether obesity has an impact on labor market outcomes, through lower cognition or related socioeconomic factors, the evidence up to this point is mixed. For instance, Amis et al. (2014) find that obesity does not affect high school graduation. On the other hand, negative effects of obesity are found on college graduates and their future earnings potential.

Using data from the National Health and Nutrition Examination Survey III, Wada and Tekin (2010) estimate wage models for respondents in the National Longitudinal Survey of Youth 1979 (NLSY79). They find that body fat is associated with decreased wages for both males and females, among whites and blacks. The researchers also show that fat-free mass is associated with increased wages.[3]

Interestingly, Fox and Hutto (2013) also use NLSY79 data and show that obesity dampens upward mobility, and increases the likelihood of downward mobility for women. In contrast, obese men have a greater likelihood of upward mobility. The implications of this study are that obesity among women in their early adulthood places them on a poor life-time income profile. This is one of the few studies that has tracked obesity, earnings, and gender gap topics within nutrition economics.

In contrast to these studies, Majumder (2013) reverses the standard claim that wages and obesity are negatively related. Using NLSY data from 1997, Majumder (2013) shows the opposite to be

[2]Zavodny (2013) shows that overweight children are more likely to get lower teacher assessments than actual test scores.
[3]Relatedly, Kosteas (2012) using NLSY data shows that regular exercise yields a 6–10% wage increase in the United States.

true, namely, that white males receive a wage premium for higher BMI. Further, this empirical examination also shows the wages of all other ethno-gender groups to be unaffected by obesity.

Lempert (2014) also makes interesting observations dealing with endogeneity that arise in this context: overweight and obesity lead to lower wages, and also low family income and low wages contribute to overweight and obesity. In fact, Lempert (2014) finds that body composition negatively affects wages at higher wage levels, and particularly more so for women. That is, at a higher wage level, the penalty of being overweight is higher for women.

The relation between body composition and wages has also been investigated for other countries. For instance, Lundborg et al. (2014) examine data for 150,000 male siblings from the Swedish military enlistment. The Swedish data also shows evidence of within-family effects of body size and cognitive skills. As a consequence, overweight and obese Swedish teenagers are at a significant disadvantage in the labor market due to poor skill acquisition (Box 14.1).

BOX 14.1 OBESE BUT UNEQUAL

Carson (2015) observes that historically, African American and white body weights decreased throughout the 19th and early-20th centuries. Southerners had taller body statures, higher BMI, and heavier body weights than workers in other parts of the country. The conclusion was that net nutrition must have been better in the South, in spite of the area being very vulnerable to disease. Carson's conclusions have serious implications for the modern obesity epidemic in the United States.

Broady and Meeks (2015) note that obesity rates have historically been higher in Southern states, due to the culture behind food preparation, consumption, and practices. The researchers also find that SNAP assistance, and lack of physical activity, are highly correlated to the obesity rate. But most importantly, the percentage of African American residents is also highly correlated to the overall level of obesity.[a]

Since World War II, the height gap between black and white women has increased by 1.95 cm, and additionally, Komlos (2010) points out that black women in the 20–39 year age range weigh 21 lbs more than their white counterparts. Komlos (2010) concludes that the obesity epidemic is partially responsible for the decline in height among this group, which leads to a double-jeopardy, since declining physical stature is reflective of systemic health deficiencies. The above finding is consistent with those under "peer effects."[b] Further, these observations are also consistent with the findings of Walker and Kawachi (2011), who note the reasons for this epidemic: overeating as a maladaptive coping strategy, occupational segregation by race, cultural practices, built-environment, and susceptibility to television-based marketing of foods.

Johnston and Lee (2011), among many other researchers, have found large gaps in the body weights of comparable black and white women. The researchers attribute this weight gap to black – white differences in energy intakes and energy expenditures. This differential is important to understand, because the overweight problem leads to potential health costs, which will fall more heavily on one sector of the population. Condliffe and Link (2014) calculate these differences in health costs: the combination of obesity and hypertension increases health care costs for whites (by 25%), Hispanics (by 48%), and substantially for African Americans (by 70%).

Further, Johnston and Lee (2011) show that the differential weight gap arises due to differences in diet, and not due to a lack of physical activity. Appropriate policies that include diet intervention and education are important to reduce future health costs facing black women. Indeed, increasing the number of healthy food outlets within a half-mile of the residence appears to be associated with lower levels of obesity, as pointed out by Broady and Meeks (2015).

[a]Relatedly, Lee and Wildeman (2013) discuss how mass imprisonment of African American women creates an environment which produces deteriorating health outcomes, including obesity. The authors also document disparities between white and African American women in these risky environments.

[b]See Fletcher (2014), Ali et al. (2014), and Forste and Moore (2012).

HEALTH COSTS OF OBESITY

The medical costs of obesity have also attracted a lot of attention among policy makers and health economists. As a motivating example, consider the studies by Hoque et al. (2010, 2013), which examine the situation of overweight adults in Texas, where this subgroup is projected to increase to 16.0 million in 2040, with the number of obese adults increasing to 14.6 million in 2040. The annual cost associated with this projected figure is equal to $40.3 billion in 2040.

Research and studies from this line of research suggest policies to confront the increasing cost by directing efforts toward Hispanics and other minorities. Nayga (2013) finds that nonwhites, lower educated individuals, and those with lower income are less likely to be aware of the link between being overweight and heart disease. From a policy point of view, this line of research implies that the design of food policy and health education campaigns about obesity and heart disease are very critical.

Several estimates and figures of health costs have been produced from this field of enquiry. Cawley and Meyerhoefer (2012), e.g., note that obesity is associated with $656 per capita higher annual medical care costs, while an Instrumental Variables result puts this figure at $2741 per capita. Thorpe et al. (2015) estimate that rising obesity levels are responsible for 11–23% of the increase in health care expenditure for several specific chronic conditions.

The combination of obesity along with other health issues has a significant effect on health expenditure. Besides the differential race effects on health costs noted by Condliffe and Link (2014), medical research now indicates that a person with diabetes and obesity has health care expenditures 14% greater than a diabetes patient without obesity (see Condliffe et al., 2013). Importantly, diabetes patients with both obesity and hypertension—the fastest growing group of diabetes patients—have health care expenditures 40% higher than others who are not obese.

MacEwan et al. (2014) estimate that a 1-unit increase in BMI for every adult in the United States is likely to increase medical expenditures by $6 billion annually. If every obese adult in the United States had a BMI of 25, then expenditures would decline by $166 billion and, in the absence of this reduction, result in a $148 billion in deadweight loss. This is also the same amount that is reported by Finkelstein et al. (2010) as direct costs of obesity, in addition to $73 billion annual indirect costs.

Finkelstein et al. (2010) note that the mandate in the 2010 Affordable Care Act (namely, that large firms offer health insurance to their employees or face a significant financial penalty) may produce incentives for businesses to invest in prevention, firms could benefit from a slimmer, healthier, and perhaps less costly, more productive work force.

Mehta and Chang (2011) note that in the United States, the mortality penalty linked with obesity has been declining, particularly with respect to Class I obesity, although Class II and III may be associated with higher mortality risks. There is a still a possibility that obese youths underuse the services available, thereby, the true costs of youth obesity are underestimated (see Wright and Prosser, 2014).

Psychological costs from stress and depression can also lead to obesity and have significant health costs. For example, Dave et al. (2011) find that among females major depression raises the probability of being overweight by 7% points. Further, this adds roughly $9 billion to the economic costs of depression.

The level of depression alongside consumption of antidepressants can also increase BMI. In order to separate the effects, Wehby and Yang (2012) use a first-difference model, and show that depression with antidepressant use could increase the BMI by about 1 point, and also increase the chances of becoming overweight or obese by about 9.2% points. Further, the study notes that this effect is larger for the unmarried and individuals with relatively low socioeconomic status. Finally, the study highlights the fact that the increase in BMI is driven mainly by the use of antidepressants, and not depression. The increasing use of antidepressants in the United States could partially explain the rise in obesity rates.

CAUSES OF OBESITY IN AMERICA

Several reasons are cited for the growing trend seen in the United States. We begin with a controversial issue: are the safety nets in the United States the reason for people gaining weight?

HOW SAFE ARE THE SAFETY NETS IN THE UNITED STATES?

Huang (2012) also notes the increase in obesity among adults in the United States, from 15% to 35% since the 1980s. Huang (2012) relates participation in SNAP to adults' BMI. Huang (2012) constructs a panel data relating participation in SNAP, and prices to individual level data spanning 1986–2006. While the least squares instrumental variable estimates indicate that SNAP increases a woman's BMI by about 1.1%, and also her probability of being obese by about 2.6% points, sophisticated models that control for participation show that for female participants, BMI reduces by 1.12%, and the likelihood of being obese reduces by 3.76%. Hence, Huang's (2012) study indicates that SNAP helps reduce obesity. Similar results are also presented in Huang et al. (2011, 2012).

As mentioned in the previous sections of this book, the National School Lunch Program provides free and reduced-cost lunches for income-eligible students, and minimally subsidizes lunches for income-ineligible students in the United States. Many researchers are interested in finding out the effect of the NSLP program, in terms of student health outcomes. Peckham (2013) compares menu offerings across school districts and relates the variation in menu choices to socioeconomic variables. Interestingly, Peckham (2013) finds that students in wealthier school districts are offered more entrees, fruit, and vegetable choices per week, possibly resulting in nutritionally superior meals, while at the same time, students receiving free lunches are more likely than students purchasing paid-price lunches to choose entrees with more fat and carbohydrates, and less protein.

Lakdawalla and Philipson (2009), and Grecu and Rotthoff (2015) hypothesize that the relationship between unearned income and weight exhibits an inverted U-shape. That is, as income increases, we see an increase in weight because of access to food. However, beyond a certain income level, households consume better quality food and, hence, weight starts to decline. Akee et al. (2013) test this hypothesis using a very interesting dataset from American Indian households in North Carolina. This study uses exogenous cash transfers to identify the effects of positive household income shocks on adolescent BMI. This study is interesting because the data also covers nonparticipants who reside in the same counties. Akee et al. (2013) show that the extra income transfers increase BMI among young children in poorer households, when compared to their counterparts in richer households.

Fan (2010) and Baum (2011), in independent enquiries, use the 1979 National Longitudinal Survey of Youth data and demonstrate that food stamps do not have any major impact on obesity, particularly for women.[4] Likewise, Burgstahler et al. (2012) show that SNAP participation and obesity among children are negatively related. This study is also very interesting because it connects the financial stress of households as a key factor in the determination of the BMI status. Similarly, Schmeiser (2012) shows that SNAP participation actually reduces the BMI percentile and the probability of being obese for boys and girls aged 5 – 11, and also for boys aged 12–18. For girls aged 12–18, SNAP participation appears to have no significant effect on these outcomes. In contrast to these findings, Robinson and Zheng (2011) show that participation in the Food Stamp Program (FSP) may contribute to obesity for older females. In a more recent and exhaustive modeling approach, Kreider et al. (2012) show that SNAP has favorable effects on child health. This study is important because it controls for endogenous participation, and also for underreporting of participation status. Both of these econometric issues create identification problems, which the authors address using latent-selection equations and auxiliary administrative data.

Jensen and Wilde (2010) provide a good synopsis of the relationship between SNAP participation, food insecurity, and obesity. First, they note that the relationship between SNAP and obesity is complex. Primarily, we observe that women in moderately food insecure households tend to have higher weights on average than women in food secure households. Jensen and Wilde (2012) note that the reason for this could be the "boom and bust" cycles in food intake. That is, food insecure households will sacrifice healthy food for cheap and unhealthy choices. Additional factors such as financial stress and lack of physical exercise can also lead to weight gain. Second, Jensen and Wilde (2012) note that SNAP benefits received only once per month are typically exhausted long before the end of the month, creating the boom and bust cycle. The study indicates that policies toward food insecurity must be more holistic and take several factors into account.

Jensen and Wilde's (2012) study is important because it indicates that the effects of safety net programs and weight are driven more by individual specific fixed-effects. Morrow (2013) also presents an exhaustive study relating safety net participation to obesity, which indirectly supports this view. Morrow's (2013) study covers several vulnerable groups, and shows that for the general population, the safety net programs help health outcomes for white and Hispanic children. However, black Head Start children are more likely to be overweight and obese at ages 5 – 6 than their non-Head Start peers. Morrow (2013) also examines the relationships between program participation, food choices, and obesity among low-income Mexican-origin women. Overall, Morrow (2013), like some of the previous researchers, finds that safety net programs are not significant drivers of BMI. This finding is in contrast to Carneiro and Ginja (2014), who find that participation in Head Start reduces the incidence of behavioral problems, health problems, and obesity in male children. Carneiro and Ginja (2014) account for program eligibility rules in their study, and allow for participation discontinuities inherent in the data.

Belfield and Kelly (2013) also examine the impact of Early Childhood Care and Education and conclude that child obesity is reduced by participation in Head Start and center-based preschool

[4]Also see Parks (2011) and Ver Ploeg (2011).

programs. Belfield and Kelly's (2013) study is interesting because it tests for self-selection on the part of healthier children into early childhood education.

Roy et al. (2012) present an excellent summary of the three most prevalent federal assistance programs in the United States that deal with nutrition issues: the Supplemental Nutrition Assistance Program (SNAP, or the Food Stamp Program), the School Breakfast Program (SBP), and the National School Lunch Program (NSLP). We have also discussed the various aspects of these programs in previous chapters.[5]

Roy et al. (2012) examine the association between program participation and the health of children, and most importantly the effect of "spillover" between programs. For example, Roy et al. (2012) point out that participation in SNAP may alter the types of food consumed at home and, consequently, may induce similar changes in food choices made at school. It is also possible that SNAP recipients eat healthier foods at home, and decide to substitute less-healthy food at the cafeteria.

Consequently, policy reforms must focus on how households view the programs and do not make unhealthy substitutions across programs. In particular, the researchers find that poorer households are in relatively poor health, and allocate more total time to child care. The time allocated to child care forces participants to join all three programs, which in turn produces beneficial effects on adolescent BMI. The researchers show that joint participation in all three programs has a beneficial effect on BMI, due to the time spent by parents on child care, which includes monitoring children's TV and movie time.

Most importantly, poorer households participate in these programs due to the need-based subsidy component of the programs, while richer households participate due to convenience. This feature has significant implications for policy. For example, the authors show that SNAP participants with an income of less than $75,000 spend less time in primary eating and drinking, more time watching television and movies, less time caring for children, and less time grocery shopping. They also have lower labor force participation, less income, and worse overall health. These households are usually nonwhite, single-parent households, which allocate less time to child health and grocery shopping. Thus, policy makers must consider family effects, and outreach efforts like SNAP-Ed must consider issues beyond "dietary guidelines."

Researchers are gradually making inroads into fully understanding the "time cost" factor involved in the individual's decision-making, particularly those who receive assistance. A good example of this subject is in Davis and You (2010), who calculate the time cost of food preparation using two methods: the market substitute approach, and the opportunity cost approach.

The market substitute approach computes the dollar value of the hours spent in food preparation at home, using the market wage in the food sector as a proxy. The opportunity cost approach estimates the benefits foregone associated with the nonmarket activity. This computation depends upon the individual's time allocation and the shadow wage.

The authors compute both costs, and show that for the general population the opportunity or time cost is about 35% of total food cost, and 21% of total cost using the substitute approach. However, the study shows that these costs are 48% and 35% for Food Stamp participants, and for those who follow the USDA Thrifty Food plan, the costs are 63% and 53%, respectively. From

[5]See Millimet et al. (2010) for further elaboration on the effects of SBP and NSLP. Also see Yin et al. (2011).

Davis and You (2010) we can conclude that the time costs are substantial for FS participants, and hence, the system creates strong incentives to eat out, thereby promoting unhealthy eating.

The "time cost allocation" problem is also present in decisions dealing with child care. Herbst and Tekin (2011) tackle the issue of childhood obesity, particularly of preschool children who are exposed to nonparental arrangements. Obviously, policy makers are concerned about the types of center care, and how the arrangements affect child health outcomes. Herbst and Tekin (2011) also note that there are many federal and state provided child care policies, such as the Child Care and Development Fund (CCDF), which provide employment-based subsidies. These subsidies allow parents to purchase child care services, thereby allowing them to enter the work force. Obviously, therefore, children's weight outcomes depend on the quality of the service, and more importantly, the subsidies increase disposable income for the parents, who might substitute healthy food for unhealthy choices.

To gain an insight into this problem, Herbst and Tekin (2011) collect data on participants who receive a subsidy the year before kindergarten on several measures of children's weight during the fall and spring of kindergarten.

The authors estimate OLS, Fixed Effects, and quantile regressions, and show that the effect of the subsidy depends upon the given distribution of BMI. Indeed, the child care subsidy is related to increases in BMI and a greater likelihood of being overweight and obese, at the top of the BMI distribution. However, subsidies have no effect on BMI at the lower end of the distribution, and show inconsistent effects in the middle of the distribution.

Further, the adoption of nonparental child care, rather than maternal employment, is the key mechanism through which the subsidy effects operate. Thus, there may be unintended consequences associated with child care subsidies, and the authors question the design features associated with states' CCDF plans. These plans may create disincentives for parents to choose high-quality care, and for providers to make costly quality-enhancing improvements. Many child care centers in the United States fail to provide children with healthy foods and sufficient opportunities for physical activity, and subsidized children perform worse on tests (Cesur et al., 2010).

Similar findings are also presented by Mandal and Powell (2014) who use the Early Childhood Longitudinal Study surveys for the years 2001–08, covering over 10,000 reports. Mandal and Powell (2014) demonstrate that child care settings affect the likelihood of childhood obesity. Indeed, if the quality of the setting is high, or if it is a paid and regulated care setting, then there is a higher consumption of fruit and vegetables.

Most importantly, among children from single-mother households, the probability of obesity increases by 15% points, due to an increase in intake of soft drinks from four to six times a week, and by 25% points, due to an increase in intake of fast-food from one to three times a week, to four to six times a week. However, among children from two-parent households, the likelihood of obesity decreases by 10% points due to eating vegetables one additional time a day.

The accessibility of services and subsidies are also important issues. Herbst and Tekin (2012) examine whether the distance to a public service agency has an effect on getting the subsidy, and ultimately an affect on obesity in low-income children. Indeed, Herbst and Tekin (2012) show that an increase in the distance to a center has a negative effect on receiving the subsidy, and that subsidized child care services lead to increases in overweight and obesity among low-income children.

In this context, the results from Denmark are rather remarkable. Greve (2011) shows that in Denmark, an increase in maternal working hours does not necessarily mean an increase in children's weight. This is because of the quality of child care in Denmark, which is superior to that in the United States. Finally, Danish fathers also contribute to children's care and health.

INCOME EFFECTS AND CHILDHOOD OBESITY

Millimet and Tchernis (2015) uncover two interesting features in the BMI of United States children. Both features have to do with a specific type of difference. The first feature arises when the authors compare males versus females, who are both in the bottom quantile of BMI distribution, at kindergarten. Given this default position, the authors find that males are more likely to gain at least 10 percentile points in BMI distribution when they reach the eighth grade.

The second feature has to do with comparisons between whites, blacks, and Hispanics. Although the whites start out from the top quantile of the BMI distribution, they are more likely to move down at least 10% points than either blacks or Hispanics.

Does family income have anything to do with childhood obesity? There is partial evidence from current research that income does have an influence on the BMI of children. For example, Chia (2013), using NLSY data, shows that there is a prevalence of childhood obesity in low-income families. As Chia (2013) correctly points out, income itself may not be the sole determinant of childhood obesity, because family income tracks unobserved household characteristics.

Hussain (2012) notes that individuals from households with financial stress and variations in net income exhibit higher BMI. This study also tracks the direction of the observed higher BMIs, from financially stressed mothers to their children.[6] In a sociological study, Wisman and Capehart (2010) note that the obesity epidemic in the United States is due to growing economic insecurity, stress, and a sense of powerlessness in modern society, where high-sugar and high-fat foods have become easily available.

Jo (2014) notes that for children in low-income families, there is a strong positive relation between income and BMI. Among high-BMI children, there is a significant positive relation between family income and BMI. Finally, Jo (2014) and Gius (2011) find that the difference in obesity rates between children from low-income and high-income families increases as children age. Indeed, as children age, they gain weight more rapidly than height. Thus, there are different drivers of obesity that vary across income groups.

In recent research using nonparametric methods, Kuku et al. (2012, 2013), we see that childhood obesity varies significantly with the level of food insecurity being experienced by the child. Moreover, this relationship differs across relevant subgroups including those defined by gender, race, ethnicity, and income.

OBESITY IN SUBURBIA

Research into obesity and its implications have proliferated in recent years, because of the above stylized facts. Mandel and Chern (2011) provide an overview of the social and environmental factors that determine BMI. Their cross-sectional data indicate that higher per capita sales in fast-food restaurants and drinking places, adjusting for the number of these restaurants, are associated with a higher prevalence of obesity and overweight among women.

[6]Pickett and Wilkinson (2012) also provide evidence that economic insecurity and income inequality generate financial and psychological stressors leading to obesity.

Lower rates of obesity, for both men and women, are found in densely populated urban areas. For the population as a whole, higher education levels and a higher consumption of fruit and vegetables are associated with lower BMI. Interestingly, increases in income lead to healthier lives for women, but increase the chances of men becoming obese. We examine several of these ideas in this section.

In recent years, researchers have connected urban sprawl to obesity. Recent suburbanization and the resulting decline in population densities in United States cities have come with their own set of problems. Since the urban sprawl and low population density development have occurred over the same time as the rise in overall obesity, researchers have become interested in exploring the empirical aspects of the subject.

It is obvious that urban sprawl forces people to travel a lot more, increases their reliance on automobiles, therefore increasing the "time cost" of travel and home-cooked meals. Commuting time also cuts into the time needed for exercise, which can be even more challenging if there are fewer sidewalks, trails, and parks.

Suburbanization also brings with it supercenters like Walmart which make foods cheaper, and hence, increase unhealthy eating habits. Economists have also linked suburban obesity to crime rates and lack of physical exercise. To what extent has urban sprawl contributed to the current obesity levels? Zhao and Kaestner (2010) investigate this issue for data between the years 1970 and 2000 that relate the obesity of metropolitan residents to population density, and show that there is a statistically significant negative association between population density and obesity. Zhao and Kaestner (2010) indicate that had populations not declined as they did, the current rate of obesity would be 13% lower.

Several studies have been generated in recent years to examine the issue of urban sprawl and its relation to obesity, from a variety of angles: remoteness, Walmart's presence, presence of food deserts, lack of parks and trails, etc. Whether any one or a combination of these factors drives obesity is still an open question, but a lot of good lessons can be gleaned from these investigations.

URBAN SPRAWL, REMOTENESS, AND THE BUILT ENVIRONMENT

Related to urban sprawl and the built environment, is the question of remoteness and its connection to obesity. Is the obesity rate in a county related to its remoteness in location? Similar to the situation with availability of parks, Guettabi and Munasib (2014) note that healthy food choices may not be easily available in remote locations. Guettabi and Munasib (2014) classify counties based on the public's access to hospitals, well-equipped gyms, chain stores, parks, and trails. The researchers hypothesize that longer distances from higher tiered centers will be closely related to the observed obesity rates within the relevant population. Indeed, Guettabi and Munasib (2014) show that the distances to urban hierarchy and county obesity are positively related, particularly in the case of remote metro-counties. The "weight penalty" from restricted access to higher order services in remote metro-counties contributes to space-obesity.[7]

[7]Levine et al. (2011) find that exercise levels are 60% greater among rural Jamaicans than for urban dwellers. Obese urban dwellers walk less than their lean urban counterparts. Obese Americans sit for almost four hours more than rural Jamaicans. Finally, urbanization is associated with low levels of exercise. Reifschneider et al. (2011) also find that Americans in excellent health are likely to spend more time exercising.

CAN HIGHER GAS PRICES REDUCE OBESITY?

Is there a connection between gas prices and obesity? Apparently, yes, according to Courtemanche (2011), who finds that increases in gas prices provide an opportunity for people to walk, and also reduce consumption of eating out, especially at restaurants. There is a strong correlation between the rise in obesity in the United States and the drop in gas prices between 1979 and 2004. Courtemanche (2011) shows that a $1 increase in gas prices will lead to a reduction in obesity in the United States by about 7–10%.

In a similar vein, Li et al. (2011) show that automobile demand in the United States is also linked to the prevalence of obesity. New vehicles demanded by consumers tend to be less fuel-efficient, because the proportion of the consumers who have become overweight has gone up. A-10% point increase in BMI reduces the MPG of new vehicles by 5%, which requires a 54 cent increase in gas prices to counteract.

Finally, Sen (2012) shows that an increase in gas prices can also lead to an increase in physical activity, such as house work, exterior cleaning, gardening, and yard work. Hence, this line of research strongly implies that the imposition of a gasoline tax is likely to have positive health and also, incidentally, positive environmental effects.

URBAN SPRAWL AND THE WALMART EFFECT

One of the important aspects of urban sprawls is the proliferation of supercenters such as Walmart. In fact, the presence of Walmart is associated with increasing levels of obesity. For example, Courtemanche and Carden (2011) use data from the Behavioral Risk Factor Surveillance System, and relate this information to the location of Walmart stores. Courtemanche and Carden (2011) show that an additional supercenter causes an increase in BMI by 0.25 units. Does this mean that Walmart supercenters are unequivocally welfare-reducing? Not necessarily, because the increase in medical costs from bad health has to be compared to savings from cheaper consumer goods. The authors show that the incremental medical costs only take away 6% of the consumers' surplus. A recent study by Bonanno and Goetz (2012a) also supports this view.

In this regard, the study by Marlow (2015) is interesting, because it shows that a slightly more nuanced interpretation is possible if stores are distinguished by their types. Marlow (2015) identifies four types of retail food outlets: supermarkets; supercenters and warehouse club stores (known in this literature as "big box" stores); convenience stores; and specialty stores. Marlow's (2015) results show that counties with more retail food stores experience a lower prevalence of adult obesity. The main driver of this inverse relationship is the greater numbers of supermarkets and specialty food stores. Interestingly, this study shows that obesity levels are positively associated with the market shares of "big box" and convenience stores.

THE BUILT ENVIRONMENT: OBESITY, PARKS, AND RECREATION

While urban sprawls, Walmart centers, and population densities are related to obesity levels, can neighborhood parks and gyms help in the reduction of the obesity problem? This question has also generated a lot of interest among researchers. Kostova (2011), using the Behavioral Risk Factor

Surveillance System (BRFSS) for 2005, shows no significant relation between physical activity and BMI. While Kostova's (2011) single equation models indicate that less urban sprawl is negatively related to obesity, and better park access is associated with lower levels of obesity, both these results disappear when participation constraint is explicitly modeled using two-stage least squares through the instrumental variable estimation.

In contrast to Kostova's (2011) findings, Fan and Jin (2014), using 2007 National Survey of Children's Health data, show that childhood obesity is significantly negatively related to the availability of neighborhood parks and playgrounds.[8]

In a very interesting study, Cawley et al. (2013) demonstrate that physical education (PE) classes among kindergarteners help in reducing obesity among 5th graders. The authors use data from the Early Childhood Longitudinal Study for 1998 – 2004, and use Instrumental Variables estimation to show lower BMI z-scores for PE participants. The PE-effect is greater for boys, because the PE-good is a complement to other physical activity that boys engage in. However, the PE-good is a substitute for girls. It is possible that while more PE activity may reduce obesity, it may also take time away from academic pursuits. However, the authors find that PE does not crowd out academic achievement for either gender.[9]

THE BUILT ENVIRONMENT: JUNK FOODS, VENDING MACHINES, VIDEO GAMES, AND GRADES

The study by Cawley et al. (2013) implies that the built environment in schools might have inadvertent effects on childhood obesity. Indeed, researchers have started looking into these effects as well. In a very interesting study from the Los Angeles Unified School District, Bauhoff (2014) finds that elimination of unhealthy foods and beverages was mainly ineffective in reducing BMI.

Although the consumption of soda and fried foods declined drastically, the students continued to consume the requisite calories from substitute products. Datar and Nicosia (2012) show that the influence of junk food on BMI is also negligible. For a national sample of fifth graders, the researchers show that junk food availability does not significantly increase BMI or obesity, despite the increased likelihood of in-school junk food purchases. Along these lines, Nakamuro et al. (2015), in a very interesting study, show that the number of hours spent on watching TV or playing video games has a very negligible effect on BMI.[10]

The importance of the in-school built environment becomes crucial if one relates the prevalence of obesity to academic achievement and grades. Gurley-Calvez and Higginbotham (2010) examine

[8]For methodological issues and a survey of this line of enquiry, see Sallis et al. (2011), and Ding and Klaus (2012). For an Obesogenic environmental critique of these studies, see Guthman (2013). The effect of the economic downturn on obesity due to lower memberships in gyms and physical fitness classes is also noted by Fitzpatrick et al. (2010).

[9]Sallis et al. (2011) and Sarma et al. (2014) note the importance of physical activity and obesity for Canada. Rashad et al. (2006) is one of the earlier studies that identified the number of restaurants per capita, the gasoline tax, the cigarette tax, indoor air pollution, household income, years of formal schooling completed, and marital status as drivers of obesity among American adults.

[10]Incidentally, Price (2012) shows that the elimination of vending machines improves the behavioral outcomes of students, such as reducing tardiness or being reported to the principal's office.

this issue for schools in West Virginia, where obesity rates among fifth graders have been near 30%. The authors find that obesity negatively affects reading proficiency in high poverty districts, but obesity rates have little effect in districts with lower poverty.

The implication of this study is that a substantial increase in instructional education spending is needed to offset the obesity effects on academic achievement, especially for students from high poverty districts.

FOOD DESERTS

The USDA defines food deserts as urban neighborhoods and rural towns without ready access to fresh, healthy, and affordable food, and estimate that roughly 23.5 million people live in food deserts (see https://apps.ams.usda.gov/fooddeserts/fooddeserts.aspx). Further, it is possible that this lack of access may contribute to unhealthy eating choices, and therefore lead to higher levels of obesity.

For instance, Chen et al. (2010) examine the issue of access to chain grocers and its relation to BMI. Using survey data on adults in Marion County, Indiana, Chen et al. (2010) note that obesity and access varies depending upon community characteristics, particularly location and income. In their simulations, the authors find that increasing access to chain grocers in low-income communities decreased average BMI for everyone.

However, extant research in this area does not provide overwhelming support for this notion. For instance, Ver Ploeg (2010) notes that access to healthy food is a problem only for a small portion of the overall population, and that low-income consumers shop where food prices are low. Bonanno and Goetz (2012b) also examine how adult obesity levels are related to food store location densities and expenditure on SNAP education, and are not able to find any significant relation between food supply environment and BMI, even after accounting for endogeniety between store locations and consumption. Rather, the authors, like other researchers, prescribe nutrition education to achieve healthy eating outcomes. Similarly, in a recent investigation, Alviola et al. (2013) use panel data covering 2007–09 for Arkansas, and fail to uncover a significant relationship between the location of food deserts and BMI among children in relevant school districts.[11]

If food deserts and lack of access are not major issues, can we then say that the availability of "locally grown food" helps in reducing obesity and related health problems? Salois (2012) uses US county-level data on obesity and diabetes, and links this information to the built environment. Salois (2012) looks at the density of farmers' markets, and local farms with direct sales, and shows that a strong local food economy helps in prevention, and suggests that community-wide interventions examine this important supply-side issue.

[11]In some instances, while prices may be significant determinants of food purchases, supermarket access may not have such a significant impact. Lin et al. (2014) point out that the price – access synergy may be an important driver of health and consumption.

FAST-FOOD AND OBESITY

The availability and consumption of cheap fast-food are also crucial factors that drive overweight and obesity. Consumption of fast-food is also related to urban sprawl and distance to highways. Dunn et al. (2012), and Dunn (2010) examine this issue in a very interesting manner, by combining different data sets on locations of fast-food chains to county patterns and individual characteristics.

Using interstate highway exits as exogenous instruments to proxy locations, Dunn et al. (2012), and Dunn (2010) find that as fast-food availability increases, so does the BMI of females in medium-density counties. Further, blacks and Hispanics also exhibit a positive BMI – location relationship. Once again, this relationship matches the observations made in the context of urban sprawl and access to supermarkets.

Chen et al. (2013) also find partial evidence of a positive relationship between BMI and the density of fast-food chains. In a very interesting combining of Natality Detail Files, Area Resource Files, and County Business Patterns, Lhila (2011) shows that greater access to fast-food restaurants is positively related to pregnant mothers' probability of excessive weight gain. However, the hypothesis does not carry through to infants' birth weights.

An important observation made from these enquiries is that most fast-food chains are located in medium-density zones, which are populated by specific demographic groups consisting of females and minorities. Consequently, Dunn (2010) questions the efficacy of "soda-fast-food-tax" solutions, because the deadweight loss of the tax may not compensate for the "time cost," or increase access to healthy diets. Anderson and Matsa (2011) also conduct similar analysis using the placement of restaurants along interstates, and conclude that restaurants are not the main drivers of the overweight problem. Consumers replace the calories from restaurant meals by eating less at other times.[12]

Qian (2014) unifies many of these concerns using data from the Arkansas public school system. First, Qian (2014) shows that the distribution of free fresh fruit and vegetables to students is associated with significant reductions in the BMI of children. Second, linking the school district data to the location of fast-food chains, the study shows that an increase in fast-food availability is closely linked to higher BMI levels for children.[13] In fact, more affluent, rural, nonminority, and female children are more strongly impacted by the location factor. Third, the presence of neighborhood parks has a significant negative effect on children's BMI, in both urban and rural areas. The effect is particularly strong for girls, and for boys in rural areas. Alviola et al. (2014) also find the same result for the Arkansas public school system.

In this context, Tomer (2011) produces an economic model which relates consumer's motivations manipulated by the "junk food" industry, which is in contrast to the standard model of rational economic behavior. Tomer (2011) relates his model to how the fast-food – industry complex exploits economically vulnerable groups to produce dysfunctional eating persistence and habit formation.

Lakdawalla and Zheng (2011) also develop an interesting theoretical model that predicts that while higher fast-food prices may decrease body weight, higher prices for fruit and vegetables may have the opposite effect when considering drivers of childhood weight. The same finding is also present in the health economics literature (Grossman et al., 2014).

---

[12] Also see Marlow and Shiers (2012, 2013).

[13] These findings are similar to Parr (2012).

Tomer's model fits very well with the observations on "portion sizes." For example, Kral and Rolls (2011) relate obesity to the portion size served by the retail food industry. Portion sizes of beverages and other energy dense foods influence energy intake, and create a sense of fullness. The authors call for policies directed toward marketing, which include promotion of consumer education and awareness, food labels and point-of-purchase information, value size pricing, package size, and portion-controlled packaging.

Interestingly, Wansink (2011) connects portion sizes to the built environment, where mindless eating creates consumption norms, juxtaposed with consumers' underestimation of the calories consumed. Redden and Haws (2013) actually demonstrate how information about "amount consumed" can help those with low self-control to monitor their intakes and overcome overconsumption. This finding merges nicely with Fan and Jin (2014), who show that information-based anti-obesity interventions have a very limited impact on improving self-control.

POLICY CHALLENGES: WILL FAT-TAXES OR SODA-TAXES WORK?

An obvious economic prescription is to impose a tax on unhealthy foods to limit consumption, and thereby reduce the negative externalities. However, as with any economic policy, there are costs and benefits associated with taxation. Economists and policy makers are divided on the role of soda-taxes or sugar-taxes, and the arguments from either side are equally compelling.[14] First, we examine the case for taxation.

Miao et al. (2013) argue that whether or not a calorie-tax would be effective depends on consumers' response to the price changes of high-calorie foods, and the availability of acceptable low-calorie substitutes. They note correctly that a tax on a nutrient will be reflected in changes in food prices, which will lead to food demand changes, and these lead to nutrient intake changes. The researchers estimate the substitution and welfare effects of taxes on added sugars and calorific sweeteners, and solid fats.

Miao et al. (2013) estimate a LINQUAD demand system that nests a CES formulation for four substitute components, constituting high- and low-added sugar and solid fats. They consider 25 aggregate food groups that are related to obesity and compute elasticities using NHANES data. An important aspect of their empirical estimation is the capture of substitution possibilities, between high- and low-fat or sugar within a food product category. This strategy allows them to compute substitution across and within food groups.[15]

The within-food groups substitution indicates that taxes induce consumers to move toward leaner and lighter choices. Consequently, the deadweight loss from the tax is also much lower. A fat-tax raises more revenue and has a lower deadweight loss per dollar of revenue than a sugar-tax does. In a related study Miao et al. (2012) show that a tax on sweeteners leads to a much smaller loss in consumer surplus than a general tax on the final sweetened goods. A similar estimation using the QUAIDS model capturing the demand for fruit consumption from two retail stores in the

[14]See Runge (2011) for an exhaustive treatment of this policy debate.

[15]Riera-Crichton and Tefft (2014) note that increases in carbohydrates are most strongly and positively correlated with the prevalence of obesity. They show that a 1% increase in carbohydrate intake yields a 1.01 point increase in obesity prevalence over five years.

Pacific Northwest by Durham and Eales (2010), also captures similar results. Durham and Eales (2010) demonstrate that the price elasticity of fresh fruit is large, and a 20% subsidy in fruit prices will increase consumption of fresh fruit by between 7% and 18%, getting the average consumer closer to the daily requirement.

Chaloupka et al. (2011a,b) argue that taxing Sugar-Sweetened Beverages (SSBs) is the best policy response to reduce SSB consumption. The researchers note the effects of soda-taxes to be similar to that of taxing tobacco. They also note that soda-taxes are likely to generate roughly $15 billion in tax revenues, which could be used toward obesity prevention efforts.[16]

A more recent study by Zhen et al. (2014) also produces similar findings from 178 beverage products using supermarket scanner data. This study shows that if a calorie-based beverage tax is applied to products purchased from all sources, then a 0.04 cent per kcal tax on sugar-sweetened beverages will reduce annual per capita beverage intake by 5800 kcal. Okrent and Alston (2012) also share the view that a tax on calories would have the lowest deadweight loss, and would even yield a net gain if the impact on public health care expenditure is also taken into account. Recent studies by Todd and Chen (2010), and Lopez and Fantuzzi (2012) also indicate the relative importance of taxes based on nutrient content, more than a general sales tax on calorifically sweetened beverages.

In a very interesting paper on tax salience, Zheng et al. (2012) show excise taxes on food and beverages to be a better policy instrument than a sales tax. The researchers point out that consumers react only to the final register price, and that the actual sales tax increase is usually never pointed out on the shelf or by the retailer. Additionally, the sales tax policies on food and beverages are not uniform across the country. Some states do not have any sales tax, and some states just tax food items, while others tax just soft drinks. Moreover, from their survey results, the researchers find that consumers are never properly informed about the exact tax rate or the corresponding price increase, which raises the question of tax salience.

Most importantly, consumers may not pay any sales tax on eligible food or beverages if they are SNAP beneficiaries. Zheng et al. (2012) demonstrate that an excise tax on unhealthy foods is a lot more effective on curbing demand than a standard increase in sales tax.[17]

While the research cited above argues by and large for calorific taxes, there are studies that indicate some of the difficulties with this policy, mainly because of substitution effects. For example, Fletcher et al. (2010a,b) and Fletcher (2011) show that a tax on soft drinks has a very small effect on BMI, partly because the tax allows for substitution of juice and whole milk which are also dense in calories. Similarly, Dharmasena and Capps (2012) show that the substitution effects between soda and other liquids such as fruit juices, low-fat milk, coffee, and tea are positive, and hence, the full effect of the tax on weight loss might be biased upward if the substitution effects are ignored. These latter studies thus expose the limitation of taxation as a policy instrument to combat obesity.

In a recent study Dharmasena et al. (2014) demonstrate that industry supply-side effects are also important, and the elasticity estimates are very sensitive to the assumptions underlying the nature of the supply curve. Indeed, the researchers use a stochastic equilibrium displacement model

[16] Also see Powell and Chriqui (2011) for related results.

[17] Using experimental evidence from 258 adults, Streletskaya et al. (2014) show that unhealthy foods tax, healthy foods advertising, and unhealthy foods tax combined with anti-obesity advertising significantly reduce the demand for calories from fat, carbohydrates, and cholesterol in meal selections. However, healthy foods subsidy and healthy foods advertising has very little effect on nutrient consumption.

(SEDM) that allows for different values of the elasticity of supply. The SEDM approach indicates that supply-side effects are also large, and that ignoring these effects can also bias the elasticity estimates upward.

After incorporating the supply-side responses, the researchers show that the decrease in the consumption of SSBs is much smaller, thus questioning the effectiveness of a soda-tax to reduce obesity. This finding is similar to the ineffectiveness of eliminating SSBs and fried foods in the Los Angeles School District found by Bauhoff (2014).

Finally, Craven et al. (2012) argue that private markets tend to provide the best possible response for obesity, and that any government intervention is likely to be inefficient. Unlike a tobacco tax, which falls only on smokers, a fat-tax or a soda-tax does not fall only on the obese, but on everyone. Inefficiency also arises because the government is not in a position to correctly evaluate the optimal tax amount, and in most cases these taxes fall regressively on the poor.[18] For example, fast-food impacts low-income women and women with children. Taxing fast-food is likely to have a greater impact on the BMI of this subgroup.

PEER EFFECTS, SOCIAL NETWORKS, DATING, AND OBESITY

Can peers, friends, colleagues, social networks, etc. influence a person's weight? Recent research into this question seems to suggest that there is indeed a "contagion effect" associated with the spread of obesity. For instance, Ali et al. (2012) use detailed data from the National Longitudinal Study of Adolescent Health and show that a person's BMI is highly influenced by that person's network. Effects of past peer weight at an early stage in a person's life linger into adulthood as well. Once again, this study implies that policy must be carefully designed to inform childhood and adolescent knowledge and preferences.

Using the same data source, Yang and Huang (2014) add an additional insight into the asymmetric nature of peer effects: weight gain is associated with an increase in the number of obese friends, but a decrease in the number of friends does not translate to weight loss.

Asirvatham et al. (2014) use data from Arkansas public schools and also find similar trends: higher BMI at the oldest grade are associated with obesity prevalence at younger grades. The effect from kindergarten lingers strongly till the fourth-grade level.[19]

Fletcher (2011) examines the data and available evidence critically and indicates that researchers should be careful in coming to conclusions about peer effects. For instance, Fletcher (2011) notes the issue of endogeneity where overweight friends may choose overweight friends, and also built environment, where the network of friends is also influenced by the prevalence of fast-food joints or lack of gym facilities.

[18]The same kind of result can also be generated from medical models of obesity as in Dolar (2010) which predict that while taxes on food will impact what people eat, the total BMI or obesity is likely to remain unaffected. See Powell and Han (2011), Han and Powell (2013), and the counterfactual simulations of Buttet and Dolar (2015).

[19]Interestingly, Zagorsky (2011) finds that first-year college students gain between 2.5 and 3.5 pounds during the freshman year, although when compared with their noncollege counterparts, this weight gain is minimal. Price and Swigert (2012) similarly do not find any matching trends among siblings, or any within-family peer effects.

Forste and Moore (2012) also find lower life satisfaction among overweight adolescents, where the negative association operates through perceptions of self, peers, parents, and school, and mostly where perceptions of body weight are strongly associated with low life satisfaction among girls compared to boys.[20]

Using survey data from 15,000 young adults, Fletcher (2014) notes that there are significant social sanctions against the overweight. But what is interesting is the lower "obesity penalties" that black women appear to receive when asked about their self-perceptions.

Interestingly, Ali et al. (2014) note that the private costs of being obese also include the costs of being left out of social life, and losing out on potential network externalities. Ali et al. (2014) find that obese white teenage girls are less likely to have been in a romantic relationship compared to their nonobese counterparts. Importantly, this group is also less likely to ever have had sex or been intimate. The researchers note that this pattern is consistent with low self-esteem, attitudes toward sex, and interviewer assessment of appearance and personality.

Incidentally, obese black teenage girls do not show this trend, once again, in conformity with Fletcher's (2014) "obesity penalty" thesis. Instrumental variables estimates and estimates from models with lagged weight status confirm the overall patterns (Box 14.2).

BOX 14.2 OBESITY OR WAR?

In a sequence of important studies, Cawley and Maclean (2012, 2013) demonstrate that the rising obesity among youth is making US military recruitment very difficult. This result has implications for US military readiness. The researchers examine the data from the National Health and Nutrition Examination Surveys, spanning over 40 years.

The authors note that as of 2008, about 5 million men and 16 million women exceeded the enlistment standards for weight and body fat. Indeed, a further rise of just 1% in weight and body fat in the overall population would further reduce eligibility for military service by over 850,000 men and 1 million women. The study indicates that the fraction of age-eligible civilians exceeding the weight and fat standards for admission has more than doubled for men and nearly quadrupled for women in the last 50 years.

Related to these findings, Maclean and Cawley (2014) examine the eligibility criteria for the United States Public Health Service Commissioned Corps (US PHSCC). The US PHSCC is the uniformed service engaged in public health. Similar to the applicants in the military, the number of eligible civilians who exceeded the required standards for weight-for-height and BMI has risen precipitously (from 9% to 18%).

The results from simulations using this data also parallel the ones derived in related studies, and find that with the PHSCC data an additional 1% increase in population body weight will result in an additional 3.42% of men and 5.08% of women exceeding PHSCC accession standards. Obviously, this finding is important for the country's capacity to deliver public health services.

Incidentally, how well do veterans of active-duty military fare, in terms of BMI and obesity? Teachman and Tedrow (2013), using extensive data from over 6000 veterans spanning over 13 years, show that veterans also exhibit higher levels of BMI and obesity, due to difficulties in transitioning to civilian life. Further, the weight gains during transition are permanent.

[20]Socioeconomic status is intertwined with health outcomes, and Corsnoe (2012) finds that obese girls at the start of high school had higher levels of internalizing symptoms, and lower levels of perceived social integration in school, only when they were exposed to high family instability.

OBESITY IN OTHER DEVELOPED COUNTRIES

CANADA

Several of the determinants that are observed for the United States regarding obesity are also noted for Canada. For instance, peer effects are also present in Canada. In a very interesting study Averett et al. (2013) use the Canadian National Public Health Survey for 1994–2008, and find that while marriage produces lots of benefits in the form of increased income, health, and longevity, there are costs as well, in the form of higher BMI and lower levels of exercise.

Further, using Canadian macroeconomic data and the Canadian Population Health Survey, Latif (2014) shows a significant relation between the unemployment rate and the probability of being overweight and obese. Besides unemployment and economic security concerns, the importance of socioeconomic determinants, e.g., in Hajizadeh et al. (2014) reveal an interesting peculiarity in the Canadian data. Hajizadeh et al. (2014) find that income-related inequality influences the risk of obesity for both the rich and the poor. For example, obesity is highly prevalent among economically disadvantaged women in the Atlantic provinces, and among better-off individuals in Alberta. The main drivers of income-inequality related obesity in Canada are demographics, income, immigration, education, drinking habits, and physical activity.[21] The authors point out that health policies should focus on poorer females and economically well-off males.

Physical activity and obesity levels are also connected in the case of Canada. Sarma et al. (2014) use an extensive data set from Canada's National Population Health Survey with individual information for about 16 years. Sarma et al. (2014) consider four measures of leisure-time physical activity (LTPA) and work-related physical activity (WRPA). Both measures of physical activity have a negative influence on BMI. For instance, at least 30 minutes of walking reduces BMI by about 0.11–0.14 points for males and almost 0.2 points for females, when compared to their inactive counterparts. Likewise, lifting loads at the workplace is associated with a reduction in BMI by 0.2–0.3 points in males and 0.3–0.4 points in females, relative to those who are reported sedentary.

OBESITY IN THE OECD

Devaux et al. (2011), Sassi (2010), Economos and Sliwa (2011), and Branca (2007, 2010) have exhaustive surveys on the causes and costs of obesity in OECD and in the European Union. Overall, the research policy here stresses the need for education to curb the impact of obesity. Both in the United Kingdom and in the United States, race, ethnicity, and migrant status are highly associated with obesity (see Martinson et al., 2012).

Many researchers have also called for a more active role for the family and particularly from women in the household. In particular, Schuring (2013) finds that the likelihood of a child being obese if the mother was employed is much higher than in the United Kingdom, implying the role of employee-friendly policies in the workplace.[22]

[21] Also see Fernando (2010) for the impact of Food Away From Home in obesity in Canada.

[22] This finding is also supported by Hong et al. (2015).

OBESITY IN LATIN AMERICA: COLOMBIA, BRAZIL, GUATEMALA, AND MEXICO

Fortich Mesa and Guitierrez (2011) note that excess weight and obesity is a big problem in Colombia, where BMI is positively related to household wealth, and negatively related to years of schooling, for both men and women. The researchers note the increasing costs of public health due to obesity, and call for adequate public action.

The experience with Food Away From Home (FAFH), observed in the United States and Canada, is also shown to be a driver of obesity in Brazil. Finocchio et al. (2015) show using data from Family Budgets Research for the years 2002 – 03 and 2008 – 09, that income and expenditure on FAFH are directly related, and as a consequence, produce a greater prevalence of overweight and obesity, especially among Brazilian men.

Jolly et al. (2013) show that food imports and per capita GNP are negatively related to BMI, while the total number of TVs influences total obesity levels positively for 25 selected Latin American and Caribbean (LAC) countries. Food importing countries within LAC countries have a lower prevalence of obesity.[23] Among the many reasons for increased levels of obesity in Latin America, Asfaw (2011) finds consumption of processed food to be a main driver of obesity in Gautemala. Asfaw (2011) shows that a 10% increase in the share of household food expenditure on partially processed food increases the BMI of family members by 3.95%. The same 10% increase in the share of highly processed food increases the BMI by 4.25%. Higher shares of processed food expenditure seem to be a high risk factor in these countries.

In an interesting study, Damon and Kristiansen (2014) test the hypothesis that increased liquidity and time allocations affect BMI outcomes. They use data on the obesity status of children who remain in Mexico, when either a male or a female from the household migrates to the United States. The study shows that urban females engage relatively more in housework, and hence are less obese. However, urban boys do not engage in similar activities and, as a consequence, become more obese.

Concerns about the impact of safety nets are equally important in Mexico. As mentioned in the other parts of the book, the conditional cash transfer program Oportunidades is very popular in Mexico, and how the participants fare in terms of BMI is an important question. Andalon (2011) examines the BMI of participating youths and finds that the provision of schooling, health information sessions, and sizable cash transfers could have significantly affected the BMI rates. Andalon (2011) notes that for female participants, a lower level of BMI was noted, probably because they were exposed to increased access to information and schooling, improved dietary quality, and increased monitoring of health outcomes and physical activity.

Prina and Royer (2014) note that information and education may not always elicit the optimal response from the affected parties. Prina and Royer (2014) note that when body weight report cards were sent to parents, so as to motivate them to address the issue of their obese children, no significant alteration in behavior was observed. In fact, the researchers observe that parents of children in the most obese classrooms were less likely to report that their obese child weighed too much relative to those in the least obese classrooms. Prina and Royer (2014) note that under these circumstances, the reference BMI measure is a moving line, and this makes policies geared toward public awareness very difficult to implement.

---

[23]Viego and Temporelli (2011) show the prevalence of double burden in Argentina.

In a compelling comparison between the United States and Mexico, Monteverde et al. (2010) find that mortality rates for obese and overweight individuals for the elderly are high in both countries. Further, the probability of experiencing obesity-related illness among the obese is larger for the elderly in the United States. However, the likelihood of dying from these illnesses is higher in Mexico.

OTHER OBESE GROUPS

The notion of acculturation was discussed in the chapter on the social determinants of nutrition. Baker et al. (2015) recently extend this to the notion of the "immigrant epidemiological paradox." This paradox refers to the observation that immigrants and their children enjoy health advantages over their US-born peers; however, these advantages diminish with greater acculturation. Baker et al. (2015) test this hypothesis, and find the opposite to be true, namely, children of US-born mothers are less likely to be obese than otherwise similar children of foreign-born mothers, and further, the children of the least-acculturated immigrant mothers are the most likely to be obese.

In another study on acculturation, Wen and Maloney (2014) note that the likelihood of illegal undocumented women gaining weight is higher than that of legal immigrant women. The odds of an illegal immigrant woman being obese is 10% points higher, and about 40% points higher of being overweight. The same pattern is not observed among immigrant men.[24] Interestingly, Ulijaszek and Schwekendiek (2013) find the incidence of overweight among adopted Koreans raised in the United States to be higher by 11%, than those adopted Koreans living in Europe.

POLICY: SHOULD WE WORRY ABOUT OBESITY?

Tackling obesity is not an easy task. As with any policy making in economics, there are many sides to argue, and many perspectives to entertain. We begin by asking whether any policy making itself is needed. Can markets generate optimal obesity levels? The external spillover costs of obesity are also a big concern, and are also highly debated among policy makers. Bailey (2013) shows that there are no external costs associated with obesity, implying that obese people pay their own costs of health in the form of lower wages. Kalist and Siahaan (2013) note that the social costs of obesity may be overstated, because the obese are less likely to be criminals, since the odds of an obese man being arrested is 64% of those who are healthy. Contrast this finding with those of Simmons and Zlatopper (2010), who demonstrate that the motor vehicle death rate has a statistically significant positive relationship with the percentage of the population that is obese.

Bhattacharya and Sood (2011) point out that the idea that the lack of external costs does not justify interventionists policies fails to consider the positive externalities that will be generated as a consequence of intervention. Bhattacharya and Sood (2011) note that public policies are justified because these will inform people about different issues, and help remove ignorance and encourage self-control.[25]

[24]Also see Zeng (2013) for a study of obesity among native Amazonians.

[25]For a conservative argument about individual responsibility and the effects of market prices to combat obesity, see Philipson and Posner (2011). Also see Muth (2010).

Roberto and Brownell (2011) note that conservative reasoning emphasizes individual responsibility, and consequently has led to weak government action. Roberto and Brownell (2011), on the other hand, stress that the environment plays a major role in determining obesity, and that suitable action that addresses school food environments, food access and cost, sugared beverage consumption, food marketing, restaurant food nutrition content and portion size, appropriate labeling, and bans on certain types of advertising are critically needed to curb the epidemic.

DOES PUBLIC OPINION SWAY POLICY?

Public opinion also plays a very important role in deciding the success of policies. Although a lot of information and education is generated to warn citizens about the consequences of obesity, the general awareness and processing of this information is not all that straightforward. Oliver and Lee (2005), using very special survey data, note that most Americans are not worried about obesity. Consequently, they do not place a lot of weight on interventionist proposals, and would rather place the burden on individual responsibility. Further, Oliver and Lee (2005) also find that the American public utilizes their secondary mental framework on smoking or the environment, and extends these intuitions on issues about obesity. In the American context, such public opinion also informs the political debates, rendering policy making to combat obesity rather difficult.[26]

Martin et al. (2010) make a very important observation concerning the future prospects of health. They note that the vast improvements in medicine, technologies, and public policies have resulted in better health outcomes for the older population, which have continued into the 2000s. However, the younger population has seen a doubling of obesity in recent decades. The authors note that information through education about the ill-effects of smoking are slowly making inroads into public forums, and the increase in obesity has slowly started to recede.[27]

In fact, Hong (2013) notes that between 1980 and 2009, the health costs of overweight individuals measured by activity limitation has declined significantly. The reason for the decline is due to technical progress in the health care sector, with affordable access. Along with increases in income, the relation between BMI and activity limitation has shifted due to technical developments in the United States, resulting in the declining burden of obesity.

POLICY ANGLE FROM BEHAVIORAL ECONOMICS

In recent years, behavioral economics has attracted a lot of attention among policy makers, and the tools from this branch of economics are now a big part of the White House Task Force on Obesity, and the Institute of Medicine (IOM). The Task Force has generated many program-based evaluation

[26]See Kersh and Morone (2011) for political incrementalism and "muddling-through" in this context. From a legalistic point-of-view, Marlow (2014) finds that state governments within the United States that enact most laws are those with relatively low obesity prevalence.

[27]Auld (2011) attributes much of the variation in body weight across time and space in the United States to individual and regional characteristics.

techniques from experimental settings in lunchrooms, grocery stores, and labs, which have a strong hold in policy design.

These newer developments have slowly started producing modest results. One of the crucial conclusions from behavioral economics is that individual rationality may breakdown, that addicts exhibit "hyperbolic discounting," and that obesity is positively related to the discount rate (Timothy and Hamilton, 2012). Hence, the best policy response may not be in taxing fast-foods or soda, but in clever information transmission regarding the long-run impacts of such behavior. Gittlesohn and Lee (2013) present three intervention studies that integrate educational, environmental, and behavioral economics to successfully tackle obesity.

Liu et al. (2014) also provide additional insights from behavioral models dealing with present-biased preferences, visceral factors, and status-quo bias that exist among consumers, which must be used to formulate regulation of restaurants and peer effects in public schools. Jones-Corneille et al. (2011) apply self-monitoring, cognitive restructuring, and stimulus control to help with weight management, aided by on-site visits, telephone, Internet, and e-mail contact.

Cash and Schroeter (2010) note that while these methods are cheap, implementable, and flexible, there is always a lingering doubt as to whether experimental results will totally match real outcomes. Many economists, such as Kenkel (2010), and Thapa and Lyford (2014), argue that changes in the agri − food sector are still badly needed, and as such, supply-side issues should not be sidelined.

THE MORAL HAZARD PROBLEM AND POLICY HEADACHES

Behavioral economics also highlights the moral hazard problem inherent in this context. Gustavsen et al. (2011) compare the visits to physicians among groups of individuals who have a similar BMI. They find that the effect of moral hazard on visits to doctors to increase in men with high BMI. In fact, moral hazard is higher among men with higher BMI.

In this context, the arguments put forth by Bhattacharya and Packalen (2012) merit our attention. Bhattacharya and Packalen (2012) indicate that Medicare-induced health insurance acts like a pooling insurance scheme, in which an individual need not take the full costs of obesity into account, thus creating a negative externality. This argument enables public policy such as "soda-taxes" and "fat-taxes" to gain traction.

However, Bhattacharya and Packalen (2012) identify another moral hazard, which may trigger a positive externality, from which obese individuals may free-ride. The positive externalities that Bhattacharya and Packlen (2012) consider are the innovations in pharmaceuticals and in the health sector, which benefit all citizens, and hence, consumers of preventive health may overinvest in preventive care.

Consequently, whether the negative externality and the moral hazard from Medicare-induced insurance is bigger than the positive spillovers from innovation is an empirical question, and when put to the test, it turns out that innovation-induced externalities offset Medicare-induced moral hazard problems. Hence, punishment strategies such as "soda-taxes" and "fat-taxes" are not justified on their original stand.

In the next few sections we examine some standard approaches to policy prescribed by many economists and researchers.

CAN NUTRITION LABELING, PACKAGING, ADVERTISING, LAW, AND EDUCATION HELP?

Can nutrition labeling affect obesity? This question has also attracted the attention of many researchers who seek to improve health outcomes through labeling requirements and consumer education. A few researchers such as Aresenault (2010) raised some important questions about the effectiveness of such programs. More evidence has been gathered since then, and many interesting observations have been made in this area (see Kiesel et al., 2011).

For example, Loureiro et al. (2012) use a switching regression model to capture the role of nutritional labels on obesity. The researchers find that the BMI for men who read nutritional labels is 0.12 point lower than men who do not read them. The effect is more significant for women users, whose BMI is lower by 1.49 points, compared to women who do not read labels. Similarly, Andreyeva (2012) discusses how a revision of food packaging in 2009 helped participants in the *Special Supplemental Nutrition Program for Women, Infants, and Children (*WIC) achieve healthy diets in Connecticut.

This line of research helps to inform policy makers about educational campaigns, and Jordan et al. (2012) show that targeted messaging based on positive feelings of nurturing and concern about child weight gain are often the most effective. Low carbohydrate information in the content is also a useful mechanism to allow consumers to substitute toward healthy foods (see Paudel et al., 2013; Box 14.3).

BOX 14.3 FOOD MARKETING AND CHILD NUTRITION

Kaur (2011) presents evidence from India that suggests that marketers use a variety of ways to reach children. The marketing channels include advertising, the Internet, in-school promotions, clever packaging, bundling with related children's goods, etc. Moreover, advertisements use cartoon characters, celebrities, animation, fast music, color effects, toys, games, and clever packaging. Unfortunately, most of these promotional ploys are from food categories such as ready-to-eat cereals, fruit snacks, candy, frozen desserts, juice, cheesy snacks, chips, and meat products.

Kaur (2011) presents two important findings. First, children are easily drawn to these types of strategies, and are hence targeted intensively by the multi-national companies (MNCs), such as Coca Cola, PepsiCo, Kellogg's, KFC, McDonalds, and Nestle. Second, and most crucially, heavy promotion and targeting takes place for foods that are nutrient poor, and high in salt, sugar, and fat.

Similar marketing techniques to attract the young audience are also found in developed countries, and Kaur (2011) points to the increasing levels of child obesity as suggestive evidence that links such marketing to the nutritional outcomes of children in these countries.

Kaur (2013) also shows that promotional packaging and promotions are very similar across countries, and that MNCs advertise with greater intensity than non-MNCs. Consequently, Kaur (2011, 2013) and others[a] call for stricter government intervention.

Regulation can enforce accurate information about nutrition content, and encourage healthier consumption. Parents and consumer groups can also form a cohesive body to demand specific regulation. Food companies must also produce and market healthier foods that contain calcium and iron, without losing the taste or the "cool" factor. Hoping that companies will self-regulate to proper ethical codes is futile. Overall, there has to be a significant increase in the awareness among all constituents regarding healthy eating, physical activity, and the advantages of a home-cooked meal.

[a]For further references that link marketing to nutrition, see Kaur (2011, 2013).

PUBLIC POLICY ANGLE 1: FOOD ADVERTISEMENT AND PROMOTION

The impact of advertising and promotion of fast-food and soft drinks on obesity among children has also attracted many empirical enquiries from the United States and Canada. Timothy and Padilla (2009) find evidence from Canada that promotion of fast-food has been a main driver of obesity. Elliott (2012) also examines data from Canada and shows that foods targeted at children under different labels such as, "better for you," are more likely to have marketing success than any addition to nutritional knowledge. Further, Dhar and Baylis (2011) is also a very good empirical study from Canada, where a ban on fast-food advertising reduced fast-food consumption by $8 million annually.

Andreyava et al. (2011) demonstrate from ECLS and Nielsen Company data that in the United States, exposure to fast-food advertising is associated with a 1.1% increase in children's consumption of fast-food. Vandewater and Wartella (2011) also provide a convincing case about the impact of advertising on children's eating habits.

The entire line of research tells us that appropriate promotion policy, labeling, and nutritional guidelines are very evasive in nature. For instance, even appropriate labeling may not be the right strategy. Indeed, for the United States, Kim et al. (2012) note that voluntary front-of-package nutritional labeling for beverages has conflicting perception outcomes among consumers. For instance, consumers' perception index about the healthiness of beverages decreased for milk and juices, and increased for soft drinks. Maher (2012) also finds similar results for children in the United States, whose information processing, as it applies to nutritional content of food products, appears to be highly confused.

Consequently, newer and more innovative advertising strategies are needed to promote healthy eating habits. In an experimental study, Liaukonyte et al. (2012) show that broad-based advertising, which is advertising the entire category of fruit and vegetables, increases consumers' willingness to pay for the product, and also reduces per capita calorific intake. The experiments show that advertisements must combine appropriate triggers to produce healthy life-styles.

Indeed, in this regard, Mello (2010) provides all the policy options that can be exercised by the Federal Trade Commission (FTC), in spite of many political and legal obstacles that are present. Particularly important is the evidence of deep capture by food industry lobbyists that prevents appropriate policy regarding obesity (Smith and Tasnadi, 2014).

Mello (2010) notes that the FTC can use its scope of authority under the legal arm of unfairness within the deception doctrine. Mello (2010) also presents strategies that the FTC can follow to boost self-regulation among industry and advertising agencies.[28]

POLICY ANGLE 2: USDA DIETARY GUIDELINES

Dharmasena et al. (2011) examine the value of the USDA guidelines for Americans on calorie intake, caffeine, and vitamin C, derived from nonalcoholic beverages. The researchers use a large panel data set and show that the dietary guidelines help in reducing calorific and nutrient intake

[28]Interestingly, Yu (2011) shows that peer effects work by fashioning parental communication styles, attitudes, and input on children. Hoy and Childers (2012) note that in recent years parent's attitude about "healthy snacks" has changed for the better, yet many parents of $6-11$ year olds note their family diet as "very healthy," even though they snack a lot. Further, parent's views on fast-food advertising and promotion influence children's attitudes toward obesity, and eventually increase the children's BMI. Also see Averett et al. (2013) for the importance of marriage in this aspect for Canadian data, where marriage is associated with a higher BMI, overweight, and obesity, and lower levels of physical exercise.

from nonalcoholic beverages. The above study is a good first step within the nonalcoholic beverages category to find some evidence that dietary guidelines may be helpful.

Other researchers have also examined the value of the Dietary Guidelines for America (DGA), and note that despite decades of these guidelines, Americans still eat a poor diet. Consequently, Duffy et al. (2012), and Knutson (2012) call for a more active multi-pronged approach.

Schuldt and Schwarz (2010) provide a very insightful angle to this aspect. They note that labeling a food as "organic" does not provide any information about its calorie content. Ironically, even when the calorie content of "organic" cookies is the same as "regular" cookies, consumers perceive that they can eat more of the former, and many of these same consumers are also highly pro-environment. Hence, not only are DGA ineffective, the may bias consumers inadvertently, in some cases.

POLICY ANGLE 3: WILL SODA-TAXES WORK?

Mellor (2011) shows that substitution effects can be somewhat counterintuitive when we start relating fat-taxes and soda-taxes to taxes on cigarettes. This is because Mellor (2011) shows that cigarette taxes increase BMI in children of smoking mothers. That is, higher cigarette costs may reduce smoking, but can increase expenses on food items, which are substitutes. Likewise a workplace ban on smoking also increases BMI (Liu et al., 2010), which is why workplace obesity preventions must be cleverly positioned (see Goetzel, 2011).

POLICY ANGLE 4: FAMILY ISSUES AND MOTHER'S ROLE

Family support and social support are very crucial to reach young boys and girls who are overweight or obese. The reason why this group really requires this kind of support-based intervention is because this group is more prone to use unhealthy weight-gain behaviors, such as skipping meals or even fasting, when compared to the actions of their healthy peers. Vander Wal (2012) observes that youngsters who had difficulty communicating with their parents, who had low levels of parent – school support, who were subject to bullying, or who had poor classmate support were more likely to engage in unhealthy weight control behaviors.

Anderson (2012) notes that as hours worked by the mother increase, several drivers of good eating habits, such as eating meals as a family or at regular times, tend to go down, and hours of television watching go up. Since many of these activities are also associated with obesity, there is a possibility that maternal employment may be an important driver of obesity. As Ben-Shalom (2010) and Miller (2011) uncover, maternal employment and childhood obesity are prevalent mostly in vulnerable families. Further, Classen (2010) documents the transmission of obesity across generations; if your parents are obese, then you are highly likely to be obese as well.[29]

Just as Averett et al. (2013) observed for Canada, Wilson (2012) notes that entry into marriage is associated with weight gain, and exit with weight loss. Wilson (2012) attributes this trend to pressures in the marriage market, along with gendered preferences regarding partner BMI.

[29]Results from experimental economics (Emke et al., 2012) show a significant relation between parent status, and parental generosity as main drivers of the obesogenic factors in the household that encourage childhood obesity.

POLICY ANGLE 5: EARLY SCREENING, MATERNAL ROLE, AND BREAST FEEDING

Early screening for potential obesity can help in reducing potential health costs. Yang et al. (2013) use US data on children's BMI, and find that adult hypertension and diabetes can be minimized through early screening among children and older adolescents.

Belfield and Kelly (2012) also note the importance of breast feeding in preventing childhood obesity. Further, breast feeding is also associated with higher motor scores.

Before concluding this section, we wish to explore the nonparametric methods in STATA to gain some insight into newer methods that are used by researchers in this field.

In the next section, we explore the same ideas through implementation in STATA. In particular, our STATA example below illustrates the importance of nonparametric regression in this context. While researchers have used latent variable models to determine the likelihood of being obese, the STATA exercise reveals that other nonparametric procedures can also be used to determine the underlying causality in the data, which the parametric models may sometimes fail to capture. The lowess procedure is also very popular among researchers, and the policy implication of the exercise is to show how data diagnostics are very important, so as to frame appropriate targets. The STATA exercise at the end of the chapter allows the students to explore this aspect in depth.

NONPARAMETRIC REGRESSION IN STATA: TESTING OBESITY AND FOOD INSECURITY

Consider the following data set on 100 children with the following information:

Obs	OBIFI		G	R	Obs	OBI	FI	G	R	Obs	OBI	FI	G	R
1	1	3	1	1	36	1	1	0	0	71	0	2	1	0
2	1	3	0	1	37	1	3	1	1	72	0	2	0	1
3	1	3	1	1	38	1	3	0	1	73	0	2	1	1
4	1	3	0	1	39	1	3	1	1	74	0	2	0	1
5	1	2	1	1	40	1	2	0	1	75	0	3	1	0
6	1	2	0	0	41	1	2	1	1	76	0	3	0	0
7	1	2	1	0	42	1	2	0	0	77	0	3	1	1
8	1	2	0	0	43	1	1	1	0	78	0	3	0	1
9	1	1	1	0	44	1	3	0	0	79	0	1	1	0
10	1	1	0	0	45	1	2	1	0	80	0	1	0	1
11	1	1	1	1	46	1	1	0	1	81	0	1	1	1
12	1	3	0	1	47	1	3	1	1	82	0	2	0	1
13	1	3	1	1	48	1	2	0	0	83	0	2	1	0
14	1	3	0	1	49	1	1	1	1	84	0	2	0	0
15	1	2	1	0	50	1	3	0	1	85	0	3	1	1
16	1	2	0	0	51	0	1	1	0	86	0	3	0	0
17	1	2	1	0	52	0	1	0	0	87	0	3	1	1
18	1	1	0	1	53	0	1	1	0	88	0	1	0	0
19	1	3	1	1	54	0	1	0	0	89	0	1	1	1

Continued

Obs	OBI	FI	G	R	Obs	OBI	FI	G	R	Obs	OBI	FI	G	R
20	1	2	0	1	55	0	1	1	0	90	0	2	0	0
21	1	1	1	0	56	0	1	0	1	91	0	2	1	1
22	1	3	0	0	57	0	2	1	1	92	0	3	0	0
23	1	2	1	1	58	0	2	0	1	93	0	3	1	1
24	1	1	0	1	59	0	2	1	1	94	0	1	0	0
25	1	3	1	0	60	0	2	0	1	95	0	1	1	1
26	1	3	0	1	61	0	2	1	0	96	0	1	0	0
27	1	3	1	1	62	0	3	0	0	97	0	2	1	1
28	1	3	0	0	63	0	3	1	0	98	0	2	0	0
29	1	3	1	1	64	0	3	0	0	99	0	2	1	1
30	1	2	0	1	65	0	3	1	1	100	0	3	0	0
31	1	2	1	1	66	0	3	0	1					
32	1	2	0	1	67	0	1	1	1					
33	1	2	1	1	68	0	1	0	1					
34	1	1	0	0	69	0	1	1	0					
35	1	1	1	0	70	0	1	0	0					

In the above table we have the following information:

OBI: A dummy variable that equals 1 if the child is obese, 0 if not.
FI: An ordinal variable reflecting food insecurity in the household (1 = not food insecure, 2 = moderate food insecurity, and 3 = food insecure).
G: A dummy variable that equals 1 if the child is a male, 0 if female.
R: A dummy variable that equals 1 if child is white, 0 if nonwhite.

We first estimate the likelihood function using a probit model in STATA:

```
. probit obi fi g r

Iteration 0:   log likelihood = -69.314718
Iteration 1:   log likelihood = -68.091602
Iteration 2:   log likelihood = -68.091332
Iteration 3:   log likelihood = -68.091332

Probit regression                               Number of obs   =        100
                                                LR chi2(3)      =       2.45
                                                Prob > chi2     =     0.4850
Log likelihood = -68.091332                     Pseudo R2       =     0.0176

------------------------------------------------------------------------------
         obi |      Coef.   Std. Err.      z    P>|z|     [95% Conf. Interval]
-------------+----------------------------------------------------------------
          fi |   .1908233    .160714     1.19   0.235    -.1241704     .505817
           g |  -.0237071   .2552404    -0.09   0.926    -.5239691     .476555
           r |   .1968171    .261163     0.75   0.451    -.3150529    .7086871
       _cons |  -.4871553   .3753415    -1.30   0.194    -1.222811    .2485005
------------------------------------------------------------------------------
```

With a standard probit model, we see that none of the variables is significant. We now generate the predicted probabilities using `predict` in STATA, and name the predicted value variable *probi*.

Based on Kuku et al. (2012, 2013) we apply the lowess procedure to derive a nonparametric regression. Using the predicted value we use the STATA command:

```
lowess probi fi, bwidth(0.5) xlabel(1(0.5)3) ylabel(0.3(0.05)0.6)
```

The lowess procedure is discussed in Fox (1997, p. 420). The procedure fits a nonparametric locally simple weighted regression, using different spans of the X values, and producing different polynomial regressions across each span. For details see Fox (1997), and also Gutierrez et al. (2003). The output from STATA produces the following lowess smoother:

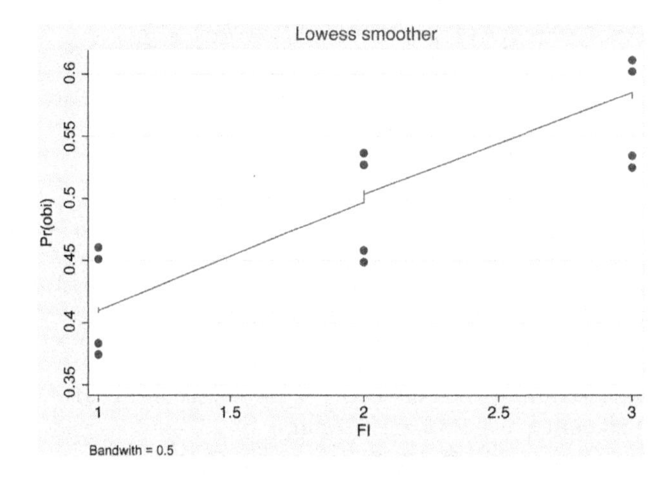

The smoothness indicates that the predicted probability of being obese increases with the level of food insecurity. We can produce more refined graphs through classifications for G and R (Problem 1).

CONCLUSIONS

The purpose of this chapter is to first come to terms with the problem of obesity in America, Europe, and also in other developing countries. While the problem of obesity and the corresponding health costs are evident, the actual causes of obesity are still very elusive. Consequently, it is very difficult for policy makers to produce targeted policies, such as soda-taxes, or fat-taxes, to combat obesity. In this chapter, we thematically discussed most of the issues ranging from diet, exercise, income effects, peer effects, fast-food, suburbia, etc., and listed the multi-faceted policy approach needed to tackle the issues.

Branca (2010) also produces a long list of policies supporting breast feeding programs in Europe.[30] Branca (2010) provides many types of intervention in Scandinavia that contributed to the high level of breast feeding, such as (1) problem-based information about breast feeding, written

[30]More detailed information about appropriate interventions is provided in Sassi (2010), Economos and Sliwa (2011), and Branca (2010). Also see Amin (2015) for improvements in SFPs in the United States.

mostly for and often by mothers, but read also by health workers; (2) how as a result of (1) more health workers also succeeded in their own breast feeding; (3) increased availability of mother-to-mother support groups, and health workers with better management skills; (4) an increase in paid maternity leave with a guaranteed return to previous employment; and (5) changing maternity ward practices to promote mother–infant contact and autonomy.

We end this section with a list of recommendations provided by Branca (2007) for Europe, which can be usefully replicated in any institutional setting (Box 14.4):

BOX 14.4 SCREENING AND PREVENTIVE MEASURES

- Develop and monitor effective physical activity interventions.
- Promote family involvement in weight control, weight maintenance, and weight-loss interventions.
- Follow-up studies in exercise-management.
- Special interventions for treating obesity in children.
- Advice on low-fat diets for obesity.
- Develop strategic health-professional management with reference to obesity.

Breast Feeding Measures and Healthy Diets

- Breast feeding promotion and infant growth.
- Baby-friendly hospitals' influence on breast feeding duration.
- Interventions for promoting the initiation of breast feeding.
- Inform mothers about formula milk versus term human milk for feeding preterm or low birth weight infants.
- Extending breast feeding duration through primary care.
- Consolidation and updating of the evidence base for the promotion of breast feeding.
- Effective interventions to promote healthy feeding in infants under one year of age.

In-School Policies

- Interventions for increasing fruit and vegetable consumption in preschool children.
- School-based interventions for primary prevention of cardiovascular disease.
- Guidelines for school health programs to promote lifelong healthy eating.
- Establishment of safe environments and opportunities for physical activity.
- Incorporate physical education to promote a physically active lifestyle.
- Health education curricula to support healthy eating.
- Parental involvement in instruction and support of physical activity.
- Adequate health services to assess and deliver physical activity.
- Community-outreach efforts to provide a range of sport and recreation programs.
- Other public health interventions.
- Workplace interventions for smoking cessation.
- Individual counseling for smoking cessation.
- Interventions for preventing tobacco sales to minors.
- School-based programs for preventing smoking.

EXERCISES

1. For the country of your choice, develop a typology of overweight and obesity and compare this with other malnutrition indicators that have been studied in Chapter 2. How does the emergence

of the overweight and obesity challenge the policy makers in their approach to malnutrition in general? Identify and discuss specific strategies and policies that are in place to address the problem of overweight and obesity in this country.

2. Use the data from the example and execute the following line:

```
lowess probi fi, bwidth(0.5) xlabel(1(0.5)3) ylabel(0.3(0.05)0.6), if g == 1
```

This produces a lowess scatter for the males in the sample. How does this result differ from the probit estimates?

PART

SPECIAL TOPICS IN NUTRITION POLICY G

CHAPTER

15 AGRICULTURE, NUTRITION, HEALTH: HOW TO BRING MULTIPLE SECTORS TO WORK ON NUTRITIONAL GOALS

We know that bio-fortification works. What is needed now is to build greater demand for bio-fortified crops within national nutrition programs. This will require addressing demand side constraints and policies that encourage the private sector to incorporate these nutritious crops in processed foods.

—Dr. Akinwumi A. Adesina, President of the African Development Bank, 2015. Address to the 3rd Annual Meeting of the Global Panel on Agriculture and Food Systems for Nutrition

INTRODUCTION

The linkages between agriculture, nutrition, and health have been on the development agenda recently, although how agriculture can contribute to nutritional outcomes is not new in development thinking (Pinstrup-Andersen, 2013). For example, the goal of the green revolution related to research and technology development, particularly in Asia, was to address the challenges of hunger many developing countries faced 50 years ago. This is a good example of addressing food insecurity and nutrition challenges through agricultural and crop improvements. While the green revolution technologies focused on macronutrients, such as calories and protein, the challenge of micronutrient deficiency or hidden hunger was grossly neglected by the green revolution approach to nutrition. This was one of the major criticisms of the introduction of high yielding rice and wheat crop varieties, which replaced other crops which had more micronutrients and resulted in reduced crop diversity, leading to a monoculture of rice and wheat.

To offset such unintended but negative consequences, scientists have embarked on setting priorities for crop improvements in international agricultural research, considering nutrition as one of its goals almost 40 years ago (Pinstrup-Andersen et al., 1976). Several attempts to understand the role of food system improvements for their contribution to better nutritional outcomes have been documented elsewhere (Kataki and Babu, 2002). Other efforts to identify opportunities for integrating nutritional goals in farming systems have also been studied: the introduction of animal production and livestock ownership (Leroy and Frangello, 2007; Azzari et al., 2015) in rice-aquaculture systems (Rajasekaran and Whiteford, 1993; Murshed-e-Jahan et al., 2010); in agroforestry systems (Babu and Rajasekaran, 1991a; Babu and Rhoe, 2002); in growing algal supplements through biotechnological advances (Babu and Rajasekaran, 1991b); and in identifying and promoting indigenous plant species (Babu, 2001). The nutritional implications of the commercialization of agriculture have also been studied (Kennedy and Von Braun, 1987; Martin and Von Braun, 1989).

Nutrition Economics. DOI: http://dx.doi.org/10.1016/B978-0-12-800878-2.00015-3

Leveraging agricultural systems to contribute to nutrition has gained revived interest in the wake of the food crisis of 2007 – 08 when the food supply at the global level failed to meet the demand for food, and food prices kept rising (Pinstrup-Andersen, 2013; Webb and Kennedy, 2014; McDermott et al., 2015). In addition to the nutritional contribution through increasing the nutrient content of crops, modifying agricultural systems is seen as a source of income and enhanced nutrition, and further, it can open up women's time spent in agriculture through technological choices. Agriculture and food production systems also affect nutrition through nutritional enhancement throughout the value chains, particularly in the context of high value crops. These opportunities, however, involve several sectors working together to achieve the nutritional goals.

In this chapter we look at selected agriculture-based intervention opportunities in the context of multi-sectoral nutrition policy making and programming. We review the literature on agriculture – nutrition–health linkages for their nutritional implications, and to identify new research areas.

CONCEPTUAL FRAMEWORK

The conceptual framework presented in Chapter 3, A Conceptual Framework for Investing in Nutrition: Issues, Challenges, and Analytical Approaches, could be reformulated to identify the theoretical pathways through which agricultural interventions can affect nutritional outcomes. Fig. 15.1 builds on an earlier effort to connect agricultural factors to the nutritional outcomes in the early 1990s (Babu and Mthindi, 1994), and on the recent development to highlight the role of women in nutrition, both as producers and providers (Herforth et al., 2012; Webb and Kennedy, 2014; Kadiyala et al., 2015). To begin with, the efficacy of agriculture-based interventions depend crucially on several broad contexts in which the agro – ecological system functions. For example, the political system and the policy making environment will determine if agriculture-based interventions could be implemented on a large scale. The agro – ecological conditions that allow growing of specific crops, and natural resource constraints such as the availability of land and water, can define the production systems and their ability to diversify and substitute one enterprise for the other. Related to this is the challenge of land use rights and water use rights of landless and share cropping households.

Socioeconomic factors such as the dominance of large-scale production systems surrounded by smallholders and landless laborers producing cash crops for the markets may not allow production of nutritious crops or diversification (Coates and Galante, 2015). In the context of policy, the price policies for inputs, such as fertilizer, and outputs will help or restrict diversification to nutritious crops. For example, in the case of India, the minimum support price given to rice and wheat—a food security oriented policy objective—severely affects the farmers' ability to diversify to other nutritious crops, such as fruit and vegetables (Jones and Moffit, 2015).

Market and price uncertainty, thus, can play a significant role in crop diversification toward nutritional objectives. When the markets function well for fruit and vegetables, farmers tend to produce for the market and may earn a better income, but whether this enhanced income results in

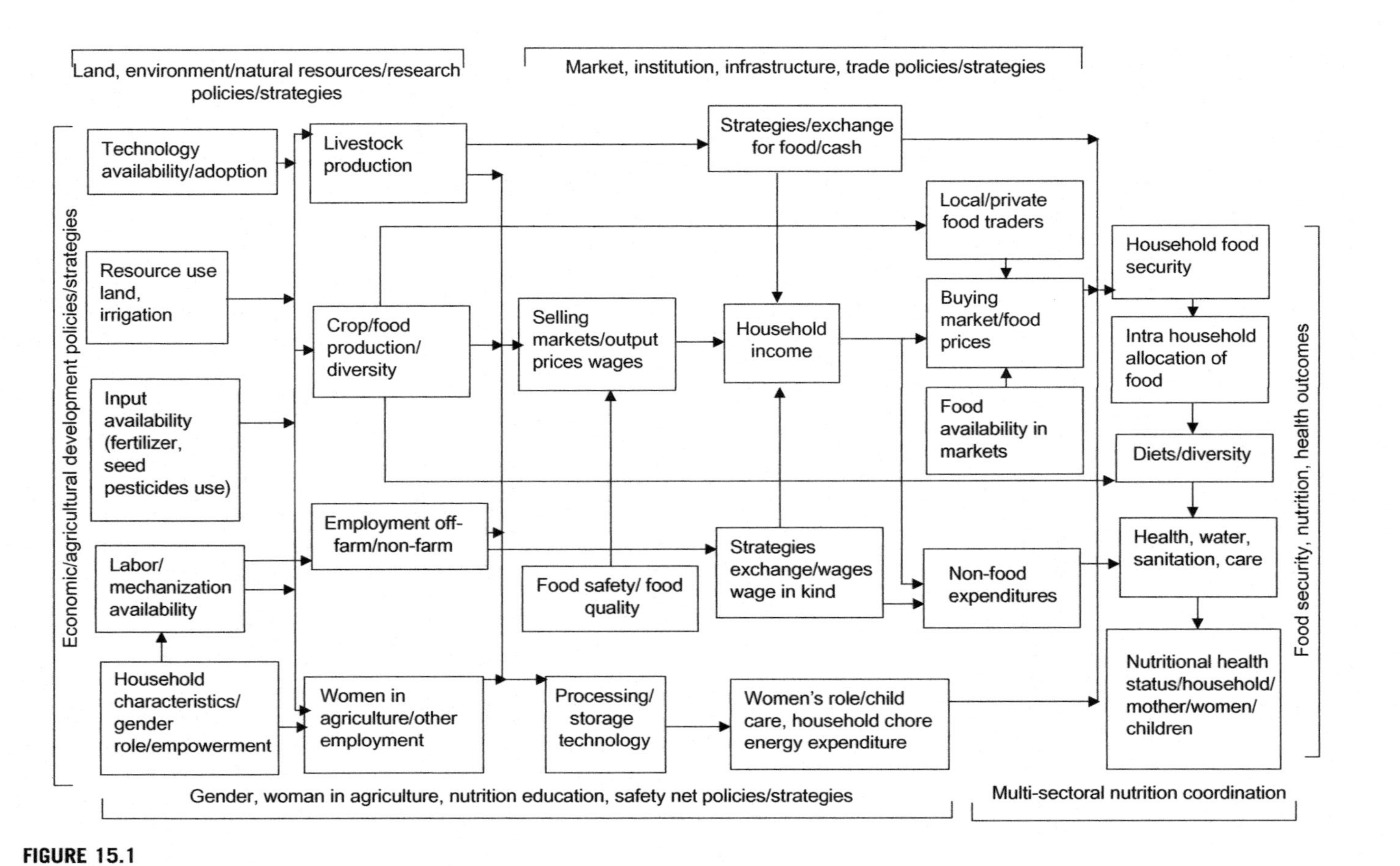

FIGURE 15.1

Agriculture – nutrition pathways: entry points for policy interventions.

Based on Babu, S.C., Mthindi, G.B., 1994. Household food security and nutrition monitoring: the Malawi approach to development planning and policy interventions. Food Policy 19 (3), 272–284; Webb and Kennedy (2014); Kadiyala et al. (2014).

improved nutritional outcomes is unclear, although studies have shown positive influences of such income increases (Murshed-e-Jahan et al., 2010). At the community level, the availability of land and the method of distribution among households will decide the success of agriculture-based nutrition interventions. For example, in the context of several African countries, the customary land tenure system does not allow long-term investments, such as allocation of land to aquaculture production.

Increasingly, agricultural sector growth is seen as one of the pathways to achieving the goal of sustainably reducing poverty, malnutrition, and hunger. Yet the goals of the agricultural sector, such as improving the productivity of food production systems, enabling better processing and storage at the local levels, identifying and expanding the markets both local and external to translate production into income, require investment in agricultural research (Webb, 2013), market infrastructure, and in local storage facilities. Recently, researchers have began to stress the development of value chains, and their potential for employment generation and poverty reduction (Pandya-Lorch et al., 2014).

From the nutritional perspective, the nutritional status of household members, particularly vulnerable groups, such as the women and children, depends on their improved access to higher quantities and qualities of food (Ruiz et al., 2015). In this context, dietary diversity plays an important role. Yet the farming systems relying on monocultures, such as the production of cereal crops, do not help much in improving the quality of diets. In addition, the consumption and use of nonfood factors such as clean water, sanitation, primary health care, and child care contribute to nutritional status. Access to and consumption of all these nonfood items are also influenced by the availability of income and resources, as well as the time and energy spent in obtaining them. This is particularly true for farming systems where women spend most to their time in the production of food and provision of nutrition (Balagamwala et al., 2015). Improving the investment in health and nutrition services, use of locally produced foods in the provision of nutritious diets, and improving nutritional behavior through education are key for nutrition to be transformed through agricultural systems (Webb and Block, 2012; Olney et al., 2015; Saaka and Larbi, 2015). Recently, food safety issues have been identified as part of providing quality food and nutrition (Leroy and Sununtnasuk, 2015). In the case of Africa, exposure to contaminants such as aflatoxin has been shown to affect the growth of infants (Turner et al., 2007).

The new paradigm that connects agriculture to nutrition focuses on the three major pathways—food production; income from agriculture and nonagricultural sources used to obtain diverse foods, along with nonfood inputs such as clean water, sanitation, and health care for all members of the family, particularly the women and children; and the gender role in agriculture, particularly how women are empowered to control their time and money given their varied roles as producer, processor, marketer, and provider of food and nutrition (McDermott et al., 2015).

Understanding the interlinkages between the agricultural factors and interventions and nutritional objectives, as depicted in Fig. 15.1, can help in the formulation of programs that are nutrition sensitive and nutrition driven. We explore this in the context of a multi-sector approach to agriculture–nutrition programming. The policies related to broad agricultural development strategies are given on the left side of the figure, as they influence several factors that affect agricultural production, marketing, and consumption. From these broad agricultural strategies, it is possible to specify key subsector policies that can influence key variables. For example, the top left corner of the figure identifies policies related to land, irrigation, natural resources such as soil and water, and issues related to the environment and climate change. These policies, along with policies related to labor and other inputs can influence cropping patterns and the crop–livestock mix.

Policies that are related to market, infrastructure, and output price are given at the right top corner of the figure. They help in translating agricultural production into a meaningful income for the farming families. Further, they enhance added-value through provision of market infrastructure, such as cold storage, for better processing and marketing of the commodities. They also help in improving the quality of food commodities through regulatory policies and procedures. Trade policies have a high level of influence in determining food availability at the macro level, and also help in managing the fluctuation of prices in the global markets. The safety net policies and policies that affect gender relations and the empowerment of women are given at the bottom left corner of the figure. Finally, the multi-sectoral nature of policy making and program implementation is given in the right bottom corner of the figure. Collectively, these policies help in influencing the causal factors responsible for attaining food security and nutrition at the household and individual level. Using the conceptual framework developed above, we describe several emerging approaches to addressing nutritional challenges throughout the agriculture–nutrition pathways.

DIETARY DIVERSITY

Dietary diversity has been recognized for its contribution to balanced nutrition for long time. Yet the initial efforts to reduce hunger and prevent famine in the context of agricultural research focused on the major cereal crops such as rice and wheat. While focused investments on increasing the productivity of these crops resulted in solving hunger problems, particularly in several countries in Asia, it also brought vast areas of land under these crops into the form of monocultures. It has also eroded biodiversity and taken land away from the crops that were contributing to diverse diets in these societies. While the real prices of the cereal crops such as rice and wheat have come down and made these more accessible to the poor, the cost of foods which contribute micronutrients have increased over the years, resulting in reduced dietary diversity. Given that more than two billion people are now afflicted with micronutrient malnutrition, there is a need to increase dietary diversity and the diversity of crops produced by rural households.

Several pathways exists to increase dietary diversity. To begin with, broad agricultural policies should take into consideration the need for diversified diets and their contribution to nutritional well-being. Policies that encourage monocultures of rice and wheat do not help in this context. Focused efforts are then needed to increase the diversity of crops, livestock, small ruminants, and aquaculture at the community level. Increasing the production of high value crops can improve the diversity of production at the farm level, but this does not guarantee the consumption of diverse diets. This will further require nutrition extension and nutrition education to guide farming households to expand the number of enterprises that can contribute to the increased availability of balanced nutrition at the household level.

The role of home gardens and growing fruit and vegetables also helps in increasing diversity. Nutrition interventions such as school feeding programs (discussed in chapter: Economics of School Nutrition: An Application of Regression Discontinuity) could be designed in such a way that the food given to the students is purchased locally, and contains diverse dietary nutrients. The social safety net programs such as conditional cash transfers can further help in increasing the diversity of diets by introducing locally produced foods (see chapter: Nutritional Implications of Social Protection: Application of Panel Data Method). Improving value chins for the high value commodities, and ensuring these are also consumed by the producing households, can increase the diversity of food consumption (we expand on this below). Increasing the nutrient content of the food already grown through bio-fortification can increase the diversity of the diets (see section Bio-Fortification).

Nutritional education is a key intervention for increasing dietary diversity and nutrient intake. As people are used to eating a small number of foods that are traditionally grown and eaten by them, breaking cultural eating patterns and taboos requires nutritional education that helps in the behavioral change toward eating a wide variety of foods. In this context, revising the university curriculum to prepare frontline professionals in agriculture, nutrition, and social work is important. The curriculum has to go from the traditional single discipline to multidisciplinary to cover nutrition, agriculture, social work, home economics, and health (Babu et al., 2016).

Finally, improving dietary diversity requires context specific interventions that take into account specific nutritional challenges faced by individuals and groups in the population. Agricultural interventions then need to take nutritional outcomes as explicit goals. Designing cropping patterns and the choice of crop varieties and technologies could then be analyzed for their nutritional contributions (see chapter: Designing a Decentralized Food System to Meet Nutrition Needs: An Optimization Approach for an example of such an approach). Designing such food production systems involves strengthening the capacity of the extension personnel. They must be trained in the monitoring and evaluation of the interventions to see how dietary diversity improves and contributes to improved nutritional status. This includes thorough knowledge about the indicators of dietary diversity and household access to food (Hoddinott and Yohannes, 2002). Increasing dietary diversity at the community level involves a multi-sectoral approach. Involvement of professionals from nutrition, agricultural extension, primary health care officials, and social welfare is required. This continues to be a challenge in developing countries and we discuss this further later in this chapter.

BIO-FORTIFICATION

Bouis (2016) summarizes the latest status of bio-fortification as an agriculture-based nutrition intervention. Since the beginning of the green revolution period the development community has been focused on the role of agriculture and food production in improving the nutritional status of the population, particularly in developing countries. The initial investment in productivity increasing technologies, such as high yielding varieties of seeds that were generated through rice and wheat breeding, have contributed to the Asian green revolution. The birth of the Consultative Group on International Agricultural Research (CGIAR), supported by the multilateral and bilateral donors, more than 40 years ago resulted in several commodity-specific breakthroughs using agriculture to increase macronutrients, such as calories and protein, at the household level. Research by internal agricultural research centers continues to address the productivity of crops, plant protection, soil and water management, and sustainability issues related to climate change and land and forest management. Yet, as we mentioned in Chapter 2, Global Nutrition Challenges and Targets: A Development and Policy Perspective, more than two billion people are suffering from one or more micronutrient deficiencies that are essential to maintain healthy living. This has led to the increased nutritional focus of the international agricultural research system. A good example is the program for bio-fortification of crops, particularly those consumed by the poor. Bio-fortification is a process of increasing the nutrient content of crops through breeding varieties that are rich in micronutrients (Bouis, 2016).

The process involves identifying crop varieties that contain high quantities of specific nutrients and minerals such as vitamin A or iron and through breeding, develop these varieties of crops that are consumed by a large number of people. Such crops include sweet potato, maize, beans, and millet. Several steps are involved in making bio-fortified crops available to meet the nutritional needs of the population (Boy, 2016). Confirmation is required that the crops and varieties from the breeding program have a sufficient amount of nutrients before they can be field tested. The varieties are then tested for their ability to retain the nutrient quality after different food processing methods commonly used by households. The next stage is to check that the consumption of these bio-fortified crops could add nutritional value, and reduce specific nutrient gaps by at least 25–50%. Further, the efficacy of the introduction of bio-fortified crops are tested to check if the specific micronutrient status is improved under controlled conditions. The bio-fortified crops are also tested for their effectiveness when introduced at a larger scale through market channels. The overall challenge in the context of the agricultural research involved in producing bio-fortified crops is that the yield levels cannot be sacrificed in the process of increasing micronutrient content (Lividini and Fiedler, 2015).

Bio-fortification has now become an accepted strategy for making agriculture nutrition-sensitive (Bouis, 2016). Initial evaluations have shown that bio-fortified crops do improve the nutritional intake of those who produce and consume them. Bio-fortified crops are grown in some 30 countries and production is expanding. Evaluation studies have shown the benefits of bio-fortified crops. Yet several policy and program challenges remain (Bouis, 2016). Investment is needed from the governments of the countries for agricultural research. Current efforts to breed crops for micronutrients depend on external assistance, and it is working. In most countries, however, the researchers do not have resources committed to continuing such research after the donor funded projects come to a close. The breeding programs have to maintain the germplasms and continue to invest in breeding crops to maintain the quality and vigor of the crop varieties. Such long-term commitment can only come from the mainstreaming of bio-fortification research in the national systems of agricultural research. The commercialization of the crops that have been bio-fortified requires regulatory mechanisms that can engage the private sector in the production and multiplication of seed varieties. The yield and micronutrient content tradeoff is still an issue at the farmer field level, particularly those who grow the crops for commercial purposes, as they do not get increased prices for growing bio-fortified crops. Understanding the pathways through which bio-fortication affects nutritional outcomes, and using this tool as one of several interventions to address nutritional challenges requires context specific and multidisciplinary approaches as discussed below (Bouis, 2016; http://www.securenutrition.org/blog-entry/financing-scale-nutritious-staple-food-crops#sthash.k7g4VUSP.dpuf).

NUTRITION VALUE CHAINS

As shown in the conceptual framework developed at the beginning of this chapter, one way to connect agricultural interventions to the nutritional status of the population is to increase the nutritive value of the commodity production chains. A recent review by Gelli et al. (2015) summarizes several issues, constraints, and challenges toward making value chains nutrition sensitive. High-value commodities such as fruit and vegetables, dairy products, milk, meat, and fish products, when produced by farming communities can help improve the nutritional content of the food consumed by the rural population. Thus, the interventions that introduce organized production of high-value commodities aim at the dual goals of increasing the value of agricultural production by farm households, and at the same time focus on improving the nutritional outcomes of the household members, particularly the women and children. Development of commodity value chains helps in producing nutrient rich commodities, making such commodities available to the households that produce them, and through the marketing process making them accessible to households that can afford to buy them in the markets.

Policy makers and program managers, however, face several policy and institutional constraints in making the traditional markets work for high value nutrient-rich commodities. For example, high value commodity market chains, when developed to meet food safety standards through investments in cold storage and other infrastructure, normally take the commodities away from the production site to urban markets. Then the challenge becomes the one, as in any intervention that introduces commercialization of agriculture—incomes generated through which may or may not result in better nutritional outcomes. Very little is known about how the development of the market chains directly or indirectly contributes to improvements in nutritional outcomes. They are trying to figure out which channel generates sufficient income to effectively impact nutrition becomes a challenge. Such policy dilemmas are common with any intervention that commercializes agriculture.

From the policy and program design perspective, it is important to understand how farming households make choices between selling what they produce to make additional income, and their consumption to increase nutritional status. Income from the sale of high value commodities is needed for nutrition and nonnutrition expenditures, but the choices households make toward nutrition could be influenced by the nutritional education and behavioral change education that will be essential to get the full benefit of using nutritive value chains as nutrition interventions.

CONNECTING THE SECTORS THROUGH MULTI-SECTOR PROGRAMMING

Enhancing the role of agriculture to meet nutritional objectives is the first step in the inter-sectoral approach to nutritional programming. In addition to agriculture, other sectors such as health, water, sanitation, and social welfare need to play their respective roles to meet nutritional objectives. While some level of coordination is possible at the planning and policy making level, the delivery of services that have nutritional objectives at the last mile level continues to be a challenge. A major question to policy makers and nutrition program managers is how one increases the nutritional impact by working with a group of professionals coming together from multiple sectors. While the agriculture sector has more proximity to the nutritional outcomes through the production of nutritious food and integrating nutritional objectives in the food system, other sectors such as water, sanitation, health, gender, and social welfare face more challenges in similar integration processes.

Innovations in the policy and planning of multi-sector nutritional planning helps in better integration of nutritional goals in various sectors, and keeps them accountable for the nutritional outcomes. Such a process starts with the highest level of policy processes. Directives from the president's office or parliamentary committees help in expressing their commitment and guiding the sectoral ministries to work together. Cross-sectoral integration has to be further followed up by vertical integration of goals and objectives through coordination and management of programs at the subregional and local levels. In each of the sectors, the stakeholders and key actors and players need to be sensitized about the nutritional objectives and outcomes. Consultations are needed among these players within each of the sectors, as well as joint meetings that bring together all the sectors for planning, setting priorities, designing implementation plans, monitoring and evaluation, and resource allocation. The development partners in each country have a key role to bring global knowledge to the table, as well as supporting such multi-sectoral approaches.

It should however, be noted that multi-sectoral approaches do not succeed without a high level of sensitization and ownership of the nutrition goals by each of the sectors. In addition, the consultative processes could consume a lot of time and may not be appreciated. The current governance structures may not fully support such multiple sectors coming together for the common cause. Even within ministries, the coordination of nutritional activities becomes difficult if appropriate leadership capacity does not exist at different levels. Major challenges also exist in terms of allocation of resources by each of the sectors and the donors who support specific nutrition interventions through a particular ministry. For example, when one development donor supports a school nutrition program through the ministry of education, and the other donor supports a similar school feeding program through the ministry of social welfare, the coordination of programs, monitoring of efforts, and resource mobilization for nutrition becomes challenging. Further, the accountability for the nutritional outcomes and monitoring and tracking the program benefits also becomes cumbersome for the nutritional policy makers at the national level (Table 15.1).

Table 15.1 Making Agriculture and Food Systems More Nutrition Sensitive
Incorporate explicit nutrition objectives and indicators into their design, and track and mitigate potential harms, while seeking synergies with economic, social, and environmental objectives. *This however, has to happen at national level policy making, providing a process for the sectoral ministries such as food, agriculture, rural development, gender, and social welfare to take nutrition seriously. Currently only a few leaders in a couple of ministries take such a multi-sectoral view about nutrition.*
Assess the context at the local level, to design appropriate activities to address the types and causes of malnutrition, including chronic or acute under-nutrition, vitamin and mineral deficiencies, and obesity and chronic disease. Context assessment can include potential food resources, agro-ecology, seasonality of production and income, access to productive resources such as land, market opportunities, and infrastructure, gender dynamics and roles, opportunities for collaboration with other sectors or programmes, and local priorities. *The challenge, however, is that such decentralized capacity hardly exists, even in countries where adequate capacity could be available at the national level. Local level context specific nutrition interventions require basic analytical capacity for identifying interventions, testing them for potential benefits, and scaling them up within the same agro-*

(Continued)

Table 15.1 Making Agriculture and Food Systems More Nutrition Sensitive *Continued*

ecological zones. Such capacity does not exist in most countries affected by nutritional challenges. Investment in such capacity would then be a first step.

Target the vulnerable and improve equity through participation, access to resources, and decent employment. Vulnerable groups include smallholders, women, youth, the landless, urban dwellers, and the unemployed. *Targeting in the context of agriculture and food systems where most of the farmers face similar challenges is easier said than done. Yet, identifying vulnerable groups and designing specific programs that would help them to effectively use agricultural programs to address nutrition would be helpful.*

Collaborate and coordinate with other sectors (health, environment, social protection, labor, water and sanitation, education, energy) and programmes, through joint strategies with common goals, to address concurrently the multiple underlying causes of malnutrition. *See the section on multi-sectoral coordination above. Such coordination requires leadership at all levels. Nutritional leadership is the most severe constraint in developing countries. Further multi-sectoral coordination requires well-articulated national directives.*

Maintain or improve the natural resource base (water, soil, air, climate, bio-diversity), critical to the livelihoods and resilience of vulnerable farmers and to sustainable food and nutrition security for all. Manage water resources in particular to reduce vector-borne illness, and to ensure sustainable, safe household water sources. *This again in itself is a multi-sectoral activity covering ministries of agriculture, land, water resources, health, and forestry to mention a few. Again this has to be coordinated at the landscape level* (Babu and Reidhead, 2000).

Empower women by ensuring access to productive resources, income opportunities, extension services and information, credit, labor and time-saving technologies (including energy and water services), and supporting their voice in household and farming decisions. Equitable opportunities to earn and learn should be compatible with safe pregnancy and young child feeding. *Agricultural policies and programs could specifically target women, giving incentives to participate in nutrition education and technology adoption for diet diversification.*

Facilitate production diversification, and increase production of nutrient-dense crops and small-scale livestock (e.g., horticultural products, legumes, livestock, and fish at a small scale, under-utilized crops, and bio-fortified crops). Diversified production systems are important to vulnerable producers to enable resilience to climate and price shocks, more diverse food consumption, reduction of seasonal food and income fluctuations, and greater and more gender-equitable income generation. *This will again require the subsectors within agriculture to come together and organize a concerted effort, keeping nutrition as a common goal. Departments of crops, livestock, horticulture, fisheries, marketing, irrigation, and mechanization for example, may function within agriculture, but coordination needs revamping approaches to research, extension, and technology dissemination. The Indian Extension model called the Agricultural Technology Management Agency is an effort in this direction* (Babu et al., 2015).

Improve processing, storage, and preservation to retain nutritional value, shelf-life, and food safety, to reduce seasonality of food insecurity and post-harvest losses, and to make healthy foods convenient to prepare. *Marketing infrastructure needs to be improved at all levels. Improvements in cold storage through solar electricity have shown some promise. Serious efforts are needed in this area. Connecting villages, towns, and urban centers through vegetable marketing can reduce the need for storage greatly.*

Expand markets and market access for vulnerable groups, particularly for marketing nutritious foods or products vulnerable groups have a comparative advantage in producing. This can include innovative promotion (such as marketing based on nutrient content), value addition, access to price information, and farmer associations. *This again requires capacity and infrastructure to collect, analyze, and disseminate real time marketing data for the producers and consumers to get connected and discover the right prices.*

Incorporate nutrition promotion and education around food and sustainable food systems that builds on existing local knowledge, attitudes, and practices. Nutrition knowledge can enhance the impact of production and income in rural households, especially important for women and young children, and can increase demand for nutritious foods in the general population. *While mass education through television and radio are the best approaches, personal communication with farmers by the farm home assistant and nutrition extension workers cannot be underestimated.*

From: FAO, 2015. Key Recommendations for Improving Nutrition through Agriculture and Food Systems. (Comments in italics added by the authors). *Available at: www.fao.org/3/a-i4922e.pdf.*

CONCLUSIONS

As the challenge of malnutrition continues to daunt the development community, the solutions for sustainable nutrition outcomes have to come from multidisciplinary approaches to solving the nutrition problems at all levels. Increase in income can enhance nutrition through provision of diverse foods through markets and trade, depending on a country's level of development. But interventions that facilitate the production of nutritious food and consumption require multiple sectors working together. Recently, the development community has begun its quest for sustainable diets that could be produced with minimal disturbance to the food production and natural resource systems that support it. Such an approach needs to recognize the role of various ecosystems, and protect biodiversity. The food consumed by societies needs to come from sources that do not unduly burden natural resources such as land, water, and landscapes. In addition, the resilience of the food production systems that are vulnerable to natural disasters and to the phenomenon of cultural change has to be proactively improved. Agricultural systems play a key role in the provision of sustainable diets.

EXERCISES

1. Consider your study country that you have been working on in several exercises. Develop an agriculture-based nutritional strategy for the country taking into account the policy system, development partners operation, their programs, and the key features that help the country to effectively use agriculture as a pathway to nutrition.
2. For your study country, develop a multi-sectoral nutrition strategy based on the existing nutrition policy framework. What linkages are currently missing in the design and implementation of a multi-sectoral nutrition strategy?

CHAPTER

DESIGNING A DECENTRALIZED FOOD SYSTEM TO MEET NUTRITION NEEDS: AN OPTIMIZATION APPROACH

16

The food system won't self-correct. We need more ambition, more innovation and more leadership to create a food system that delivers affordable, healthy diets to everyone in the world.

—Marc Van Ameringen, Executive Director, Global Alliance for Improved Nutrition (GAIN) in 2015 (www.Huffingtonpost.com)

Recently there has been much rhetoric about nutrition sensitive food systems (Pinstrup-Andersen, 2013). Yet, little documented evidence exists on how to modify existing food systems to meet the nutritional needs of the community. While few attempts have been made on a pilot basis to design nutrition-sensitive food systems, much less has been documented about their potential for scaling-up (FAO, 2015). Pilot intervention projects do not scale-up, partly because they do not get into the mainstream in public service delivery systems. As a result, many of the pilot interventions close down soon after the donor funding dries up. Knowledge on food system-based nutrition interventions needs to be made simple for decentralized decision-makers in the sectors responsible for nutrition to understand, assimilate, and modify by themselves with little or no help from the outside, or dependency on technical assistance (Glendenning et al., 2010; Babu et al., 2016). Innovations and technological developments linking agriculture and nutrition can bear fruit only when the delivery mechanisms are well equipped to reach out to the intended groups (Kawtrakul, 2012).

In this chapter, we look at how changing the components of a local food production system can help produce improvements in nutritional availability and intake. The question is how to design and implement nutritionally directed local food systems? As seen in the last chapter on agriculture and nutrition, explicit nutritional objectives; context-based local level solutions; targeting of vulnerable segments of the society; empowering women; facilitating diversification for dietary diversity, including growing nutrient dense crops; improved processing, storage, and marketing; local production use in social safety nets; and behavioral change through nutrition education are all important for addressing nutrition goals through local food system interventions (FAO, 2015).

In order to guide the farming communities to grow better crops to achieve nutritional goals, researchers need to understand the interlinkages between farming objectives, resource constraints, the quality of diets consumed, nutritional deficits of the farm families, and the agro-ecological conditions under which the crops are grown. Such understanding can be particularly useful for the chronically food deficient and nutritionally vulnerable regions of a country. Modeling the farm

Nutrition Economics. DOI: http://dx.doi.org/10.1016/B978-0-12-800878-2.00016-5

households for their nutritional needs and incorporating nutritional considerations in farm level recommendations for crop enterprises, technological innovations, and food processing can help in optimizing the meagre resources that resource poor farmers have, and can help in improving nutritional outcomes (Kataki and Babu, 2002). Further, effectively using such knowledge in the development of farm level advice and communication from the extension professionals and nutritionists could be an effective way to reduce malnutrition (Babu et al., 2016). We demonstrate one such practical approach in this chapter.

Recently, there has been high level recognition of the food system as a source of nutritional well-being, and leveraging agriculture for nutritional outcomes is seen as a priority intervention (McIntyre et al., 2001; Gillespie et al., 2012). Agricultural strategies with a focus on attaining household food and nutrition security have been among the prime goals of national development in developing countries for some time (Pinstrup-Andersen and Caicedo, 1978; Delgado, 1995; Pinstrup-Anderson, 2013). Low productivity resulting in reduced income, along with high dependency on limited land resources, continues to affect food and nutrition security outcomes (Cleaver and Schreiber, 1994; Leather and Foster, 2009; Babu et al., 2015). The result has been chronic low food security levels that contribute to high levels of under-nutrition (Webb and Kennedy, 2014). Further, food systems face external shocks, both natural and economic, that make them vulnerable (Babu and Blom, 2013). How can food systems contribute to sustainable nutritional outcomes? What role can agriculture play in easing nutritional security at the household and community levels?

As seen in Chapter 5, Macroeconomic Aspects of Nutrition Policy, macroeconomic policies play a crucial role in determining the sectoral policy outcomes. The agriculture – nutrition disconnect has been highlighted in several country case studies (Gillespie et al., 2012; Babu et al., 2016). They suggest the following: agriculture contributes to food availability, and through that pathway helps in increased nutrition at the household level. Agriculture can increase the income of farmers, which in turn can be used to access nutrition. Policies that help agriculture or distort incentives in agriculture thus can have a positive or negative influence on food and nonfood prices. Women's role in agriculture can have an impact on their nutritional status through intra-household allocation of resources, as seen in Chapter 9, Intrahousehold Allocation and Gender Bias in Nutrition: Application of Heckman Two-Step Procedure. Further, women's involvement in agriculture can take away the needed input to child care and feeding, thus affecting child nutrition. It can also affect the nutritional and health status of women themselves (Quisumbing and Meizen-Dick, 2012).

Gillespie et al. (2012) identify three major sets of challenges in leveraging agriculture to improve nutritional status: politics and governance; knowledge and evidence; and capacity and financial resources. Based on a set of country case studies conducted in East Africa, they conclude that in order to better leverage agriculture for improved nutritional outcomes, high-level coordination mechanisms are needed to facilitate multi-sectoral collaboration, and to design and implement nutrition sensitive agricultural programs and policies (Gillespie et al., 2013, in *Lancet*). These studies also show that there is insufficient evidence on how agriculture can enhance nutritional well-being.

A major challenge identified by these studies on the agriculture – nutrition disconnect is the capacity to collect timely data on nutrition and agriculture in order to develop potential interventions at the community and household levels. This has been a long-standing issue in the decentralized programming of food and nutritional interventions (Babu and Pinstrup-Andersen, 1994). Further, educating policy makers to design nutrition-sensitive agricultural interventions, enhancing the skills of extension workers to incorporate nutritional goals in their curriculum for training farmers, and strengthening nutrition field workers to develop programs that include agricultural and food system planning remain major challenges in translating agriculture and food system interventions into effective nutritional outcomes (Babu and Rhoe, 2002; Babu et al., 2016).

It is well-known that farm households that are not connected to markets continue to be vulnerable. Policies and programs affecting the subsistence nature of food and agricultural systems can affect their food and nutrition security (Abdulai and Delgado, 1995). Production-oriented policies affect the type of food produced, seasonal food availability, and the intake of such foods. This further affects their nutritional outcomes (von Braun et al., 1992; Babu et al., 1993). Thus, understanding the production systems, resources, and sustainability challenges in the context of achieving nutritional outcomes is essential for decentralized program development. Increasing household food security and nutrition thus requires an in-depth understanding of their production conditions, resource constraints, and physical and social infrastructure. Such understanding can help in designing programs that help in evolving appropriate research strategies, input provision, setting priorities for research, and designing farming systems (Cleaver and Schreiber, 1994).

In recent years there has been a revival of interest in connecting nutritional and health objectives to agricultural interventions (McDermott et al., 2015). Additionally, designing cropping and food systems for achieving self-sufficiency in food production and nutritional and health objectives at household, community, and national levels has also been discussed (Pinstrup-Andersen, 1985; World Bank, 1994; Babu, 2002b; Ruel and Alderman, 2013). Designing decentralized cropping systems depends on resource endowments, farming objectives, and other agro-ecological characteristics of the food systems (Babu and Rajasekaran, 1991a,b). Yet the policy and programs designed to leverage agriculture for nutritional objectives and to focus on the national level objectives often ignore the context and specific interventions that may vary among different agro-ecological regions in a country.

In this chapter, we show that with simple tools of analysis, field extension and nutrition professionals can design and implement decentralized policies that meet specific conditions and characteristics of agro-ecological regions. Such decentralization of approaches to rural development and nutrition interventions have been emphasized elsewhere (Staatz et al., 1990; Binswanger, 1994; Fiorella et al., 2016). The use of such tools can help in decentralized interventions based on food system modifications. With the advent of information and communication technologies, and the availability of personal computers at local levels of administration, approaches to modifying local food systems become much more practicable (Peterson, 1991; Gruber, 1995). Increasingly, the tools of ICT and knowledge portals have helped to reach out to the knowledge professional to design and implement intervention programs. Thus, with appropriate tools and skills development, new ICT methods could be effectively deployed for designing and implementing institutional interventions (De Silva et al., 2012; Walisadeera et al., 2014).

In what follows we develop a practical tool for designing interventions at the local food system level to meet the nutritional needs of the population. The chapter is based on Babu (1999), which presents a case study of one of the poorest rural subsistence farming systems in Africa to illustrate the model and its applications.

A PRACTICAL TOOL FOR DESIGNING DECENTRALIZED FOOD NUTRITION INTERVENTIONS

Computer-based tools are becoming common for guiding agricultural program development (Balachandran et al., 1989; Babu and Hassan, 1995; Walisadeera et al., 2013). Knowledge-based systems are ICT-based applications that can undertake analysis and research to reach conclusions about a challenge with an expert level equivalent or close to the human level. Once developed, the knowledge-based systems can be used by field level professionals with little or no help from the subject matter expert. Knowledge-based systems are successfully employed in decision-making about technology choices in agricultural production (McKinion and Lemmon, 1985, 1992). They have been effectively used in guiding farmers choices related to varietal selection, input use, crop mix, and farm plans that ensure food and nutrition security (Babu et al., 1990; Babu, 2009).

Fig. 16.1 describes the use of a knowledge-based system in making choices in the context of a food system related to nutritional outcomes (Babu and Rhoe, 2002). Farm households begin with the basic decision related to achieving food security through production or purchase, or a

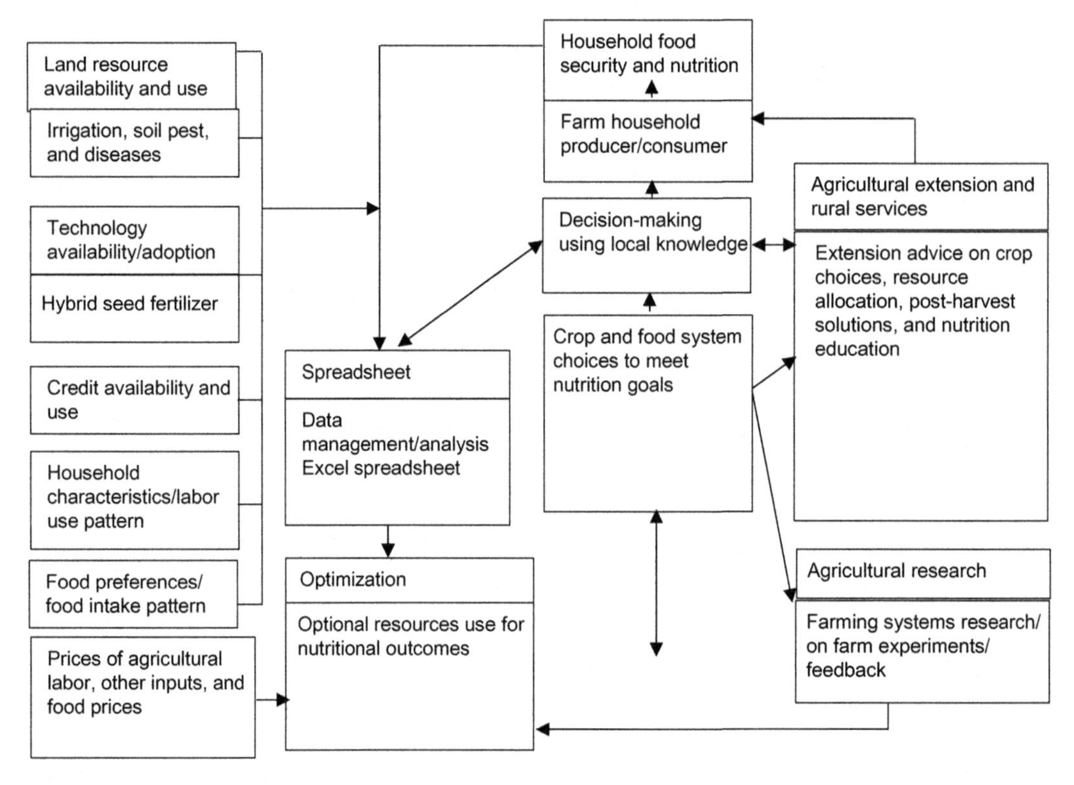

FIGURE 16.1

Designing food systems for nutritional outcomes.

Adapted Based on Babu (1999).

combination of both. Such decisions have implications for the use of farm level resources, such as land and labor allocation to farming activities. For example, landless labor households depend primarily on selling their labor to other farming households to earn income or food in kind. Further, the level of income thus obtained determines their nutritional status, based on the consumption pattern of the household members. For households that solely depend on their land to produce their food, the quantity of food produced depends on the choice of technology, availability of inputs to produce the food, and the productivity of the land. In addition, the availability of water for irrigation, timely and affordable seeds of crop varieties, and the accessibility of credit to obtain inputs, in turn affects the ability of the farmers to use improved seed varieties and chemical fertilizers. Data collected on these characteristics of farm households, as well as the food system in general, are captured using farm level surveys that are available in most countries.

The knowledge and evidence developed by researchers on the contribution of various inputs to the productivity of crop and livestock enterprises provides the database for the decision-making process in a knowledge-based system. In addition, data on resource availability; water use efficiency; soil fertility; weather conditions during crop growth including temperature, humidity, and rainfall pattern; the occurrence of diseases and pests; and weed incidence help in organizing the decision systems. Further, external factors such as the availability of markets for the producers, and prices of inputs and outputs, help the farmers to decide the land allocation to various agricultural production enterprises. In the context of leveraging agriculture and food systems for nutritional outcomes, food security and nutritional requirements need to be considered along with the information needs discussed above.

Once the database for decision-making is assembled, it can be used in the farm level optimization framework to develop decisions and recommendations related to farming systems crop and livestock enterprise choices, levels of input use related to fertilizer and water, technology choices related to mechanization and labor use, resource conservation decisions, and choice and timing of the marketing of the produce (Rajasekaran et al., 1995). The results of the optimization models could be used by the agricultural extension and nutrition professionals to recommend the appropriate and farm specific choices farmers have to make to achieve their nutritional goals. Further, feedback from the farmers after executing the recommendations could be incorporated to update the results of the optimization models.

The knowledge-based system we describe below has two components: the knowledge base and the influence engine. The knowledge-base, as described above, involves information from the farmers, the decision-making rules, and additional experience of the farm households in solving a specific problem. The influence engine, on the other hand, uses logical deductions and manipulations to interpret the rules of the knowledge-base to provide specific recommendations depending on the information or situation provided by the farmers (Liebenow, 1987).

In the context of leveraging agriculture for nutrition, and in designing food system-based decentralized decision-making for nutritional outcomes, information about the following is useful: the nature of farming systems, the food and nutritional needs of the family, food and nonfood consumption patterns, seasonality of food availability in the region, the prices of various food commodities and their substitutes, and the solutions to the farm household optimization model. In addition, the experience of the farmers and the experts after implementing specific recommendations from the knowledge base are also useful (Babu, 2000).This knowledge-base is stored in the form of conditional statements. A conditional statement specifies an action that will result from meeting a set of conditions. Below we demonstrate using the case study of Malawian smallholding farmers, how a knowledge-based system could be used to develop recommendations for decentralized planning and programming food system-based nutritional interventions (Babu et al., 2014).

Fig. 16.1 presents a schematic view of various inputs that go into the designing of food systems for nutritional outcomes. Food and nutrition security is determined by the household's ability to generate income, either through food/crop production or through directly consuming the food they produce on their farms. In the context of food production, resource availability including land, labor, and capital for purchased inputs, along with the use of improved technology and the availability of irrigation, determine food availability at the household level. Collecting farm level data on these characteristics helps in analyzing various options farm households have in meeting food and nutritional requirements. The optimization models can help in guiding the choice of resources to meet the nutritional outcomes.

Food production and availability of food at the household level can be improved through the adoption of advanced technologies; such technological advances are available through the national agricultural research systems and are disseminated by extension workers. The technological information and the opportunities for increasing yields and for using farm level resources in a sustainable manner can be obtained through the research extension linkages. Such information could be used along with the resource constraints and the nutritional goals of the households to generate technological options available for farmers at the local level (Rajasekaran et al., 1995). Agricultural extension systems can then use such options in their extension messages (Babu et al., 2015).

As a practical tool, the computer-based recommendations use the solutions of the optimization model. The knowledge-base with a series of solutions for a range of possibilities of farm level conditions can then be used as the basis for designing a rule-based decision system. The knowledge-base is then driven by the inference engine which will interpret the decision rules and provide solutions for extensions based on logical deductions (Liebowitz, 1996). The information on farm level characteristics becomes the starting knowledge-base, and the outputs of the optimization becomes the knowledge-base for advising the farmers to achieve nutritional goals.

APPLICATION OF THE MODEL TO THE FOOD SYSTEM IN MALAWI

Malawi continues to be a rural-based economy, with more than 80% of the population living in rural areas. A majority of the poor live in rural regions unconnected to markets, and depend on subsistence farming to meet nutritional goals. Agriculture continues to contribute to more than 30% of the gross domestic product, and more than 60% of the value of national exports comes from agriculture. Maize, cassava, and beans are the major food crops. Smallholders also grow other cash crops, such as groundnuts, tobacco, cotton, and rice.

The landholding size continues to go down with the increase in population. More than 50% of the farming households cultivate less than 1.0 hectare (2.4 acres) of land, and depend on the food produced from such land as their major source of diet and nutrition. Population pressure on the land and the low productivity of the agricultural systems combine to limit household access to food and nutrition. The farming systems in many other African and Asian countries are similar. Meeting the nutritional needs of such farming households who form the majority of the poor, malnourished, and vulnerable population in developing countries requires that they grow the right crops, and in the right mixture to meet their nutritional goals. Although the Malawian farming system once supported subsistence agriculture, and made most farmers self-sufficient in maize, the major staple food (Liebenow et al., 1989; Babu et al., 2014), a major food crisis from climatic and price shocks affected the food and nutrition security of the population in the 1990s and 2000s. However, through high-level investment in fertilizer subsidies, Malawi attained food self-sufficiency in the mid-2000s, and was able to export maize to neighboring countries (Sofranko and Fliegel, 1989; Chirwa, 2005). However, recurring climate shocks continue to keep the majority of smallholders vulnerable to nutritional insecurity (Babu and Chapasuka, 1996).

In addition, the marketing and trade policies over the years, in the name of liberalization and structural adjustment, have reduced the farm households' access to organized markets. In this context, food and nutrition security largely depends on designing innovative interventions for the food system that can ensure optimum nutritional outcomes. Local level intervention programs following national policy directives have higher chances of yielding results. However, such interventions have to take into consideration the local agro-ecology, the resource constraints of the farmers, the nature of training objectives such as subsistence, semi-subsistence, and market-oriented farming, tastes and cultural food preferences, and market access for the output produced. The model presented below allows the design of such locality specific interventions for the output produced (Babu and Khaila, 1996).

THE OPTIMIZATION PROCEDURE

Designing policies and strategies for agricultural development requires a national level approach to align the sector level objectives to national development goals. Yet, in implementing these policies at decentralized levels, context specific approaches are needed. In this section we develop a farm level model to identify farm enterprises that can achieve nutritional objectives in a specific food system. Such models have been used successfully to guide policy makers, and to bring together extension and nutrition professionals to increase the nutritional impact of food system interventions (Calkins, 1981; Babu and Hallam, 1989; Babu and Rajasekaran, 1991a).

As mentioned above, agro-ecological and farm level information form the basis for analysis leading to decentralized nutrition solutions for farming systems. Modeling choices depend on the policy options to be analyzed. A farm level linear programming model presented below can provide information needed by local level professionals and guide them in designing an appropriate crop-mix for the farm households to achieve nutritional goals. Various studies used linear programming for increasing nutritional outcomes by least cost rations (Calkins, 1981); by policies for household consumption pattern changes (Babu, Hallam, and Rajasekaran, 1990); by understanding the role of taste on food intake (Silberberg, 1996); by evaluating the implications of new food introduction (Babu and Rajasekaran, 1991b); and by increasing the quality of food (Weiske, 1981).

Babu (1999) and Babu et al. (2014) presented a simple optimization model for identifying the best crop combinations:

Maximize $J = C'X$

Subject to the following constraints:

$AX \leq B;$
$DX \geq N;$
and $X \geq O$
where X is a vector of potential crop choices,
C is a vector of coefficients of net return of crops,
A is a matrix of input requirement coefficients,
B is a vector of resources and input availability,
D is a matrix of nutrient content of crop activities,
N is a vector of food security and nutrition requirements, and
$X \geq 0$ is a nonnegativity constraint also encompassing the food security and nutrition considerations of the household.

In the context of developing practical recommendations for the farmers to meet their nutritional goals, higher level modeling may not be fully understood by the nutrient and agricultural development professional with basic designs in their own fields. A significant share of the rural households in Malawi are female headed households, and the model can accommodate such demographic characteristics. The food produced in one season is stored for the rest of the year, and this is incorporated into the model by considering the annual needs of the nutrition requirements over the years. Applied in the context of various types of farmers, the model allows for both selling and consumption activities, and still aims to meet the nutritional objectives. The model considers varying requirements of nutrition for different gender and age groups who may have different patterns of energy spending, based on their occupation and levels of activity.

The household and cultural preferences for various food types and tastes can be incorporated as part of the constraints set. In general, farmers will maximize their net farm income over consumption and nutritional needs. The farm level data for the modeling exercise comes from various sources. The input/output coefficients come from farm household surveys (MOALD, 1984). Annual surveys of agriculture in Malawi provided the data on crop yields. The market prices of inputs and outputs were gathered through various rounds of data collection (Govindan and Babu, 1996).

The crop planning model was solved for different categories of farm households to reflect both subsistence and market-oriented farming systems. The nutritional needs of the farm households are met solely by production of food crops in a subsistence scenario. In the market-oriented scenario, farmers maximize the net farm income. Tables 16.1 and 16.2 summarize the solutions of the optimization problem for the above two scenarios, respectively. They provided information on the solutions related to three farm sizes in terms of dietary preferences and technology needs for the farm households. The set of 72 solutions that were generated using various combinations helped decision-makers to guide farming systems and good system development to meet the nutritional needs of the local areas.

The decision-making system called FOODEXPERT is then used to develop a decision tree, as shown in Fig. 16.2. This decision provides recommendations on the optimal crop plans for different food systems characterized by agro-ecological regions. The optimization process is as follows. The agro-ecological region is identified first. Then the farmer type is chosen, based on nutritional and food security needs. Based on whether the farmer is a subsistence farmer or a market-oriented farmer, they are further classified according to their farm size. Their technological preferences and adoption in terms of hybrid maize seed use and chemical fertilizer use further stratifies them.

Table 16.1 Summary of Optimal Solutions for Choosing Crop Mixes for Food Security and Nutrition Planning (Subsistence Farming)

Farm Type	Farm Size	Food Preference (Sample Staple)	Technology Choice		Maximum Profit (K)	Optimal Farm Plan	
			Hybrid (a) Local & Hybrid (b)	Fertilizer (1) Org. Manure (2)		Crop Mix Area/Allocated (HA)	
Subsistence Farming	Small farm (≤0.75 ha)	Maize		1	315.29	HM,B	0,39,0,36
				2	286.17	HM,B	0,52,0,23
					296.37	LM,HM,B	0,28,033,0,14
					273.82	LM,HM,B	0,32,0,25,0,18
		Cassava	a	1	366.68	CA,B	0,47,0,29
				2	316.14	CA,B	0,39,0,36
			b	1	366.69	CA,B	0,47,0,29
				2	316.14	CA,B	0.31.0.44
		Maize and Cassava	a	1	376.74	HM,CA,B	0,11,0,29,0,26
				2	383.65	HM,CA,B	0,19,0,26,0,36
			b	1	383.65	HM,CA,B	0,16,0,28,0,31
				2	396.43	HM,CA,B	0,12,0,30,0,33
	Medium farm (≥75$ ≤1.50 ha)		a	1	977.80	HM,LM,B	0,94,0,26,0,40
				2	945.06	HM,LM,B	0,85,0,30,0,35
			b	1	988.08	HM,LM,B	0,70,0,21,0,59
				2	985.04	HM,LM,B	0,79,0,18,0,56
		Cassava	a	1	1108.46	CA,B,GN	0,47,0,36,0,72
				2	1106.08	CA,B,GN	0,46,0,32,0,72
			b	1	1197.81	CA,B,GN	0,40,0,30,0,80
				2	1086.43	CA,B,GN	0,43,0,38,0,69
		Maize and Cassava	a	1	1206.67	HM,CA,B,GN	0,33,0,24,0,21,0,72
				2	1199.72	HM,CA,B,GN	0,38,0,26,0,23,0,63
			b	1	1206.67	HM,CA,B,GN	0,33,0,24,0,21,0,72
				2	1199.72	HM,CA,B,GN	0,38,0,26,0,23,0,63

(*Continued*)

Table 16.1 Summary of Optimal Solutions for Choosing Crop Mixes for Food Security and Nutrition Planning (Subsistence Farming) ***Continued***

Farm Type	Farm Size	Food Preference (Sample Staple)	Technology Choice		Maximum Profit (K)	Optimal Farm Plan	
			Hybrid (a) Local & Hybrid (b)	Fertilizer (1) Org. Manure (2)		Crop Mix Area/Allocated (HA)	
	Large farm (>1.5 ha)	Maize	a	1	1841.45	HM,B,CA,GN	0,88,0,29,0,41,1,42
				2	1632.78	HM,B,CA,GN	0,76,0,31,0,86,1,07
			b	1	1743.66	HM,B,CA,GN	0,76,0,39,0,66,1,20
				2	1521.46	HM,B,CA,GN	0,66,0,47,0,83,1,04
		Cassava	a	1	1816.40	CA,B,GN	0,81,0,43,1,76
				2	1705.06	CA,B,GN	0,93,0,51,1,56
			b	1	2018.43	CA,B,GN	0,56,0,55,1,87
				2	1917.15	CA,B,GN	0,61,0,61,1,78
		Maize and Cassava	a	1	2078.19	HM,CM,CA,GN	0,42,0,45,0,24,1,99
				2	1996.04	HM,CM,CA,GN	0,48,0,41,0,29,1,82
			b	1	2066.52	HM,CM,CA,GN	0,46,0,50,0,21,1,83
				2	1997.06	HM,CM,CA,GN	0,47,0,42,0,34,1,77

HM, *Hybrid Maize;* CA, *Cassava;* B, *Beans;* GN, *Groundnut.*
From: *Babu (1999).*

Table 16.2 Summary of Optimal Solutions for Choosing Crop Mixes for Food and Nutrition Planning (Market-Oriented Farming)

Farm Type	Farm Size	Food Preference (Sample Staple)	Technology Choice Maize Fertilizer Variety Use		Maximum Profit (K)	Optimal Farm Plan	
			Hybrid (a) Local & Hybrid (b)	Fertilizer (1) Org. Manure (2)		Crop Mix Area/Allocated (HA)	
Market-oriented farming	Small farm (≤0.75 ha)	Maize		1	502.86	HM,B,GN	0.28,0.36,0.11
				2	481.08	HM,B,GN	0.26,0.31,0.19
					497.88	HM,B,GN	0.26,0.28,0.21
					463.78	HM,B,GN	0.29,0.38,0.09
		Cassava	a	1	535.38	CA,B,GN	0.16
				2	510.03	CA,B,GN	0.32,0.26,0.17
			b	1	490.29	CA,B,GN	0.33,0.29,0.41
				2	462.81	CA,B,GN	0.31,0.31,0.13
		Maize and Cassava	a	1	585.46	CA,HM,B,GN	0.21,0.20,0.12,0.22
				2	548.64	CA,HM,B,GN	0.25,0.18,0.16,0.16
			b	1	563.52	CA,HM,B,GN	0.23,0.18,0.19,0.16
				2	530.14	CA,HM,B,GN	0.27,0.11,0.16,0.21
	Medium farm (≥75 & ≤ 1.50 ha)	Maize	a	1	1041.51	HM,B,GN	0,56,0.36,0.59
				2	910.50	HM,B,GN	0.66,0.31,0.53
			b	1	1041.51	HM,B,GN	0.56,0.36,0.59
				2	910.50	HM,B,GN	0.66,0.31,0.53
		Cassava	a	1	1107.25	HM,CA,B,GN	0.47,0.36,0.67
				2	1032.78	HM,CA,B,GN	0.33,0.24,0.36,0.57
			b	1	1107.25	HM,CA,B,GN	0.47,0.36,0.67
				2	1032.78	HM,CA,B,GN	0.33,0.24,0.36,0.57
		Maize and Cassava	a	1	1075.00	HM,CA,B,GN	0.28,0.24,0.35,0.63
				2	1075.00	HM,CA,B,GN	0.28,0.24,0.35,0.63
			b	1	1075.00	HM,CA,B,GN	0.28,0.24,0.35,0.63
				2	1075.00	HM,CA,B,GN	0.28,0.24,0.35,0.63

(*Continued*)

Table 16.2 Summary of Optimal Solutions for Choosing Crop Mixes for Food and Nutrition Planning (Market-Oriented Farming) ***Continued***

Farm Type	Farm Size	Food Preference (Sample Staple)	Technology Choice Maize Fertilizer Variety Use		Maximum Profit (K)	Optimal Farm Plan	
			Hybrid (a) Local & Hybrid (b)	Fertilizer (1) Org. Manure (2)		Crop Mix	Area/Allocated (HA)
	Large farm (> 1.5 ha)	Maize	a	1	2494.24	HM,B,GN	0.28.0.94,1.78
				2	2321.48	HM,B,GN	0.36,0.89,1.75
			b	1	2494.24	HM,B,GN	0.28.0.94,1.78
				2	2321.48	HM,B,GN	0.36,0.89,1.75
		Cassava	a	1	2537.35	CA,B,GN	0.47,0.67,1.86
				2	2537.35	CA,B,GN	0.47,0.67,1.86
			b	1	2537.35	CA,B,GN	0.47,0.67,1.86
				2	2537.35	CA,B,GN	0.47,0.67,1.86
		Maize and Cassava	a	1	2545.93	HM,CA,B,GN	0.28,0.23,0.65,1.83
				2	2545.93	HM,CA,B,GN	0.28,0.23,0.65,1.83
			b	1	2545.93	HM,CA,B,GN	0.28,0.23,0.65,1.83
				2	2545.93	HM,CA,B,GN	0.28,0.23,0.65,1.83

HM, *Hybrid Maize;* CA, *Cassava;* B, *Beans;* GN, *Groundnut.*
From: *Babu (1999).*

Ecological Region	Farmer Type (Goals)	Farm Size	Maize Variety	Fertilizer Use	Food Preference (Staple food)	- Crop Mixes Maize+Beans
Region 1	Subsistence Farming	Small	Hybrid	Chemical Fertilizer	Maize	- Maize+Cassava+Bean - Bean+Groundnuts
Region 2		Medium	Composite		Cassava	- Groundnuts only - Cassava+Groundnuts - Cassava+Maize only
Region 3	Market-oriented Farming	Large	Local	Organic Manicure	Maize and Cassava	- Maize only

FIGURE 16.2

Decision tree for choosing crop mix for food security and nutrition planning at farm level.

Based on Babu (1999).

Finally, based on their food preferences for staple foods, maize, cassava, or a combination of both, the crop mix they should adopt is recommended. The farm level knowledge-based system program is described in Fig. 16.3. The program has three major components. The input from the farmer is recorded in the "user interactive section," and the recommendations are shown in this section. In the decision system section (component 2), the raw information from the user is coded for developing the rules of decision-making. The farm types are coded with numerical values which are then used in decision-making rules.

Section 3 selected the crop choices as decisions based on the forward chain procedure in knowledge-based system programming. The program selects the optimal mix of crops to be grown by the farmer given the inputs on various characteristics such as farm size, nutritional needs, food preferences, and adoption of modern technology by the farmers. In section 1, as shown in Fig. 16.3, when the inputs are shown as: Region is 1; number of children is 1; size of farm is 3; soil type is 2; number of pairs of oxen owned is 2; the crop growing season is 2; the farm type is 1; the preference for staple food is 1; the variety of crop chosen is 2; and the choice of fertilizer is 2; then the recommendation for the farm household is to "grow maize and beans in an area of 0.5 ha and 0.25 ha, respectively."

Based on such feedback on recommendations, the agricultural extension and nutrition professionals are able to recommend the types of crops to be grown by farmers, and the allocation of land to these crops to meet the nutritional goals. The knowledge-base will have to be changed based on the changes in the parameters of the model by incorporating them as other parameters which can also come from research institutions that develop new technologies from various locational trials that they conduct to test their technologies. Modifying these parameters based on locality-specific information can enhance the realistic nature of the recommendations.

The program described above can also help in incorporating field experimental information conducted at local levels by the researchers, and can improve the optimal solutions at the decentralized levels.

Section 1

User Interaction and Recommendations

ALTP	To Start	
REGION	1	RECOMENDATION
ADULT MALE:	1	Crops: grow maize and beans
ADULT FEMALE:	1	Area: 0.5ha and 0.25ha
CHILDREN:	1	
FARM SIZE:	3	
SOIL TYPE:	2	
OXEN:	2	
SEASON:	2	
FARM TYPE:	1	
STAPLE FOOD:	1	
VARIETY:	2	
FERTLIZER USE:	2	

```
/wgra/agra
(goto)B5~/xiENTER THE REGION:~~
(goto)B6~/xnENTER THE NUMBER OF ADULT MALES:~~
(goto)B7~/ xnENTER THE NUMBER OF ADULT FEMALES:~~
(goto)B8~/ xnENTER THE NUMBER OF CHILDREN:~~

(goto)B9~/ xnENTER THE SIZE OF THE FARM IN ACRES:~~

(goto)B10~/xiENTER THE TYPE OF SOIL:~~

(goto)B11~/xiENTER THE NUMBER OF OXEN IN PAIRS:~~

(goto)B12~/xiENTER THE SEASON:~~
(goto)B13~/xiENTER THE FARM TYPE 1=SUBSISTANCE 2=MARKET:~~
/x1FARMTYPE=~1~~~
/xgi
/x1FARMTYPE=~2~~~
/xgi28~
```

Section 2

The Decision System

DECISION: USE YOUR OWN OXEN
RENT A PAIR OF OXEN
RENT A PAIR OF OXEN AND HIRE SOME LABOR

HIRE LABOR (MAN DAYS)

GROW MAIZE ONLY
GROW MAIZE AND BEANS
GROW MAIZE AND GROUNDNUTS
GROW GROUNDNUTS AND BEANS
GROW GROUNDNUTS ONLY

```
/xiFARMSIZE<=3#AND#OXEN>2~/CC45~F6~/CC47~F7~(goto)A3~/xq
/xiFARMSIZE<=3#AND#OXEN=2~/CC45~F6~/CC47~F7~/xq
/xiFARMSIZE<=3#AND#OXEN=1~/CC41~F6~/CC47~F7~/xq
/xiFARMSIZE<=3#AND#OXEN=0~/CC45~F5~/CC46~F7~/xq
/xiFARMSIZE<=3#AND#FARMSIZE<=7#AND#ODEN>2~/CC44~F5~/CC4
7~F7~/Xq
/xiFARMSIZE<=3#AND#FARMSIZE<=7#AND#ODEN=2~/CC44~F6~/CC4
7~F7~/Xq
/xiFARMSIZE<=3#AND#FARMSIZE<=7#AND#ODEN=1~/CC43~F6~/CC4
7~F7~/Xq
/xiFARMSIZE<=3#AND#FARMSIZE<=7#AND#ODEN00~/CC43~F6~/CC4
8~F7~/Xq
/xiFARMSIZE<=7#AND#OXEN>2~/CC44~F5~/CC49~F7~/xq
/xiFARMSIZE<=7#AND#OXEN=2~/CC43~F5~/CC48~F7~/xq
/xiFARMSIZE<=7#AND#OXEN=1~/CC43~F5~/CC48~F7~/xq
/xiFARMSIZE<=7#AND#OXEN=0~/CC43~F5~/CC47~F7~/xq
```

Section 3

Selection of Crops Choice using Forward Chain Procedure

```
/xiFARMSIZE<=3#and OXEN>2~/CC45~F6~/CC47~F7~(goto)A3~/xq
/xiFARMSIZE<=3#and OXEN>2~/CC45~F6~/CC47~F7~/xq
/xiFARMSIZE<=3#AND#OXEN>1~/CC41~F6~/CC47~F7~/xq
/xiFARMSIZE<=3#AND#OXEN>0~/CC45~F6~/CC46~F7~/xq
/xiFARMSIZE<=3#AND#OXEN>0~/CC45~F5~/CC46~F7~/xq
/xiFARMSIZE<=3#and#FARMSIZE<=7#AND#ODEN>2~/CC44~F5~/CC47~
F7~/Xq
/xiFARMSIZE<=3#and#FARMSIZE<=7#AND#ODEN=2~/CC44~F6~/CC47~
F7~/Xq
/xiFARMSIZE<=3#and#FARMSIZE<=7#AND#ODEN>1~/CC43~F6~/CC47~
F7~/Xq
/xiFARMSIZE<=3#and#FARMSIZE<=7#AND#ODEN>0~/CC43~F6~/CC48~
F7~/Xq
/xiFARMSIZE<=7#AND#OXEN>2~/CC44~F5~/CC49~F7~/xq
/xiFARMSIZE<=7#AND#OXEN=2~/CC43~F5~/CC48~F7~/xq
/xiFARMSIZE<=7#AND#OXEN=1~/CC43~F5~/CC48~F7~/xq
/xiFARMSIZE<=7#AND#OXEN=0~/CC43~F5~/CC47~F7~/xq
/xiFARMSIZE<=3#AND#OXEN>2~/CC45~F6~/CC47~F7~(goto)A3~/xq
/xiFARMSIZE<=3#AND#OXEN=2~/CC45~F6~/CC47~F7~/xq
/xiFARMSIZE<=3#AND#OXEN=1~/CC41~F6~/CC47~F7~/xq
/xiFARMSIZE<=3#AND#OXEN=0~/CC45~F5~/CC46~F7~/xq
```

FIGURE 16.3

Knowledge-based system for farm level recommendations on crop selection to achieve household food security in Malawi.

Based on Babu (1999).

CONCLUSIONS

Recently, emphasis has been placed on a food system-based approach to achieving nutritional objectives. New food systems need to move toward the principle (McClaferty and Zukerman, 2014; Fan, 2016) that more emphasis needs to be given to nutrition per unit of input or resources used in food and agricultural production. Importance needs to be given to production, processing, storage, transportation, and marketing of food with due attention to food losses and food waste. Consumers and the private sector cannot be ignored in developing new food systems. Moving away from addressing the food supply problem, food accessibility needs to be addressed. While recognizing that small farmers matter, there is a need to recognize that context specificity matters. The role of women, and designing food systems that reduce the drudgery of women and increase gender equality of food and nutrition consumption will be a key outcome of new food systems.

In this chapter we presented an illustrative but practical tool that will help frontline extension and nutrition professionals to design new local food systems with decentralized nutrition interventions that are agro-ecologically sustainable. Extensions of the model presented here can include the introduction of risk that the farmers face in terms of climatic risks, price variability, seasonal availability of foods in the market, and the changes in farm level solutions for nutritional outcomes can be analyzed. This will require better understanding of the attitudes of farm households toward risk, and the nature of uncertainty they face in the context of their food system. One way to address such risk is to introduce risk in the linear programming model presented in this section. This is usually done by reformulating the model as a quadratic programming model, in order to minimize the variances in the net returns as objective functions. In many models, the target income becomes a parameterized constraint (Hazell and Norton 1986; Babu and Rajasekaran, 1991b; McCarl and Spreen, 2004). Such models require additional data in terms of time services information on yields of crops and prices which are increasingly available in most developing countries. In homogenous farming systems where few crops are grown in major areas for food security purposes, constraining acreage under these crops to meet the food security and nutritional challenges, particularly under stable yield conditions, can reflect the risk aversion levels of subsistence farming households (Calkins, 1981).

Achieving nutritional outcomes with food system interventions can be a practical and decentralized strategy to increase the nutritional status of the rural poor. While such an integrated multidisciplinary approach to bringing economists, nutritionists, and agronomists together has been attempted (Donaldson et al., 1995), practical implementation requires the development of extension tools that are based on the analysis of real world data. Teaching such tools, starting with a farm level knowledge-based system, is the objective of this chapter. Such tools are needed in the context of integrating agriculture, nutritional, and health objectives. This chapter demonstrated that, using a spreadsheet and its macro-programming tools, a rule-based system can be developed and used at decentralized levels to make farm level recommendations. This requires no additional expertise in computer programming. Borrowing the results of the published case strongly from Malawi, this chapter introduced the programming approach to achieving nutritional goals through food system-based interventions. The solutions to the nutritional optimization models could be stored as a knowledge-base for field level extension and nutrition professionals to use to recommend crop choices. This considerably reduces the need for highly trained optimization modelers, if such knowledge-based data could be developed and stored for various agro-ecological zones (Balachandran et al., 1989; Babu et al., 2016).

Once developed at decentralized levels, the decision-making systems such as the one presented in this chapter can be effective in addressing locality specific problems related to water management, nutrient management, and optimizing pest control decisions (Weiske, 1981; McKinion and Lemmon, 1985; Edward-Jones, 1992; Hochman et al., 1995). With the increasing use of ICT in agriculture, decision support systems to address nutritional challenges and to bring agricultural interventions that are nutrition sensitive into real programming activities are possible at the community level. Developing and operationalizing such tools are important for finding decentralized solutions for nutritional challenges facing the poor and malnourished.

It should be noted, however, that while the optimization model presented in this chapter can be a useful tool in guiding crop choices to obtain optimal nutrition, farmers' motivation and awareness about nutritional choices needs to be given due attention. Interventions that involve the introduction of nutritionally optimal crops and livestock enterprises should consider community, institutional, and market factors that will affect farmers' choices (Babu, 2000). Further, systems that back up the farmers' choices to grow more nutritious crops will be needed. These include the effective functioning of the input supply systems, particularly quality seeds for new crops, and agricultural and nutrition extension systems to maintain sustained interest for farmers to invest in the new enterprises suggested by the optimization models (Babu, 2002b). Finally, the importance of the monitoring and evaluation systems that capture the benefits and costs of the new crops or combinations of crops introduced cannot be overemphasized in modifying and refining the intervention to make the food systems nutritionally sustainable.

EXERCISES

1. Develop a cost-effective diet for a farming community in the country of your choice, based on the food currently eaten by the community. Could the dietary choices be changed to improve the nutritional availability and yet reduce the cost of nutrition further?
2. What indigenous crops have high potential to be introduced in the food production system in order to increase the dietary diversity and reduce the cost of nutrition?
3. Develop a knowledge-sharing strategy that the local extension workers and nutrition workers could use to guide the farming households toward improved nutrition access and intake from the local farming systems.

PART

H

CONCLUSION

CHAPTER

FUTURE DIRECTIONS FOR NUTRITION POLICY MAKING AND IMPLEMENTATION

17

The historical evolution of food policy analysis ... raises several questions going forward: who will do the analysis and where will they be trained; what is the appropriate institutional base for food policy analysts; and why do this difficult analysis if "Politics is in command?"

—Peter Timmer (2013)

Peter Timmer's quote above applies equally well for nutrition policy making. Answering his questions in the context of nutrition raises more fundamental questions. Why does the study of the economics of nutrition matter? What have we learned in nutrition policy making and what needs to be done?

Malnutrition in all forms can have economic implications both in the short run and in the long run to the individual affected and the country where he or she lives. A malnourished child can have lifelong developmental and productivity challenges. For example, a stunted child can have lower human capacity due to low cognitive development, low educational attainment, low adult productivity and income earning potential resulting from poor childhood nutrition, and this can have social and economic implications all through her or his life (Hoddinott et al., 2013; Hoddinott, 2016). An additional challenge is that nutritional damage caused during early childhood is not reversible during the lifetime. Nutritional development is therefore fundamental for economic development. Thus, addressing malnutrition is an economic challenge which goes beyond human rights, moral, and ethical imperatives (IFPRI—Global Nutrition Report, 2015). The chapters of this book aim to present various facets of the nutritional challenges countries face, and to help develop analytical skills to address them through policy and program interventions.

The nutritional challenges that humanity faces can be summarized as follows (FAO 2015; IFPRI—Global Nutrition Report, 2015; WHO, 2015a,b; UNICEF/WHO/World Bank, 2015): about 800 million people eat a calorie-deficient diet every day; about two billion people face hidden hunger in the form of micronutrient deficiencies; among children under 5 years of age, 161 million are stunted for their age, 51 million are wasted, and 42 million are overweight; and about two million people are overweight or obese. In addition, under-nutrition is a major causal factor contributing to 45% of all child deaths in developing countries (Black et al., 2013). While one set of countries suffer mostly from under-nutrition (mostly in sub-Saharan Africa and South Asia), another set of countries are afflicted by over-nutrition related health challenges (mostly in North America, Europe, Latin America, the Middle East, North Africa, and the Caribbean), although both under-nutrition and over-nutrition can coexist in the same household, community, and country.

Nutrition Economics. DOI: http://dx.doi.org/10.1016/B978-0-12-800878-2.00017-7

Some of the key lessons emerging from nutrition planning and policy making in the last three decades toward which the contents of this book attempt to contribute include the following:

1. Economic growth is essential for reduction in poverty and malnutrition. However, the rate of reduction in malnutrition is less than the rate of poverty reduction.
2. Malnutrition is a multi-sectoral challenge and needs to go beyond the health and agriculture sectors. Intervention strategies have to be coordinated with key ministries such as water, sanitation, gender, education, social protection, and food and agriculture.
3. Policy environment, leadership, governance, coordination, financing and sustainability of interventions are key for such synchronization of multi-sectoral activities.
4. In addition to food and nutrition intake, issues related to primary health care, immunization, breast feeding, mothers' education, child spacing, and other socioeconomic and cultural determinants that are context-specific need to be fully understood.
5. Intervention strategies that help to improve service delivery in sanitation, child care, clean water, and nutrition education for behavioral change are needed to increase the effectiveness of nutritional investments.
6. Empowering women through education and interventions that increase their decision-making power at the household level is key for improving women's and children's nutrition. Continued understanding of intra-household dynamics in resource allocation and utilization of nutrition and health services are needed.
7. Monitoring and evaluation of programs implemented, both for their process lessons and impact of the benefits, are needed.
8. Social safety net programs require context-specific approaches, and the nutritional benefits from them can only be realized if they are specifically addressed to nutritional goals during the design stage.
9. School nutrition programs continue to be a most popular intervention to attract children to school, to keep them in school, and to increase their learning abilities. However, the results differ depending on the context and program design.
10. Overweight and obesity are increasing even in developing countries. Strategies to address over-nutrition and under-nutrition occurring in the same community and households require innovations in all sectors.
11. Micronutrient deficiencies continue to be major set of nutrition challenges. Continued multipronged interventions are needed for iodine, iron, vitamin A, and other micronutrients.
12. Agriculture and food systems have a major role to play in solving nutritional problems. Designing interventions through research and innovation for increasing dietary diversity, bio-fortification, food safety, and nutritional enhancement throughout the food value chains is critical.
13. Nutritional challenges are widespread, and yet the solutions require locality specific interventions. Designing and implementing decentralized context-specific nutrition interventions that bring several sectors together in a coordinated manner requires local capacity at all levels.

At the global level, the problem of malnutrition is not just a challenge for the poor developing countries. More than half of the world's poor people still live in middle income countries, and poor people in these countries suffer most from nutrition problems. Nutrition problems also affect an increasing share of the population in developed countries, mostly in the form of over-nutrition, resulting in high costs associated with chronic noncommunicable diseases and high health care costs. Thus, malnutrition in all forms is a global and economic challenge and, as seen in Chapter 2, Global Nutritional Challenges and Targets: A Development and Policy Perspective, remains on top of the global development agenda (WHO and FAO, 2014; IFPRI—Global Nutrition Report, 2015).

The global community has set nutrition targets to tackle this challenge and they include: by the year 2015, a 40% reduction in the global number of children under five who are stunted; a 50% reduction of anemia in women of reproductive age; a 30% reduction in low birth weight; no increase in childhood overweight; increase the rate of exclusive breast feeding in the first six months up to at least 50%; and a reduction and maintaining of reduced childhood wasting to less than 5% (World Health Assembly, 2012). Focused attention in about 35 most affected developing countries in South Asia and sub-Saharan Africa and scaling up the specific interventions is one of the strategies suggested to achieve these global goals (Ruel and Alderman, 2013).

At the country level, translating the above global nutrition targets will require country and context-specific analysis, and policy and program development. For example, while nutritional problems may be recognized by the policy makers, countries may not have specific plans of action to address the nutrition problems. Even when the nutrition policies and strategies may be available, countries may lack resources for implementing these strategies. It is possible that the evidence for designing intervention programs may be lacking.

Translating economic growth benefits into better nutritional outcomes requires appropriate nutrition policies and intervention strategies at national levels. Designing context-specific programs to address the nutrition challenges will involve bringing evidence generated using the methods discussed in the chapters of this book to policy discussions at national, regional, and community levels. In order to achieve this, the process of nutrition policy making needs to be fully understood (Resnick et al., 2015; Babu et al., 2016; Haggblade et al., 2016). This will help understanding about the role of key players and actors in the nutrition policy arena and their influence on the policy making process. Actors and players in the nutrition policy process—frequently called the nutrition community in a country—often lack the capacity to meaningfully engage in policy debates and dialogues, and as a result are not able to hold the policy makers accountable to the global and national nutritional goals. Strengthening the capacity of the nutrition community, in terms of its leadership, research and analysis, advocacy, and influencing budget allocation, matters for designing and implementing appropriate nutrition policies and programs (Haddad et al., 2014).

Nisbett et al. (2015) summarize the key characteristics of leadership that are needed at the country level. According to them, nutrition leadership depends on the individual capacity, the knowledge-base available, and the political economy of the country. Yet without nurturing local leadership for nutrition policy making, research, outreach, and implementation, externally designed initiatives are likely to fail or become unsustainable.

Thus, designing policies and programs to address the context-specific nutrition challenges requires well equipped professionals who have a good understanding of a common set of nutrition issues, and challenges, and have the tools required to address them. Lack of such capacity at the national, regional, and community levels has been one of the primary causes for the failure of multidisciplinary approaches to nutrition problem solving. The foregoing chapters of this book were intended to impart thematic and analytical skills for analyzing nutrition challenges as a multisectoral problem to be addressed by a multidisciplinary team of professionals who could come together from various disciplines and sectors. While each of these professionals may have in-depth training in their own fields, solving nutrition problems requires a common understanding of the issues, the progress made in theoretical development, and quantitative approaches to generate evidence for policy and program interventions. The chapters of this book set out to achieve this goal.

In the chapters of this book we reviewed key issues, constraints, and challenges facing policy makers in designing and implementing nutrition intervention programs and policies that can help in achieving the nutrition targets of the global development community. The chapters of this book bring together current evidence on various thematic issues, their policy implications, analytical methods to address them, and also demonstrate how to implement empirical studies to generate evidence for policy debate and dialogue. The main focus of the chapters is to enable a multidisciplinary team of nutritionists, agriculturalists, anthropologists, economists, policy researchers, and analysts to develop their capacity to solve nutrition challenges in their own countries, in the context in which they need to be addressed. This book's contents are motivated by the conviction that without such local multidisciplinary capacity, policy makers will continue to lack evidence to act on nutritional challenges and goals. Further, without such capacity, future progress in addressing nutritional challenges will continue to be frustratingly slow. While this challenge is well recognized, the academic and research community until recently has done little to help develop such capacity, mainly due to its single disciplinary approach to problem solving. This book is an attempt to fill this void.

We presented analytical methods in the context of nutrition and economics, the two major disciplines in development studies. Bringing together these two disciplines is crucial as a first step to begin to integrate other related sectors for nutrition policy making and implementation of intervention programs. The rest of the related disciplines, such as physical anthropology, sociology, agronomy, and public health, could be accommodated in a similar fashion in the context of nutritional policy development and implementation. Thus, throughout this book we emphasize an applied and multidisciplinary approach to generating evidence and understanding the significance of the findings for policy analysis.

While individual professions have a definitive role to play in nutrition policy making and program interventions, recognizing that professionals have to work in an integrated multidisciplinary manner has become inevitable. While this is often talked about at almost every major conference, to the extent that it has become a cliché (similar to the terms thrown around such as "political will" and "enabling environment"), there has not been any serious effort to build the multidisciplinary capacity needed to tackle nutrition problems at various levels of decision-making in the developed and developing world. Academic institutions in developed or developing countries are not fully geared toward developing such interdisciplinary capacity through the curriculum they teach or the research methods they employ. Employing nutrition researchers in economics and public policy departments, and offering faculty positions to economists in food science and nutrition departments and home economics colleges will be a good start. Encouraging nutrition graduates to go on to study public policy and social sciences is key for generating future multidisciplinary problem solvers. Some progress has been made in this direction, and this has been rewarded highly by bringing nutrition issues to the forefront of development policy making in the last 20 years.

As mentioned in the beginning of this chapter, the nutritional well-being of a population in a country is at the core of its development process. It is becoming more of a universal challenge to achieve optimal nutrition. While one set of countries may be addressing under-nutrition as a priority, another set of countries may be giving emphasis to over-nutrition in its policy agenda. Nevertheless, the approach to developing evidence on what works and why, and applying such evidence to design and implement nutrition interventions, evaluate them and refine to scale-up, requires a common set of tools.

Nutritional challenges could be treated as being on a continuum, with under-nutrition on one end of the spectrum and over-nutrition on the other. Yet, due to the challenges countries are facing with both extremes, and also a combination of these extremes called the double burden of nutrition, strategic approaches are needed for research, advocacy, and outreach activities. Added to this is hidden hunger in the form of micronutrient deficiencies, making it a triple burden to some countries. Depending on the level of development of the country, one or more forms of malnutrition could coexist at varying levels, and this has implications for all segments of society. Local capacity for addressing these challenges will require changes in the curriculum of the universities in developing and developed countries.

Income growth of the countries does have a positive impact on under-nutrition, but it also has a negative impact on over-nutrition levels. A 10% increase in national income results in a 6% reduction in stunting, while the same level of income growth also results in a 7% increase in overweight and obesity in women (Ruel and Alderman, 2013). The economic transformation of the countries is one of the important determinants of the nutrition transition that countries undergo (Webb and Black, 2013). Moving from under-nutrition focused policies and programs to over-nutrition challenges, and finally achieving a nutritional status of the population where optimal nutrition is provided to all citizens of the world, will be part of the development challenge no matter where the country stands in the path of its economic transformation. While this may seem an ambitious goal, this is certainly achievable, and at least countries should be striving to achieve this goal—for the economic, moral, ethical, human rights, and health reasons mentioned in the introductory chapters, and to improve the overall quality of life of their populations.

Rural and urban populations differ in their access to food, nutrition, and health. Countries have to increasingly tackle nutritional challenges in both geographical settings. They need to be concerned not only about how to feed the growing population, but also how to feed them correctly to achieve a high quality of life. Measured in terms of optimal nutrition and health, this will be a major development challenge for the future. The chapters of this book looked at various factors that affect food security, nutrition, and health, and those factors that play interconnecting roles. The role of these factors needs to be measured and analyzed in the country, and from a local community perspective. Having the evidence about the indicators and causal factors of nutrition is a fundamental prerequisite for nutrition policy making and program intervention (Babu and Pinstrup-Andersen, 1994). The chapters of this book have demonstrated how one could translate such data and analytical information into policies and programs.

Making sectors responsible for nutritional outcomes more "nutrition sensitive" is probably not enough. Saying nutrition will be everywhere in development programs, but not necessarily at the center and at the core of the development process, has not worked in the past. This also makes it convenient for program managers to make nutrition a cross cutting issue at best, and transfer the responsibility to other sectors. Nutrition sensitive agriculture is a good example of this approach. Food is a major input to nutrition. Agriculture and food systems need to play a critical role in providing nutrition both in terms of quantity and quality. How do we tackle the problem of malnutrition—both the under-nutrition and over-nutrition challenges—from a food systems perspective?

The capacity for identifying the nutrition gaps in current food systems, designing modified food systems through innovation to fill the gaps, and implementing new innovations, are needed for such approaches. Agriculturalists, nutritionists, extension and advisory services, credit institutions, processing units, and agribusiness specialists have to work together toward this goal.

In the past two decades, global research systems such as the Consultative Group for International Agricultural Research (CGIAR) have been successful in bringing together a multidisciplinary team of researchers to address the nutritional challenges from the food system perspective (Kataki and Babu, 2007; McDermott et al., 2015). Solutions for under-nutrition could be found at local levels if the food systems are designed to meet the nutritional needs of local communities. However, operationalizing this approach again requires multidisciplinary scientists coming together to understand the specific nutritional problems, agro-ecological conditions, resource constraints of the farming households, market availability, and opportunities for storing and processing at local levels. Global programs have been seeking solutions to the question: how can agriculture help in improving nutrition and health? Agriculture will not solve the problem by itself (McDermott et al., 2015). Policy coordination at the national and local levels is important as well.

How to scale-up remains an unanswered question. In most cases we know what to do. But the issue of how to do it at scale continues to be a development challenge. Gillespie et al. (2015) identify several factors that contribute to the successful scaling up of a nutrition program. High impact nutrition program interventions have clear cut goals and specific targets to achieve. Often multiple and mixed up goals compete with each other, and nutrition goals are often overlooked.

The policy system and the process by which the programs are designed, adopted, implemented, monitored, and evaluated for further refinement needs to be not only transparent, but also accountable. This can ensure the solid governance structures needed for program delivery. The committed presence of local nutrition leaders who are credible, and political champions who have long-term interests helps in keeping nutrition on the policy agenda. Identifying local contexts and the opportunities for scaling up by designing programs that can help in addressing the specific needs of the local communities is a key to successful scaling up of nutrition interventions. While adequate funding is a prerequisite for program implementation, local capacity is needed at all levels to deliver the program benefits. Finally, the institutional frameworks of effective monitoring of food and nutrition program interventions are fundamental for refinement and sustainability of the program interventions (Gillespie et al., 2015).

The global nutrition community has been striving hard to make nutrition a development priority for countries through various declarations and global goal-setting processes (FAO and WHO, 2014; WHO, 2014). However, what the countries do after signing the international declarations, and how they use the support given by the international community remains to be further studied. Translating such external assistance into national level impact will also require strengthening the policy process, improving the research and analytical capacity for generating evidence, and strengthening implementation capacity. For example, ICN2 has been another process through which countries jointly take the pledge of addressing their nutritional challenges. Its predecessor ICN1 was successful in highlighting the nutritional challenges of the developing countries, but until MDG were formulated, it did not have the needed momentum in developing countries. During ICN2, higher emphasis has been given to agriculture for a nutrition framework and to increase the scale-up of nutrition programming, funding, and evidence-based interventions. The recently developed and agreed upon Sustainable Development Goals recognize nutrition directly or indirectly through several of its individual goals (IFPRI—Global Nutrition Report, 2015). However, the key challenge of designing and implementing intervention programs that will reduce the nutritional problems at a country level remains.

How can nutrition be improved by building the capacity of those who are responsible for designing the policies, adopting the policies in the national policy systems, and implementing them on the ground? How can the needed resources be mobilized to address nutrition challenges; how much countries are contributing to this cause; and how to guide the process of investment toward the right set of problems, and maximizing the returns on such investments is crucial. A recent call for investing in nutrition estimates that an additional annual investment of US$7 billion will be needed on top of what is being spent to address stunting, wasting, anemia, and breast feeding targets by 2025 (Shekar et al., 2016). In the context of mobilizing resources and investing in nutrition, leadership at the country level is important.

There is a general consensus that country ownership and partnership is critical for developing the evidence for identifying the problem, generating evidence, and monitoring performance and accountability. Assessment of the SUN in 2014 showed that results were not fully achieved in terms of nutrition goals, but they have done a good job in sensitizing the developing country governments to put nutrition on the national political, policy, and strategy agenda of the countries. Yet questions related to what has been learned in terms of the policy process, knowledge management and sharing, and the governance of nutrition policies and programs remain.

In order to improve nutrition action on the ground, there is a need to improve accountability for the investments made. Collecting data on the results through effective monitoring, evaluation, and assessment systems provides such feedback. However, there is a need for strengthening food security and nutrition monitoring systems where data on the program implementation from various sources are brought together, harmonized, and analyzed in an open data context. But this requires data collecting, processing, and analysis capacity at national levels.

An assessment of the in-country capacity to develop a plan for implementing the programs and policies, monitoring them, and further refinement of the interventions is needed as a first step. Food security and nutrition monitoring systems need to be inclusive, and be accessible in real time. But reducing the time between data collection and decision-making requires capacity at all levels of the information value chains—from data to information based on policy analysis of the data, and from information to data for decision-making (Babu 2015). In this book we provided the necessary tools for analysis in a multidisciplinary manner.

Finally, current efforts to address nutritional challenges at the global level require a multisectoral approach. Such an approach will recognize that the resources for social sector investments compete with other long-term development priorities. For example, poverty reduction, food security, and nutrition goals have to be addressed through possible ways of benefiting each other. Health, nutrition, and rural development goals require common investments such as improving rural roads, and one set of goals becomes a prerequisite for achieving the other. For example, severely malnourished people are often located in remote areas where there are no roads for the nutrition and health workers to reach them. They are also isolated and disconnected from the markets for the produce they cultivate. Service delivery for agriculture is also affected by the absence of rural roads. Thus, infrastructure development with a focus on delivering nutritional objectives provides opportunities to develop other spheres of the economy.

There has been a call for action toward nutrition from various global reports recently, including the SDGs. They will remain just "calls" if there is no capacity to adopt and apply them in the country context. Nutrition-related funding has been on the rise at the global level, although it is still not nearly adequate to address the enormous problem at hand. Linking the goals of poverty reduction, health, nutrition, and ending hunger in a multidisciplinary manner can help in achieving them in a cost-effective and efficient manner. Sadly, such an approach to development problem solving is still missing at all levels. The higher education and research community has just begun to reflect on these development challenges in the context of developing countries. But investments in producing high quality multidisciplinary nutrition capacity, and the challenges in the design of appropriate curriculum and methods of developing skills for achieving SDGs are not yet taken seriously.

Public provision of health and nutritional services presents a significant set of opportunities in a more synergistic manner (Wage et al., 2015). Such synergies begin at international levels where organizations that work on related issues come out of their silos and address them holistically, so that similar efforts can be undertaken at the national and subnational levels. In such efforts, the role of in-country capacity—that the contents of this book aims to strengthen—to address the multi-sectoral nutritional challenges can hardly be overemphasized.

References

Abubakar, A., Uriyo, J., Msuya, S.E., Swai, M., Stray-Pedersen, B., 2012. Prevalence and risk factors for poor nutritional status among children in the Kilimanjaro Region of Tanzania. Int. J. Environ. Res. Public Health 9, 3506–3518.

Abuya, B.A., Onsomu, E.O., Kimani, J.K., Moore, D., 2011. Influence of maternal education on child immunization and stunting in Kenya. Matern. Child Health J. 15 (8), 1389–1399.

Adair, L.S., Fall, C.H., Osmond, C., Stein, A.D., Martorell, R., Ramirez-Zea, M., et al., 2013. Associations of linear growth and relative weight gain during early life with adult health and human capital in countries of low and middle income: findings from five birth cohort studies. Lancet 382 (9891), 525–534.

Adhau, B.P., 2011. The problem of malnutrition in tribal society (with special reference to Melghat Region of Amravati District). Int. J. Res. Commerce Manag. 2 (9), 109–111.

Afridi, F., 2010. Child welfare programs and child nutrition: evidence from a mandated school meal program in India. J. Dev. Econ. 92 (2), 152–165.

Afridi, F., 2011. The impact of school meals on school participation: evidence from rural India. J. Dev. Stud. 47 (11), 1636–1656.

Agoramoorthy, G., Hsu, M.J., 2009. India needs sanitation policy reform to enhance public health. J. Econ. Policy Reform 12 (4), 333–342.

Ahmed, A.U., Babu, S.C., 2007. The impact of food for education programs in Bangladesh. Education 3, 8.

Akee, R., Simeonova, E., Copeland, W., Angold, A., Costello, E.J., 2013. Young adult obesity and household income: effects of unconditional cash transfers. Am. Econ. J. Appl. Econ. 5 (2), 1–28.

Alderman, H., 1988. An analysis of food demand in Pakistan using market price aggregates. Pak. Dev. Rev. 27, 89–108.

Alderman, H., 2010. The economic cost of a poor start to life. J. Dev. Origins Health Dis. 1 (1), 19–25.

Alderman, H., Bundy, D., 2011. School Feeding Programs and Development: Are We Framing the Question Correctly? The World Bank Research Observer, lkr005.

Alderman, H., Gertler, P., 1997. Family resources and gender differences in human capital investments: the demand for children's medical care in Pakistan. Intrahousehold Resource Allocation in Developing Countries: Models, Methods, and Policy. Johns Hopkins University Press for the International Food Policy Research Institute, Baltimore and London, pp. 231–248.

Alderman, H., Mustafa, M., 2013. Social Protection and Nutrition. FAO, Rome, Note prepared for the technical panel discussions on What are the policy lessons learned and what are the success factors for the ICN2.

Alderman, H., Timmer, C.P., 1980. Food policy and food demand in Indonesia. Bull. Indonesian Econ. Stud. 16, 83–93.

Alderman, H., Yemtsov, R., 2013. How can safety nets contribute to economic growth? World Bank Econ. Rev.

Alderman, H., Behrman, J.R., Lavy, V., Menon, R., 1999. Child Nutrition, Child Health, and School Enrollment: A Longitudinal Analysis. <http://dx.doi.org/10.1596/1813-9450-1700>.

Alderman, H., Behrman, J.R., Lavy, V., Menon, R., 2001. Child health and school enrollment: a longitudinal analysis. J. Hum. Resour. 185–205.

Alderman, H., Hoogeveen, H., Rossi, M., 2006. Reducing child malnutrition in Tanzania: combined effects of income growthand program interventions. Econ. Hum. Biol. 4 (1), 1–23.

Ali, M.M., Amialchuk, A., Gao, S., Heiland, F., 2012. Adolescent weight gain and social networks: is there a contagion effect? Appl. Econ. 44 (22–24), 2969–2983.

Ali, M.M., Rizzo, J.A., Amialchuk, A., Heiland, F., 2014. Racial differences in the influence of female adolescents' body size on dating and sex. Econ. Hum. Biol. 12, 140–152.

Allais, O., Bertail, P., Nichele, V., 2010. The effects of a fat tax on French households' purchases: a nutritional approach. Am. J. Agric. Econ. 92 (1), 228–245.

Alston, M.J., Rickard, J.B., Okrent, M.A., 2010. Farm policy and obesity in the United States. Choices, 3rd Quarter 25 (3).

Alviola IV, A.P., Nayga Jr, M.R., Thomsen, M., 2013. Food deserts and childhood obesity. Appl. Econ. Perspect. Policy 35 (1), 106–124.

Alviola, A.P. IV, Nayga, M.R. Jr., Thomsen, R.M., Danforth, D., Smartt, J., 2014, The effect of fast-food restaurants on childhood obesity: a school level analysis. Econ. Hum. Biol. Jan (12), 110–119.

Amaral, S., Bandyopadhyay, S., Sensarma, R., 2015. Employment programmes for the poor and female empowerment: the effect of NREGS on gender-based violence in India. J. Interdiscip. Econ. 27 (2), 199–218.

Amin, S.A., Yon, B.A., Taylor, J.C., Johnson, R.K., 2015. Impact of the National School Lunch Program on fruit and vegetable selection in Northeastern Elementary Schoolchildren, 2012–2013. Public Health Rep. 130 (5).

Amis, J.M., Hussey, A., Okunade, A.A., 2014. Adolescent obesity, educational attainment and adult earnings. Appl. Econ. Lett. 21 (13–15), 945–950.

Andalon, M., 2011. Oportunidades to reduce overweight and obesity in Mexico? Health Econ. 20 (Suppl. 1), 1–18.

Anderson, L.M., Matsa, A.D. 2011. Are restaurants really supersizing america? Am. Econ. J.: Appl. Econ. 3 (1), 152–188.

Anderson, M.P., 2012. Parental employment, family routines and childhood obesity. Econ. Hum. Biol. 10 (4), 340–351.

Andres, L.A., Briceno, B., Chase, C., Echenique, J.A. 2014. Sanitation and Externalities: Evidence From Early Childhood Health in Rural India. The World Bank, Policy Research Working Paper Series: 6737.

Andreyeva, T., 2012. Effects of the revised food packages for women, infants, and children (WIC) in Connecticut. Choices, 3rd Quarter 27, 3.

Andreyeva, T., Kelly, I.R., Harris, L.J., 2011. Exposure to food advertising on television: associations with children's fast food and soft drink consumption and obesity. Econ. Hum. Biol. 9 (3), 221–233.

Angrist, J., Kruegger, A., 1991. Does compulsory school attendance affect schooling and earnings? Q. J. Econ. 106 (4), 979–1104.

Angrist, J.D., 1990. Lifetime earnings and the Vietnam era draft lottery: evidence from social security administrative records. Am. Econ. Rev. 80 (3), 313–336.

Arifeen, S., Black, R., Caulfield, L., Antelman, G., Baqui, A., 2001. Determinants of infant growth in the slums of Dhaka: size and maturity at birth, breastfeeding and morbidity. Eur. J. Clin. Nutr. 55, 167–178.

Arnold, C., Conway, T., Greenslade, M., 2011. DFID Cash Transfers Literature Review. GFID, London.

Asfaw, A., 2011. Peer-effects in obesity among public elementary school children: a grade-level analysis. Health Econ. 20 (2), 184–195.

Asirvatham, J., Nayga Jr., R.M., Thomsen, M.R., 2014. Race and gender differences in the cognitive effects of childhood overweight. Appl. Econ. Perspect. Policy 36 (3), 438–459.

Aturupane, H., Deolalikar, A.B., Gunewardena, D., 2008. The Determinants of Child Weight and Height in Sri-Lanka: A Quantile Regression Approach. World Institute for Development Economic Research (UNU-WIDER).

Audsley, B., Halme, R., Balzer, N., 2010. Comparing cash and food transfers: a cost-benefit analysis from rural Malawi. In: Omamo, S.W., Gentilini, U., Sandström, S. (Eds.), Revolution: From Food Aid to Food Assistance. Innovations in Overcoming Hunger. World Food Programme, Rome, pp. 89–102.

Augsburg, B., Rodriguez-Lesmes, P., 2015, Sanitation Dynamics: Toilet Acquisition and Its Economic and Social Implications. Institute for Fiscal Studies, IFS Working Papers: W15/15.

Auld, M.C., 2011. Effect of large-scale social interactions on body weight. J. Health Econ. 30 (2), 303–316.
Averett, S.L., Smith, J.K., 2014. Financial hardship and obesity. Econ. Hum. Biol. 15, 201–212.
Averett, S.L., Stifel, D.C., 2013. The Applied Economics of Weight and Obesity. Taylor and Francis, Routledge Collective Volume Article, London and New York, pp. 68–74.
Averett, S.L., Argys, L.M., Sorkin, J., 2013. In sickness and in health: an examination of relationship status and health using data from the Canadian National Public Health Survey. Rev. Econ. Household 11 (4), 599–633.
Babu, S.C., 1989. Challenges facing agriculture in southern Africa. In: A Conference Report: Inter-Conference Symposium of the International Association of Agricultural Economists, Badplass, South Africa, 10–16 August 1998.
Babu, S.C., 1997a. Rethinking training in food policy analysis: how relevant is it for policy reforms. Food Policy 22 (1), 1–9.
Babu, S.C., 1997b. Facing donor community with informed policy decisions – lessons from food security and nutrition monitoring in Malawi. Afr. Dev. 22 (2), 5–24.
Babu, S.C., 1997c. Multi-disciplinary capacity strengthening for food security and nutrition policy analysis: lessons from Malawi. Food Nutr. Bull. 18 (4), 363–375.
Babu, S.C., 2001. Food and nutrition policies in Africa: capacity challenges and training options. Afr. J. Food Nutr. Sci. 1 (1), 19–28.
Babu, S.C., 2002a. Food systems for improved human nutrition: linking agriculture, nutrition and productivity. J. Crop Prod. 6 (1/2), 7–30.
Babu, S.C., 2002b. Designing nutrition interventions with food systems: planning, communication, monitoring and evaluation. J. Crop Prod. 6 (1/2), 365–373.
Babu, S.C., 2009. Hunger and Food Security, World at Risk: A Global Issues Sourcebook. second ed. CQ Press, Washington, DC.
Babu, S.C., 2011. Developing multi-disciplinary capacity for agriculture, health, and nutrition – challenges and opportunities. Afr. J. Food Agric. Nutr. Dev. 11 (6), 1–3.
Babu, S.C., 2015a. Policy processes and food price crises: a framework for analysis and lessons from country studies. In: Pinstrup-Andersen, P. (Ed.), Food Price Policy in an Era of Market Instability: A Political Economy Analysis, Part II: Syntheses of Findings from Country Studies. Oxford University Press, Oxford, UK, pp. 76–101. Chapter 4.
Babu, S.C., 2015b. Private sector extension with input supply and output aggregation: case of sugarcane production system with EID-Parry in India. In: Zhou, Y., Babu, S.C. (Eds.), Knowledge Driven Development: Private Extension and Global Lessons. Academic Press, London, UK, Chapter 4.
Babu, S.C., Andersen, P.P., 1994. Food security and nutrition monitoring: a conceptual framework, issues and challenges. Food Policy 19 (3), 218–233.
Babu, S.C., Chapasuka, E., 1997. Mitigating the effects of drought through food security and nutrition monitoring: lessons from Malawi. U.N. Univ. Food Nutr. Bull. 18 (1), 71–81.
Babu, S.C., Gajanan, S.N., Sanyal, P., 2014. Food Security Poverty Nut Policy, 2nd Edition. http://dx.doi.org/10.1016/B978-0-12-405864-4.00035-1 © 2014 Elsevier Inc. All rights reserved.
Babu, S.C., Hallam, A., 1989. Socio-economic impacts of school feeding programmes: empirical evidence from a south Indian village. Food Policy 14 (1), 58–66.
Babu, S.C., Mthindi, G.B., 1994. Household food security and nutrition monitoring: the Malawi approach to development planning and policy interventions. Food Policy 19 (3), 272–284.
Babu, S.C., Pinstrup-Andersen, P., 1994. Food security and nutrition monitoring: a conceptual framework, issues and challenges. Food Policy 19 (3), 218–233.
Babu, S.C., Sanyal, P., 2008. Persistent food insecurity in Malawi and policy options. In: Pinstrup-Andersen, P. (Ed.), Globalization and Food Security. Cornell University Press, Ithaca.

Babu, S.C., Subramanian, S.R., 1988. Nutritional poverty — distribution and measurement. Indian J. Nutr. Diet. 25 (3), 75–81.

Babu, S.C., Thirumaran, S., Mohanam, T.C., 1993. Agricultural productivity, seasonality, and gender bias in rural nutrition: empirical evidence from South India. Soc. Sci. Med. 37 (11), 128–1413.

Babu, S.C., Singh, M., Hymavathi, T.V., Rani, U., Kavitha, G.G., Karthik, S., 2016. Improved Nutrition Through Agricultural Extension and Advisory Services: Case Studies of Curriculum Review and Operational Lessons From India (English). The World Bank, Washington, DC.

Babu, S.C., Manvatkar, R., Kolavalli, S., 2016. Strengthening capacity for agribusiness development and management in Sub-Saharan Africa. Afr. J. Manag. 2 (1), 1–30. Available from: <http://dx.doi.org/10.1080/23322373.2015.1112714>

Bailey J., 2013. Who pays for obesity? Evidence from health insurance benefit mandates. Econ. Lett. 121 (2), 287–289.

Baker, E.H., Rendall, M.S., Weden, M.M., 2015. Disease prevalence, disease incidence, and mortality in the United States and in England Banks. Demography 52 (4), 1295–1320.

Ban, R., Das, G., Monica, R.V., 2008. The Political Economy of Village Sanitation in South India: Capture or Poor Information? The World Bank, Policy Research Working Paper Series: 4802, 48 pp.

Banerjee, A.V., Duflo, E., 2007. The economic lives of the poor. J. Econ. Perspect. 21 (1), 141–167.

Banerjee, A.V., Duflo, E., 2008. What is middle class about the middle classes around the world? J. Econ. Perspect. 22 (2), 3–28.

Banerjee, A.V., Duflo, E., 2009. The experimental approach to development economics. Annu. Rev. Econ. 1 (1), 151–178.

Bauhoff, S., 2014. The effect of school district nutrition policies on dietary intake and overweight: a synthetic control approach. Econ. Hum. Biol. 12, 45–55.

Baum, C.L., 2011. The effects of food stamps on obesity. South. Econ. J. 77 (3), 623–651.

Becker, G.S., 1960. An economic analysis of fertility. In: Becker, G. (Ed.), Demographic and Economic Change in Developing Countries. Princeton University Press, Princeton, NJ.

Becker, S., Ichino, A., 2002. Estimation of average treatment effects based on propensity scores. Stata J. 2 (4), 358–377.

Beesley, A., Ballard, R., 2013. Cookie cutter cooperatives in the KwaZulu-Natal school nutrition programme. Dev. South. Afr. 30 (2), 250–261.

Begum, S., Ahmed, M., Sen, B., 2011. Do water and sanitation interventions reduce childhood diarrhoea? New evidence from Bangladesh. Bangladesh Dev. Stud. 34 (i3), 1–30.

Behrman, J., 1988. Nutrition and Incomes: Tightly Wedded or Loosely Meshed? PEW/Cornell Lecture Series on Food and Nutrition Policy. Cornell University Food and Nutrition Policy Program, Ithaca.

Behrman, J.R., Alderman, H., 2004. Estimated Economic Benefits of Reducing Low Birth Weight in Low-Income Countries. HNP Discussion Paper, World Bank, Washington, DC.

Behrman, J.R., Deolalikar, A., 1987. Will developing countries' nutrition improve with income? J. Polit. Econ. 95, 492–507.

Behrman, J.R., Wolfe, B.L., 1984. More evidence on nutrition demand: income seems overrated and women's schooling underemphasized. J. Dev. Econ. 14 (1), 105.

Behrman, J.R., Alderman, H., Hoddinott, J., 2004. Malnutrition and hunger. Global Crises Global Solutions. pp. 363–420.

Behrman, J.R., Alderman, H., Hoddinott, J., 2004. Hunger and Malnutrition. Challenge Paper for Copenhagen Consensus 2004.

Belfield, C.R., Kelly, I.R., 2012. Early Education and Health Outcomes of a 2001 U.S. Birth Cohort 6 (3), 251–277.

Belfield, C.R., Kelly, I.R., 2013. Early Education and Health Outcomes of a 2001 U.S. Birth Cohort. Econ. Hum. Biol. 11 (3), 310–325.

Ben-Shalom, Y., Moffitt, R.A., Scholz, J.K., 2011. An Assessment of the Effectiveness of Anti-poverty Programs in the United States. The Johns Hopkins University, Economics Working Paper.

Bertail, P., Caillavet, F., 2008. Fruit and vegetable consumption patterns: a segmentation approach. Am. J. Agric. Econ. 90 (3), 827–842.

Bhargava, A., 2008. Food, Economics and Health. OUP, Oxford.

Bhattacharya, J., Packalen, M., 2012. The other ex ante moral hazard in health. J. Health Econ. 31 (1), 135–146.

Bhattacharya, J., Sood, N., 2011. Who pays for obesity? J. Econ. Perspect. 25 (1), 139–158.

Bhattacharya, J., Currie, J., Haider, S.J., 2006. Breakfast of champions? The school breakfast program and the nutrition of children and families. J. Hum. Resour. 41 (3), 445–466.

Bhutta, Z.A., Das, J.K., Rizvi, A., Gaffey, M.F., Walker, N., Horton, S., et al., 2013. Evidence-based interventions for improvement of maternal and child nutrition: what can be done and at what cost? Lancet 382 (9890), 452–477.

Black, R.E., Allen, L.H., Bhutta, Z.A., et al., 2008. Maternal and child under nutrition: global and regional exposures and health consequences. Lancet 371 (9608), 243–260.

Black, R.E., Victora, C.G., Walker, S.P., Bhutta, Z.A., Christian, P., De Onis, M., et al., 2013. Maternal and child undernutrition and overweight in low-income and middle-income countries. Lancet 382 (9890), 427–451.

Bonanno, A., Goetz, S.J., 2012a. Wal-Mart and local economic development: a survey. Econ. Dev. Q. 0891242412456738.

Bonanno, A., Lopez, R.A., 2012b. Wal-Mart's monopsony power in metro and non-metro labor markets. Reg. Sci. Urban Econ. 42 (4), 569–579.

Bouis, H., Haddad, L., 1992. Are estimates of calorie-income elasticities too high?: a recalibration of the plausible range. J. Dev. Econ. 39 (2), 333–364.

Branca, F., Nikogosian, H., Lobstein, T., 2007. The Challenge of Obesity in the WHO European Region and the Strategies for Response. WHO Publication, Europe Community Interventions for the Prevention of Obesity, xiv + 59 pages.

Briceno, B., Coville, A., Martinez, S., 2015. Promoting Handwashing and Sanitation: Evidence From a Large-Scale Randomized Trial in Rural Tanzania. The World Bank, Policy Research Working Paper Series: 7164.

Broady, K.E., Meeks, A.G., 2015. Rev. Black Polit. Econ. 42 (3), 201–209.

Brody, A., Spieldoch, A., Aboud, G., 2014. Gender and Food Security: Towards Gender-Just Food and Nutrition Security. IDS, UK.

Broussard, H.N., 2012. Food aid and adult nutrition in rural Ethiopia. Agric. Econ. 43, 45–59.

Brown, L., 2015. Rebalancing Agriculture Will Deliver for Nutrition and Gender Equality.

Brown, L., Deshpande, C., Hill, C.L.M. et al., 2009. Module 1: Gender and Food Security. In: Gender and Agriculture Sourcebook: Investing in Women as Drivers of Agricultural Growth. World Bank; Washington DC.

Browning, M., Chiappori, P.A., 1998. Efficient Intra-household allocations: a general characterization and empirical tests. Econometrica 66 (6), 1241–1278.

Bundy, D., Burbano, C., Grosh, M., Gelli, A., Jukes, M., Drake, L., 2009. Rethinking School Feeding: Social Safety Nets, Child Development, and the Education Sector. World Bank, Washington, DC.

Burchi, F., 2010. Child nutrition in Mozambique in 2003: the role of mother's schooling and nutrition knowledge. Econ. Hum. Biol. Elsevier 8 (3), 331–345.

Burgstahler, R., Gundersen, C., Garasky, S., 2012. The supplemental nutrition assistance program, financial stress, and childhood obesity. Agric. Resour. Econ. Rev. 41 (1), 29.

Buttenheim, A.M., Alderman, H., Friedman, J., 2011. Impact Evaluation of School Feeding Programs in Lao PDR. World Bank Policy Research Working Paper, 5518.

Buttet, S., Dolar, V., 2015. Toward a quantitative theory of food consumption choices and body weight. Econ. Hum. Biol. 17, 143–156.

Calonico, S., Cattaneo, D.M., Farrell, H.M., 2016. rdrobust: Software for regression discontinuity designs. Stata J. (2), 1–30, forthcoming. <http://www-personal.umich.edu/~cattaneo/papers/Calonico-Cattaneo-Farrell-Titiunik_2016_Stata.pdf>.

Cameron, C.A., Pravin, K.T., 2010. Microeconometrics Using Stata. Stata Press, College Station, Texas.

Cameron, L., Shah, M., Olivia S., 2013. Impact Evaluation of a Large-Scale Rural Sanitation Project in Indonesia. The World Bank, Policy Research Working Paper Series: 6360.

Carneiro, P., Ginja, R., 2014. Long-term impacts of compensatory preschool on health and behavior: evidence from head start. Am. Econ. J. Econ. Policy 6 (4), 135–173.

Carson, S.A., 2015. A weighty issue: diminished net nutrition among the U.S. working class in the nineteenth century. Demography 52 (3), 945–966.

Cash, S.B., Schroeter, C., 2010. Behavioral economics: a new heavyweight in Washington? Choices Mag. 25 (3), 38.

Cawley, J., Maclean, J.C., 2012. Unfit for service: the implications of rising obesity for US military recruitment. Health Econ. 21 (11), 1348–1366.

Cawley, J., Maclean, J.C., 2013. The consequences of rising youth obesity for US military academy admissions. Appl. Econ. Perspect. Policy 35 (1).

Cawley, J., Meyerhoefer, C., 2012. The medical care costs of obesity: an instrumental variables approach. J. Health Econ. 31 (1), 219–230.

Cawley, J., Frisvold, D., Meyerhoefer, C., 2013. The impact of physical education on obesity among elementary school children. J. Health Econ. 32 (4), 743–755.

Cesur, R., Herbst, C.M., Tekin, E., 2010. Chapter 3 Child Care Choices and Childhood Obesity. Current Issues in Health Economics (Contributions to Economic Analysis, Volume 290), 290. Emerald Group Publishing Limited, pp. 37–62.

Chaloupka, F.J., Powell, L.M., Chriqui, J.F., 2011a. Sugar-sweetened beverage taxation as public health policy-lessons from tobacco. Choices 26 (3), 1–6.

Chaloupka, F.J., Powell, L.M., Chriqui, J.F., 2011b. Sugar-sweetened beverages and obesity: the potential impact of public policies. J. Policy Anal. Manag. 30 (3), 645–655.

Chang, K.-L., Zastrow, M., Zdorovtsov, C., Quast, R., Skjonsberg, L., Stluka, S., 2015. Do SNAP and WIC programs encourage more fruit and vegetable intake? A household survey in the northern great plains. J. Family Econ. Issues 36 (4), 477–490.

Charman, A.J.E., 2008. Empowering Women Through Livelihoods Orientated Agricultural Service Provision: A Consideration of Evidence from Southern Africa. UNU-WIDER, Helsinki, Finland.

Chen, S., Florax, R.J., Snyder, S., Miller, C.C., 2010. Obesity and access to chain grocers. Econ. Geogr. 86 (4), 431–452.

Chen, S.E., Florax, R.J., Snyder, S.D., 2013. Obesity and fast food in urban markets: a new approach using geo-referenced micro data. Health Econ. 22 (7), 835–856.

Chia, Y.F., 2013. Dollars and pounds: the impact of family income on childhood weight. Appl. Econ. 45 (14), 1931–1941.

Chiappori, P.A., 1988. Rational household labor supply. Econometrica 56 (1), 63–89.

Chiwaula, L.S., Kaluwa, B.M., 2008. Household consumption of infant foods in two low-income districts in Malawi. J. Int. Dev. 20 (5), 686–697.

Chowhan, J., Stewart, M.J., 2014. While mothers work do children shirk? Determinants of youth obesity. Appl. Econ. Perspect. Policy 36 (2), 287–308.

Classen, T.J., 2010. Measures of the intergenerational transmission of body mass index between mothers and their children in the United States, 1981–2004. Econ. Hum. Biol. 8 (1), 30–43.

Cleveland, G., Krashinsky, M., 1998. The Benefits and Costs of Good Child Care. University of Toronto at Scarborough Report.

Coady, P.D., 2004. Designing and Evaluating Social Safety Nets: Theory, Evidence and Policy Conclusions. International Food Policy Research Institute, Washington, DC, Food Consumption and Nutrition Division Discussion Paper No. 172.

Cohen, R.J., Brown, K.H., Canahuati, J., Rivera, L.L., Dewey, K.G., 1995. Determinants of growth from birth to 12 months among breast-fed Honduran infants in relation to age of introduction of complementary foods. Pediatrics 96, 504–510.

Condliffe, S., Link, C.R., 2014. Racial differences in the effects of hypertension and obesity on health expenditures by diabetes patients in the US. Appl. Econ. Lett. 21 (4–6), 280–283.

Conner, D.S., Izumi, B.T., Liquori, T., Hamm, M.W., 2012. Sustainable school food procurement in large K-12 districts: prospects for value chain partnerships. Agric. Resour. Econ. Rev. 41 (1), 100–113.

Courtemanche, C., 2011. A silver lining? The connection between gasoline prices and obesity. Econ. Inquiry 49 (3), 935–957.

Courtemanche, C., Carden, A., 2011. Supersizing supercenters? The impact of Walmart Supercenters on body mass index and obesity. J. Urban Econ. 69 (2), 165–181.

Craven, B.M., Stewart, G.T., 2013. Economic implications of socio-cultural correlates of HIV/AIDS: an analysis of global data. Appl. Econ. 45 (13–15), 1789–1800.

Craven, B.M., Marlow, M.L., Shiers, A.F., 2012. Fat taxes and other interventions won't cure obesity. Econ. Aff. 32 (2), 36–40.

Crosnoe, R., 2012. Obesity, family instability, and socioemotional health in adolescence. Econ. Hum. Biol. 10 (4), 375–384.

Crutchfield, S., Kuchler, F., Variyam, J.N., 2001. The economic benefits of nutrition labeling: a case study for fresh meat and poultry products. J. Consum. Policy 24 (2), 185–207.

Cuesta, J., 2007. Child malnutrition and the provision of water and sanitation in the Philippines. J. Asia Pac. Econ. 12 (2), 125–157.

Damon, A., Kristiansen, D., 2014. Childhood obesity in Mexico: the effect of international migration. Agric. Econ. 45 (6), 711–727.

Dasgupta, P., 1995. The population problem: theory and evidence. J. Econ. Lit. 33 (4), 1879–1902.

Dasgupta, S., 2012. Sex-Selective Abortions, Gender Discrimination in Child Health and Nutrition, and Marriage Patterns: Empirical Evidence From India. University of Colorado.

Datar, A., Nicosia, N., Datt, G., Jolliffe, D., Sharma, M., 2001. A profile of poverty in Egypt. Afr. Dev. Rev. 13 (2), 202–237.

Datta, U., 2015. Socio-economic impacts of JEEViKA: a large-scale self-help group project in Bihar, India. World Dev. 68, 1–18.

Dave, D.M., Tennant, J., Colman, G.J., 2011. Isolating the Effect of Major Depression on Obesity: Role of Selection Bias (No. w17068). National Bureau of Economic Research.

Davis, G.C., You, W., 2010. The time cost of food at home: general and food stamp participant profiles. Appl. Econ. 42 (19–21), 2537–2552, World Bank, DC, USA.

Davis, G.C., You, W., 2013. Estimates of returns to scale, elasticity of substitution, and the thrifty food plan meal poverty rate from a direct household meal production function. Food Policy 43, 204–212.

Dawson, P.J., Tiffin, R., 1998. Is there a long-run relationship between population growth and living standards? The case of India. J. Dev. Stud. 34 (5), 149–156.

de Brauw, A., Hoddinott, J., 2011. Must conditional cash transfer programs be conditioned to be effective? The impact of conditioning transfers on school enrollment in Mexico. J. Dev. Econ. 96 (2), 359–370.

de Onis, M., Dewey, K.G., Borghi, E., Onyango, A.W., Blössner, M., Daelmans, B., et al., 2013. The World Health Organization's global target for reducing childhood stunting by 2025: rationale and proposed actions. Matern. Child Nutr. 9 (Suppl. 2), 6–26.

Deaton, A., 2010. Instruments, randomization, and learning about development. J. Econ. Lit. 48 (2), 424–455.

Deaton, A., Drèze, J., 2009. Food and nutrition in India: facts and interpretations. Econ. Polit. Wkly 14, 42–65.

Dehejia, R., 2013. The Porous Dialectic. WIDER Working Paper No. 2013/11, United Nations University.

Dehejia, R., Wahba, S., 2002. Propensity score matching methods for. Rev. Econ. Stat. 84, 151–161.

Deininger, K., Liu, Y., 2009. Longer-Term Economic Impacts of Self-Help Groups in India. The World Bank, Policy Research Working Paper Series: 4886.

Dekker, L.H., Mora-Plazas, M., Marín, C., Baylin, A., Villamor, E., 2010. Stunting associated with poor socioeconomic and maternal nutrition status and respiratory morbidity in Colombian school children. Food Nutr. Bull. 31, 242–250.

Delpeuch, F., Traissac, P., Martin-Prével, Y., Massamba, J., Maire, B., 2000. Economic crisis and malnutrition: socioeconomic determinants of anthropometric status of preschool children and their mothers in an African urban area. Public Health Nutr 3, 39–47.

Deng, Z. (Ed.), 2009. China's economy: rural reform agricultural development. In: Series on Developing China – Translated Research from China, vol. 1. World Scientific, Hackensack, NJ and Singapore.

Devaux, M., Sassi, F., Church, J., Cecchini, M., Borgonovi, F., 2011. Exploring the relationship between education and obesity. OECD J. Econ. Stud. 5 (1), 121–159.

Dewey, K.G., Cohen, R.J., 2007. Does birth spacing affect maternal or child nutritional status? A systematic literature review. Matern. Child Nutr. 3, 151–173.

Dhar, T., Baylis, K., 2011. Fast-food consumption and the ban on advertising targeting children: the Quebec experience. J. Market. Res. 48 (5), 799–813.

Dharmasena, S., Capps, O., 2012. Intended and unintended consequences of a proposed national tax on sugar-sweetened beverages to combat the US obesity problem. Health Econ. 21 (6), 669–694.

Dharmasena, S., Capps, O., Clauson, A., 2011. Ascertaining the impact of the 2000 USDA dietary guidelines for Americans on the intake of calories, caffeine, calcium, and vitamin C from at-home consumption of nonalcoholic beverages. J. Agric. Appl. Econ. 43 (1), 13–27.

Dharmasena, S., Davis, G.C., Capps Jr., O., 2014. Partial versus general equilibrium calorie and revenue effects associated with a sugar-sweetened beverage tax. J. Agric. Resour. Econ. 39 (2), 157–173.

Díaz-Bonilla, E., 2015. Macroeconomics, Agriculture, and Food Security: A Guide to Policy Analysis in Developing Countries. International Food Policy Research Institute.

Dickinson, K.L., Patil, S.R., Pattanayak, S.K., Poulos, C., Yang, J.-H., 2015. Nature's Call: Impacts of Sanitation Choices in Orissa. Econ. Dev. Cult. Change 64 (1), 1–29.

Ding, D., Klaus, G., 2012. Built environment, physical activity, and obesity: what have we learned from reviewing the literature? Health Place 18 (1), 100–105.

Doepke, M., Tertilt, M., 2011. Does Female Empowerment Promote Economic Development? NBER Working Paper No. 19888.

Dolar, V., 2010. Assessing the Effect of Changes in Relative Food Prices and Income on Obesity Prevalence in the United States (Doctoral dissertation). University of Minnesota.

Doss, C.R., 1996. Testing among models of intrahousehold resource allocation. World Dev. 24 (10), 1597–1609.

Doss, C.R., 2001. Designing agricultural technology for African women farmers: lessons from 25 years of experience. World Dev. 29 (12), 2075–2092.

Duffy, P., Yamazaki, F., Zizza, C.A., 2012. Can the dietary guidelines for Americans (2010) help trim America's waistline? Choices, 1st Quarter 27 (1).

Duflo, E., 2003. Grandmothers and granddaughters: old-age pensions and intra-household allocation in South Africa. World Bank Econ. Rev. 17 (1), 1–25.

Duflo, E., Rachel, G., Michael, K., 2007. Using randomization in development economics research: a toolkit. In: Schults, T.P., Strauss, J. (Eds.), Handbook of Development Economics, vol. 4. Elsevier Science Ltd., North Holland, pp. 3862–3895.

Duflo, E., Kremer, M., Robinson, J., 2011. Nudging farmers to use fertilizer: theory and experimental evidence from Kenya. Am. Econ. Rev. 101 (6), 2350–2390.

Duflo, E., Greenstone, M., Guiteras, R., Clasen, T., 2015. Toilets Can Work: Short and Medium Run Health Impacts of Addressing Complementarities and Externalities in Water and Sanitation. National Bureau of Economic Research, Inc, NBER Working Papers: 21521.

Dunn, R.A., 2010. The effect of fast-food availability on obesity: an analysis by gender, race, and residential location. Am. J. Agric. Econ. aaq041.

Dunn, R.A., Sharkey, J.R., Horel, S., 2012. The effect of fast-food availability on fast-food consumption and obesity among rural residents: an analysis by race/ethnicity. Econ. Hum. Biol. 10 (1), 1–13.

Durham, C., Eales, J., 2010. Demand elasticities for fresh fruit at the retail level. Appl. Econ. 42 (11), 1345–1354.

Ecker, O., Qaim, M., 2011. Analyzing nutritional impacts of policies: an empirical study for Malawi. World Dev. 39 (3), 412–428.

Economos, D.C., Sliwa, A.S., 2011, Social Science Insights into Prevention, Treatment, and Policy: Community Interventions, in The Oxford Handbook of the Social Science of Obesity, pp. 713–40, Collective Volume Article by John Cawley, Oxford Handbooks Series. Oxford University Press, Oxford and New York.

Engle, P.L., Menon, P., Haddad, L., 1999. Care and nutrition: concepts and measurement. World Dev. 27 (8), 1309–1337.

Fan, M., 2010. Do food stamps contribute to obesity in low-income women? Evidence from the National Longitudinal Survey of Youth 1979. Am. J. Agric. Econ. aaq047.

FAO, 2006. Food Security. Policy Brief Issue 2, June 2006. United Nations Food and Agriculture Organization (FAO), Rome. Available from: <http://www.fao.org> (accessed October 2007.).

FAO, 2009. Designing CCT Programs to Improve Nutrition Impact: Principles, Evidence and Examples: Working Paper #06. Research organized by the Hunger-Free Latin America and the Caribbean Initiative.

FAO, 2015. The State of Food Insecurity in the World. FAO, Rome.

FAOSTAT, 2015. <http://faostat3.fao.org/home/E> (accessed 23.01.13.).

Fernando, J., 2010. Three Essays on Canadian Household Consumption of Food Away From Home With Special Emphasis on Health and Nutrition (Doctoral dissertation). University of Alberta.

Figlio, D.N., Joshua, W., 2005. Food for thought: the effects of school accountability plans on school nutrition. J. Public Econ. 89 (2–3), 381–394.

Finkelstein, E.A., Strombotne, K.L., Popkin, B.M., 2010. The costs of obesity and implications for policy-makers. Choices, 3rd Quarter 25 (3).

Finocchio, C.P.S., Dewes, H., 2015. Food away from home and obesity in Brazil. J. Agribusiness Dev. Emerg. Econ. 5 (1), 44–56.

Fischer, E., Qaim, M., 2012. Gender, agricultural commercialization and collective action in Kenya. In: Paper Presented at: the International Association of Agricultural Economists Triennial Conference, Foz do Iguaçu, Brazil, 18–24 August, 2012.

Fitzpatrick, D.J., Toner, E.A., Sommers, P.M., 2010. The skinny on obesity rates and the US economy. Atl. Econ. J. 38 (1), 119.

Fletcher, M.J., 2011. Peer effects and obesity. In: The Oxford Handbook of the Social Science of Obesity, pp. 303–312, Collective Volume Article by John Cawley, Oxford Handbooks Series. Oxford University Press, Oxford and New York.

Fletcher, M.J., 2014. The interplay between gender, race and weight status: self perceptions and social consequences. Econ. Hum. Biol. 14, 79–91.

Fletcher, J.M., Frisvold, D.E., Tefft, N., 2010a. The effects of soft drink taxes on child and adolescent consumption and weight outcomes. J. Public Econ. 94 (11), 967–974.

Fletcher, J.M., Frisvold, D., Tefft, N., 2010b. Can soft drink taxes reduce population weight? Contem. Econ. Policy 28 (1), 23–35.

Forste, R., Moore, E., 2012. Adolescent obesity and life satisfaction: perceptions of self, peers, family, and school. Econ. Hum. Biol. 10 (4), 385–394.

Fortich, M., Roberto, G., Juan, D., 2011. Los determinantes de la obesidad en Colombia. (Determinants of obesity in Colombia. With English summary). Technological U Bolivar; Economia y Region 5 (2), 155–182.

Fox, J., 1997. Applied Regression Analysis, Linear Models, and Related Methods. Sage Publications, Inc, node/3608/3501/397795.

Fox, L., Hutto, N., 2013. The effect of obesity on intergenerational income mobility. Appl. Demogr. Public Health. pp. 33–44.

Gaiha, R., Jha, R., Kulkarni, V., 2013. Demand for nutrients in India: 1993 to 2004. Appl. Econ. 45 (14), 1869–1886.

Gavan, J.D., Chandrasekara, I.S., 1979. The Impact of Public Food Distribution on Food Consumption Welfare in Sri Lanka. IFFPRI Research Report 13. International Food Policy Research Institute, Washington, DC.

Gelli, A., Hawkes, C., Donovan, J., Harris, J., Allen, S., et al., 2015. Value Chains and Nutrition – A Framework to Support the Identification, Design and Evaluation of Interventions. 01413. IFPRI Discussion Paper. Washington, DC.

Gelli, A., Masset, E., Folson, G., Kusi, A., Arhinful, D.K., Asante, F., et al., 2016. Evaluation of alternative school feeding models on nutrition, education, agriculture and other social outcomes in Ghana: rationale, randomised design and baseline data. Trials 17, 37. Available from: <http://dx.doi.org/10.1186/s13063-015-1116-0>

Gentilini, U., 2014. Our Daily Bread: What Is the Evidence on Comparing Cash Versus Food Transfers? Social Protection and Labour Discussion Paper No. 1420. World Bank, Washington, DC.

Gentilini, U., 2016. Revisiting the "cash versus food" debate: New evidence for an old puzzle? World Bank Res. Obs. 31 (1), 135–167.

Geruso, M., Spears, D., 2015. Neighborhood Sanitation and Infant Mortality. NBER Working Paper No. 21184.

Gibson, J., Kim, B., 2013. Quality, quantity, and nutritional impacts of rice price changes in Vietnam. World Dev. 43, 329–340.

Gibson, J., Rozelle, S., 2011. The effects of price on household demand for food and calories in poor countries: are our databases giving reliable estimates? Appl. Econ. 43 (25–27), 4021–4031.

Gillespie, S., 2013. Myths and realities of child nutrition. Econ. Polit. Wkly XLVII (34), 64–67.

Gillespie, S., Haddad, L., 2000. Attacking the double burden of malnutrition in Asia: a synthesis of findings from the ADB-IFPRI Regional Technical Assistance Project 5824 on Nutrition Trends, Policies and Strategies in Asia and the Pacific.

Gillespie, S., Haddad, L., 2002. Food security as a response to AIDS. AIDS Food Secur. 10–16.

Gillespie, S., Haddad, L.J., 2002. Food Security as a Response to AIDS: IFPRI 2001–2002 Annual Report Essay (No. 2002 Essay2). International Food Policy Research Institute (IFPRI).

Gillespie, S., Harris, J., Kadiyala, S., 2012. The Agriculture–Nutrition Disconnect in India: What Do We Know? IFPRI Discussion Paper 01187. International Food Policy Research Institute, Washington, DC.

Gillespie, S., Menon, P., Kennedy, A.L., 2015. Scaling up impact on nutrition: what will it take. Adv. Nutr. Int. Rev. J. 6 (4), 440–451.

Gilligan, D.O., Hoddinott, J., Taffesse, A.S., 2008. The Impact of Ethiopia's Productive Safety Net Program and Its Linkages. IFPRI Discussion Paper 00839. International Food Policy Research Institute, Washington, DC.

Gilligan, D.O., Hoddinot, J., Kumar, N., Taqsesse, A.S., 2009. An Impact Evaluation of Ethiopia'a Productive Safety Nets Programme. IFPRI, Washington, DC.

Gittelsohn, J., Lee, K., 2013. Integrating educational, environmental, and behavioral economic strategies may improve the effectiveness of obesity interventions. Appl. Econ. Perspect. Policy 35 (1), 52–68.

Gius, P.M., 2011. The prevalence of obesity and overweight among young adults: an analysis using the NLSY. Int. J. Appl. Econ. 8 (1), 36–45.

Goetzel, Z.R., 2011. Workplace obesity prevention programs. In: The Oxford Handbook of the Social Science of Obesity, pp. 683–712, Collective Volume Article by John Cawley, Oxford Handbooks Series. Oxford University Press, Oxford and New York.

Gopalan, C., Ramasastri, B.V., Balasubramanian, S.C., 1977. Nutritive Value of Indian Foods. National Institute of Nutrition, Hyderabad, India.

Grecu, M.A., Rotthoff, W.K., 2015. Economic growth and obesity: findings of an obesity Kuznets curve. Appl. Econ. Lett. 22 (7-9), 539–543.

Greve, J., 2011. New results on the effect of maternal work hours on children's overweight status: does the quality of child care matter? Labour Econ. 18 (5), 579–590.

Grosh, M., Del Ninno, C., Tesliuc, E., Ouerghi, A., 2008. For Protection and Promotion: The Design and Implementation of Effective Safety Nets. World Bank, Washington, DC.

Grosh, M., Del Nino, C., Tesliuc, E., Ouerghi, A., 2008. The Design and Implementation of Effective Safety Nets for Protection and Promotion. World Bank, Washington, DC.

Grossman, M., Tekin, E., Wada, R., 2014. Food prices and body fatness among youths. Econ. Hum. Biol. 12, 4–19.

Guettabi, M., Munasib, A., 2014. "Space obesity": the effect of remoteness on county obesity. Growth Change 45 (4), 518–548.

Gunther, I., Fink, G., 2013. Saving a life-year and reaching MDG 4 with investments in water and sanitation: a cost-effective policy? Eur. J. Dev. Res. 25 (1), 129–153.

Gurley-Calvez, T., Higginbotham, A., 2010. Childhood Obesity, academic achievement, and school expenditures. Public Finance Rev. 38 (5), 619–646.

Gustavsen, G.W., Nayga Jr, M.R., Wu, X., 2011. Obesity and moral hazard in demand for visits to physicians. Contem. Econ. Policy 29 (4), 620–633.

Guthman, J., 2013. Too much food and too little sidewalk? Problematizing the obesogenic environment thesis. Environ. Plann. 45 (1), 142–158.

Gutierrez, R.G., Linhart, J.M., Pitblado, J.S., 2003. From the help desk: local polynomial regression and Stata plugins. Stata J. 3 (4), 412–419.

Habib, R.R., 2012. Understanding water, understanding health: the case of Bebnine, Lebanon. In: Charron, D. F. (Ed.), Insight and Innovation in International Development, Ecohealth Research in Practice, Innovative Applications of an Ecosystem Approach to Health. International Development Research Centre Ottawa, Ontario, Canada, pp. 203–215. Springer, Chapter 19.

Haddad, L., Kanbur, R., 1990. How serious is the neglect of intra-household inequality? Econ. J. 100 (402), 866–881.

Haddad, L., Hoddinott, J., Alderman, H., 1997. Intrahousehold Resource Allocation: Policy Issues and Research Methods. Johns Hopkins University Press, Baltimore, MD.

Haddad, L., Nisbett, N., Barnett, I., Valli, E., 2014. Maharastra's Child Stunting Declines: What Is Driving Them? Findings of a Multidisciplinary Analysis. Institute of Development Studies, UNICEF.

Haggblade, S., Babu, S.C., Harris, J., Mkandawire, E., Nthani, D., Hendriks, S.L., 2016. Drivers of Micronutrient Policy Change in Zambia: An Application of the Kaleidoscope Model. Innovation Lab for Food Security Policy Working Paper No.C3-3. Michigan State University; International Food Policy Research Institute (IFPRI); and University of Pretoria, Lansing, MI, <http://ebrary.ifpri.org/cdm/ref/collection/p15738coll2/id/130253>.

Hajizadeh, M., Campbell, M.K., Sarma, S., 2014. Socioeconomic inequalities in adult obesity risk in Canada: trends and decomposition analyses. Eur. J. Health Econ. 15 (2), 203–221.

Hall, A., 2008. Brazil's Bolsa Familia: a double-edged sword. Dev. Change 39 (5), 799–822.

Hammer, J., Spears, D., 2013, Village Sanitation and Children's Human Capital: Evidence From a Randomized Experiment by the Maharashtra Government. The World Bank, Policy Research Working Paper Series: 6580.

Han, E., Powell, M.L., 2013. Fast food prices and adult body weight outcomes: evidence based on longitudinal quantile regression models. Contem. Econ. Policy 31 (3), 528–536.

Hanbury, M.M., 2013. The US Safety Net and Obesity. University of California, Davis.

Handa, S., King, D., 2003. Adjustment with a human face? Evidence from Jamaica. World Dev. 31 (7), 1125–1145.

Harding, M., Lovenheim, M., 2014. The Effect of Prices on Nutrition: Comparing the Impact of Product- and Nutrient-Specific Taxes. National Bureau of Economic Research, Inc, NBER Working Papers: 19781.

Harvey, P., 2007. Cash-Based Responses in Emergencies. Humanitarian Policy Group Report 24. Overseas Development Institute, London.

Harvey, P., Bailey, S., 2011. Good Practice Review: Cash Transfer Programming in Emergencies. Good Practice Reviews No. 11. Humanitarian Practice Network. Overseas Development Institute, London.

Hathi, P., Haque, S., Pant, L., Coffey, D., Spears, D., 2014. Place and Child Health: the Interaction of Population Density and Sanitation in Developing Countries. The World Bank, Policy Research Working Paper Series: 7124.

Hawkes, C., Ruel, M., Babu, S., 2007. Agriculture and health: overview, themes, and moving forward. Food Nutr. Bull. 28 (2 Suppl.), S221–S226.

Heckman, J.J., Urzúa, S., 2009. Comparing IV with Structural Models: What Simple IV Can and Cannot Identify.

Heflin, C., Mueser, P.R., 2013. Aid to Jobless Workers in Florida in the Face of the Great Recession: The Interaction of Unemployment Insurance and the Supplemental Nutritional Assistance Program.

Herbst, M.C., Tekin, E., 2011. Child care subsidies and childhood obesity. Rev. Econ. Household 9 (3), 349–378.

Herbst, M.C., Tekin, E., 2012. The accessibility of child care subsidies and evidence on the impact of subsidy receipt on childhood obesity. J. Urban Econ. 71 (1), 37–52.

Hernandez-Diaz, S., Peterson, K., Dixit, S., Hernandez, B., Parra, S., Barquera, S., et al., 1999. Association of maternal short stature with stunting in Mexican children: common genes vs common environment. Eur. J. Clin. Nutr. 53, 938–945.

Herrmann, H. 2009. An Introduction and Review of Cash Transfer Experiences and Their Feasibility as a Food Security Tool for World Food Programme in Bolivia. Consultancy Report. Swiss Agency for Development and Cooperation and the World Food Programme, Bolivia. HLPE (High Level Panel of Experts of the United Nations Committee on World Food Security).

Himes, L.C., 2011. The demography of obesity. In The Oxford Handbook of the Social Science of Obesity, pp. 35–47, Collective Volume Article by John Cawley, Oxford Handbooks Series. Oxford University Press, Oxford and New York.

Himes, L.C., Episcopo, V., 2011, Obesity: a sociological examination. In: Handbook of Sociology of Aging, pp. 513–531, Collective Volume Article, by Settersten A. Richard Jr, and Angel L. Jacqueline, Handbooks of Sociology and S1995aocial Research. Springer, New York and Heidelberg.
Hoddinott, J., et al., 2013. The economic rationale for investing in nutrition. John Wiley & Sons Ltd, Matern. Child Nutr. 9 (Suppl. 2), 69–82.
Hoddinott, J., 2016. The Economics of Reducing Malnutrition in Sub-Saharan Africa. Global Panel on Agriculture and Food Systems for Nutrition.
Hoddinott, J., Skoufias, E., 2004. The impact of PROGRESA (Programa de Educación, Salud, y Alimenación) on food consumption. Econ. Dev. Cult. Change 53 (1), 37–61.
Hoddinott, J., Wiesmann, D., 2010. The impact of conditional cash transfer programs on food consumption. In: Adato, M., Hoddinott, J. (Eds.), Conditional Cash Transfers in Latin America. Johns Hopkins University Press, Baltimore, pp. 258–283.
Hoddinott, J., Maluccio, J.A., Behrman, J.R., Flores, R., Martorell, R., 2008. Effect of a nutrition intervention during early childhood on economic productivity in Guatemalan adults. Lancet 371, 411–416.
Hoddinott, J., Berhane, G., Gilligan, D., Kumar, N., Taffesse, A.S., 2012. The impact of Ethiopia's productive safety net programme and related transfers on agricultural productivity. J. Afr. Econ. 21 (5), 761–786.
Hoddinott, J., Behrman, J.R., Maluccio, J.A., Melgar, P., Quisumbing, A.R., Ramirez-Zea, M., et al., 2013. Adult consequences of growth failure in early childhood. Am. J. Clin. Nutr. 98 (5), 1170–1178.
Hoddinott, J., Gilligan, D., Hidrobo, M., Margolies, A., Roy, S., Sandström, S., et al., 2013. Enhancing WFP's Capacity and Experience to Design, Implement, Monitor, and Evaluate Vouchers and Cash Transfer Programs: Study Summary. International Food Policy Research Institute, Washington, DC.
Hong, H.G., Yue, Y., Ghosh, P., 2015. Bayesian estimation of long-term health consequences for obese and normal-weight elderly people. J. R. Stat. Soc. Ser. A (Stat. Soc.) 178 (3), 725–739.
Hong, S.C., 2013. Has the burden of obesity declined in America since 1980? Korea World Econ. 14 (2), 343–379.
Hoque, N., Howard J., 2013. The Implications of Aging and Diversification of Population on Overweight and Obesity and the Cost Associated With Overweight and Obesity in Texas, 2000–2040.
Hoque, N., McCusker, M.E., Murdock, S.H., Perez, D., 2010. The implications of change in population size, distribution, and composition on the number of overweight and obese adults and the direct and indirect cost associated with overweight and obese adults in Texas through 2040. Popul. Res. Policy Rev. 29 (2), 173–191.
Horrell, S., Oxley, D., 2015. Gender Discrimination in 19th c England: Evidence From Factory Children. University of Oxford, Discussion Papers in Economic and Social History, Number 133, February.
Horton, S., Steckel, R., 2013). Global Economic Losses Attributable to Malnutrition 1900–2000 and Projections to 2050. Assessment Paper for Copenhagen Consensus on Human Challenges.
Horton, S., Ross, J., 2003. The economics of iron deficiency. Food Policy 28 (1), 51–75.
Hoy, G.M., Childers, C.C., 2012. Trends in food attitudes and behaviors among adults with 6–11 year-old children. J. Consum. Aff. 46 (3), 556–572.
Huang, Y., 2012. An Econometric Study of the Impact of Economic Variables on Adult Obesity and Food Assistance Program Participation in the NLSY Panel.
Huang, Y., Huffman, W., Tegene, A., 2012. Impacts of economic and psychological factors on women's obesity and food assistance program participation: evidence from the NLSY panel. Am. J. Agric. Econ. 94 (2), 331–337.
Hussain, Z., 2012. Three Essays on Health and Labor Economics.
Imbens, G.W., 2000. The role of propensity score in estimating dose–response functions. Biometrika 87 (3), 706–710.

Imbens, G.W., 2010. Better LATE than nothing: some comments on Deaton (2009) and Heckman and Urzua (2009). J. Econ. Literature 399–423.

IFPRI, 2014. Global Nutrition Report 2014: Actions and Accountability to Accelerate the World's Progress on Nutrition. International Food Policy Research Institute (IFPRI), Washington, DC. Available from: <http://dx.doi.org/10.2499/9780896295643>

International Food Policy Research Institute, 2015. Global Nutrition Report 2015: Actions and Accountability to Advance Nutrition and Sustainable Development. Washington, DC. <http://10.2499/10.2499/9780896298835>.

Ishdorj, A., Crepinsek, M.K., Jensen, H.H., 2013. Children's consumption of fruits and vegetables: do school environment and policies affect choices at school and away from school? Appl. Econ. Perspect. Policy 35 (2), 341–359.

James, M., Alastair, S.J.P., 2015. The effect of school district nutrition policies on dietary intake and overweight: a synthetic control approach. Demography 47 (Suppl), S211–S231.

Jegasothy, K., Duval, Y., 2003. Food demand in urban and rural Samoa. Pac. Econ. Bull. 18 (2), 50–64.

Jehn, M., Brewis, A., 2009. Paradoxical malnutrition in mother-child pairs: untangling the phenomenon of over- and under-nutrition in underdeveloped economies. Econ. Hum. Biol. 7 (1), 28–35.

Jensen, H.H., Wilde, E.P., 2010. More than just food: the diverse effects of food assistance programs. Choices, 3rd Quarter 25 (3).

Jensen, R., 2011. Do labor market opportunities affect young women's work and family decisions? Experimental evidence from India. Q. J. Econ. 127 (2), 753–792.

Jensen, R., 2012. Another mouth to feed? The effects of (in)fertility on malnutrition. CESifo Econ. Stud. 58 (2), 322–347.

Jensen, R.T., Nolan, H.M., 2011. Do consumer price subsidies really improve nutrition? Rev. Econ. Stat. 93 (4), 1205–1223.

Jha, R., Bhattacharyya, S., Gaiha, R., Shankar, S., 2009. "Capture" of anti-poverty programs: an analysis of the National Rural Employment Guarantee Program in India. J. Asian Econ. 20, 456–464.

Jha, R., Bhattacharyya, S., Gaiha, R., 2011. Social safety nets and nutrient deprivation: an analysis of the National Rural Employment Guarantee Program and the Public Distribution System in India. J. Asian Econ. 22 (2), 189–201.

Jha, R., Gaiha, R., Pandey, M.K., 2013. Body Mass Index, participation, duration of work and earnings under the National Rural Employment Guarantee Scheme: evidence from Rajasthan. J. Asian Econ. 26, 14–30.

Jha, R., Gaiha, R., Deolalikar, A.B. (Eds.), 2014. Handbook on Food: Demand, Supply, Sustainability and Security. Elgar, Cheltenham, U.K. & Northampton, MA.

Jha, R., Gaiha, R., Pandey, M.K., Shankar, S., 2015. Determinants and persistence of benefits from the National Rural Employment Guarantee Scheme – panel data analysis for Rajasthan, India. Eur. J. Dev. Res. 27 (2), 308–329.

Johnston, W.D., Lee, W.-S., 2011. Explaining the female black-white obesity gap: a decomposition analysis of proximal causes. Demography 48 (4), 1429–1450.

Jolly, C.M., Namugabo, E., Nguyen, G., Diawara, N., Jolly, P., Ovalle, F., 2013. Net food imports and obesity in selected Latin American and Caribbean countries. Adv. Manag. Appl. Econ. 3 (6), 159–177.

Jones, A.D., Allison M., 2015. Examining the relationship between farm production diversity and diet diversity across subsistence- and market-oriented farms in Malawi. In: Paper Presented at the 5th Annual Leverhulme Centre for Integrative Research on Agriculture and Health (LCIRAH) Conference, London, England, June 3, 2015. <http://lcirah.ac.uk/sites/default/files/FINAL_Abstract_Bookletv2.pdf>.

Jones-Corneille, R.L., Stack, M.R., Wadden, A.T., 2011. Behavioral treatment of obesity. In: The Oxford Handbook of the Social Science of Obesity, pp. 771–791, Collective Volume Article by John, Oxford Handbooks Series. Oxford University Press, Oxford and New York.

Jones, W.P., Vavra, M., von Lampe, A., Fournier, L., Fulponi, C., Giner, P., et al., 2010. OECD-FAO Agricultural Outlook 2010–2019 Highlights, Food and Agricuture Organization and Organizaiton for the Economic Cooperaiton and Development, Rome and Geneva.

Jordan, A., Piotrowski, J.T., Bleakley, A., Mallya, G., 2012. Developing media interventions to reduce household sugar-sweetened beverage consumption. Ann. Am. Acad. Polit. Soc. Sci. 640, 118–135.

Kabeer, N., 1994. Reversed Realities: Gender Hierarchies in Development Thought. Verso.

Kalist, E.D., Siahaan, F., 2013. The association of obesity with the likelihood of arrest for young adults. Econ. Hum. Biol. 11 (1), 8–17.

Kandpal, E., 2011. Beyond average treatment effects: distribution of child nutrition outcomes and program placement in India's ICDS. World Dev. 39 (8), 1410–1421.

Kaur, P., 2011. Food promotion to children: understanding the need of responsibility in marketing to children. Inf. Manag. Business Rev. 2 (4), 133–137.

Kaur, P., 2013. Promoting foods to Indian children through product packaging. J. Competitiveness 5 (4), 134–146.

Kaushal, N., Muchomba, F.M., 2015. How consumer price subsidies affect nutrition. World Dev. 74, 25–42.

Kavitha, G., Lal, B.S., 2013. Economic impact of inadequate sanitation on women's health: a study in Warangal District. Int. J. Environ. Dev. 10 (2), 209–220.

Kenkel, D., 2010. Are health behaviors driven by information? In: Obesity and the Economics of Prevention: Fit Not Fat, pp. 141–145, in Collective Volume Article by Sassi Franco, in association with the Organisation for Economic Co-operation and Development. Elgar, Cheltenham, U.K. and Northampton, MA.

Kennedy, E., Peters, P., Haddad, L., Biswas, M.R., Gabr, M., 1994. Effects of Gender of Head of Household on Women's and Children's Nutritional Status. Oxford University Press, pp. 109–124.

Kersh, R., Morone, J., 2011. Obesity politics and policy. In: The Oxford Handbook of the Social Science of Obesity, pp. 158–172, Collective Volume Article by John, Oxford Handbooks Series. Oxford University Press, Oxford and New York.

Khera, R., 2006. Mid-day meals in primary schools: achievements and challenges. Econ. Polit. Wkly 4742–4750.

Kiesel, K., McCluskey, J.J., Villas-Boas, B.S., 2011. Nutritional labeling and consumer choices. Annu. Rev. Resour. Econ. 3 (1), 141–158.

Kim, H., House, A.L., Rampersaud, G., Gao, Z., 2012. Front-of-package nutritional labels and consumer beverage perceptions. Appl. Econ. Perspect. Policy 34 (4), 599–614.

Kim, S., Nguyen, P.H., 2016. A follow-up study on the sustained impacts of A&T's behavior change communication interventions on infant and young child feeding (IYCF) practices in Bangladesh. Paper presented at IFPRI on September 27, 2016.

Knutson, D.R., 2012. Potential impacts of (2010) dietary guidelines for Americans. Choices, 1st Quarter (1), 27.

Komlos, J., 2010. The recent decline in the height of African-American women. Econ. Hum. Biol. 8 (1), 58–66.

Kostova, D., 2011. Can the built environment reduce obesity? The impact of residential sprawl and neighborhood parks on obesity and physical activity. East. Econ. J. 37 (3), 390–402.

Kosteas, D.V., 2012. The effect of exercise on earnings: evidence from the NLSY. J. Labor Res. 33 (2), 225–250.

Kral, V.E.T., Rolls, J.B., 2011. Portion size and the obesity epidemic. In: The Oxford Handbook of the Social Science of Obesity, pp. 367–384, in Collective Volume Article by John Cawley, Oxford Handbooks Series. Oxford University Press, Oxford and New York.

Kreider, B., Pepper, V.J., Gundersen, C., Jolliffe, D., 2012. Identifying the effects of SNAP (food stamps) on child health outcomes when participation is endogenous and misreported. J. Am. Stat. Assoc. 107 (499), 958–975.

Kuku, O., Garasky, S., Gundersen, C., 2012. The relationship between childhood obesity and food insecurity: a nonparametric analysis. Appl. Econ. 44 (19–21), 2667–2677.

Kuku, O., Garasky, S., Gundersen, C., 2013. The Relationship between childhood obesity and food insecurity: a nonparametric analysis. In: The Applied Economics of Weight and Obesity, pp. 33–43, Collective Volume Article, by Taylor, Mark P. Taylor and Francis, Routledge, London and New York.

Kumar, D., Goel, N., Mittal, P.C., Misra, P., 2006. Influence of infant-feeding practices on nutritional status of under-five children. Indian J. Pediatr. 73, 417–421.

Kumar, S., Vollmer, S., 2013. Does access to improved sanitation reduce childhood diarrhea in rural India? Health Econ. 22 (4), 410–427.

Lakdawalla, D., Philipson, T., 2009. The growth of obesity and technological change. Econ. Hum. Biol. 7 (3), 283–293.

Lakdawalla, D., Zheng, Y., 2011. Food prices, income, and body weight. In: The Oxford Handbook of the Social Science of Obesity, pp. 463–479, Collective Volume Article by John, Oxford Handbooks Series. Oxford University Press, Oxford and New York.

Lamba, S., Spears, D., 2013. Caste, "cleanliness" and cash: effects of caste-based political reservations in Rajasthan on a sanitation prize. J. Dev. Stud. 49 (11), 1592–1606.

Latif, E., 2014. The impact of macroeconomic conditions on obesity in Canada. Health Econ. 23 (6), 751–759.

Lee, D.S., Lemieux, T., 2010. Regression discontinuity designs in economics. J. Econ.Lit. 48, 281–355.

Lee, H., Munk, T., 2008. Using regression discontinuity design for program evaluation. In: Proceedings of the 2008 Joint Statistical Meeting, pp. 3–7.

Lee, H., Wildeman, C., 2013. Things fall apart: health consequences of mass imprisonment for African American women. Rev. Black Polit. Econ. 40 (1), 39–52.

Lee, L.-F., Rosenzweig, M.R., Pitt, M.M., 1997. The effects of improved nutrition, sanitation, and water quality on child health in high-mortality populations. J. Econ. 77 (1), 209–235.

Leibenstein, H.A., 1957. Economic Backwardness and Economic Growth. Wiley, New York.

Lempert, D.A., 2014. The Economic Causes and Consequences of Overweight and Obesity in the United States. CUNY Academic Works, <http://academicworks.cuny.edu/gc_etds/246>.

Levine, A.J., McCrady, K.S., Boyne, S., Smith, J., Cargill, K., Forrester, T., 2011. Non-exercise physical activity in agricultural and urban people. Urban Stud. 48 (11), 2417–2427.

Levinson, F.J., 1974. Morinda: An Economic Analysis of Malnutrition Among Young Children in Rural India. Cornell/MIT International Nutrition Policy Series, Cambridge, MA.

Lhila, A., 2011. Does access to fast food lead to super-sized pregnant women and whopper babies? Econ. Hum. Biol. 9 (4), 364–380.

Li, S., Liu, Y., Zhang, J., 2011. Lose some, save some: obesity, automobile demand, and gasoline consumption. J. Environ. Econ. Manag. 61 (1), 52–66.

Liaukonyte, J., Rickard, J.B., Kaiser, M.H., Okrent, M.A., Richards, J.T., 2012. Economic and health effects of fruit and vegetable advertising: evidence from lab experiments. Food Policy 37 (5), 543–553.

Lin, B.-H., Ver, P.M., Kasteridis, P., Yen, T.S., 2014. The roles of food prices and food access in determining food purchases of low-income households. J. Policy Model. 36 (5), 938–952.

Linden, L.L., Shastry, G.K., 2012. Grain inflation: identifying agent discretion in response to a conditional school nutrition program. J. Dev. Econ. 99 (1), 128–138.

Linnemayr, S., Alderman, H., Ka, A., 2008. Determinants of malnutrition in senegal: individual, household, community variables, and their interaction. Econ. Hum. Biol. 6 (2), 252–263. UNICEF, 2008. Division of Communication. Tracking Progress on Child and Maternal Nutrition: A Survival and Development Priority. UNICEF, <http://www.unicef.org/publications/index_51656.html> (accessed March, 2015).

Liu, M., Kasteridis, P., Yen, S.T., 2013. Breakfast, lunch, and dinner expenditures away from home in the United States. Food Policy 38, 156–164.

Liu, P.J., Wisdom, J., Roberto, C.A., Liu, L.J., Ubel, P.A., 2014. Using behavioral economics to design more effective food policies to address obesity. Appl. Econ. Perspect. Policy 36 (1), 6–24.

Lividini, K., Fiedler, J.L., 2015. Assessing the promise of biofortification: a case study of high provitamin a maize in Zambia. Food Policy 54 (July), 65–77.

Logan, D.T., 2009. The transformation of hunger: the demand for calories past and present. J. Econ. Hist. 69 (2), 388–408.

Lokshin, M., Radyakin, S., 2012. Month of Birth and Children's Health in India. J. Human Resources 47 (1), 174–203.

Lokshin, M., Das Gupta, M., Gragnolati, M., Ivaschenko, O., 2005. Improving child nutrition? The integrated child development services in India. Dev. Change 36 (4), 613–640.

Long, S.K., 1991. Do the school nutrition programs supplement household food expenditures? J. Hum. Resour. 26 (4), 654–678.

Lopez, A.R., Fantuzzi, L.K., 2012. Demand for carbonated soft drinks: implications for obesity policy. Appl. Econ. 44 (22–24), 2859–2865.

Loureiro, L.M., Yen, T.S., Nayga Jr., M.R., 2012. The effects of nutritional labels on obesity. Agric. Econ. 43 (3), 333–342.

Lundberg, S., Pollak, R., Wales, T., 1997. Do husbands and wives pool their resources? Evidence from the United Kingdom child benefit. J. Hum. Resour. 32 (3), 463–480.

Lundborg, P., Nystedt, P., Rooth, D.-O., 2014. Body size, skills, and income: evidence from 150,000 teenage siblings. Demography 51 (5), 1573–1596.

Lundh, C., 2013. Was there an urban-rural consumption gap? The standard of living of workers in Southern Sweden, 1914–1920. Scand. Econ. Hist. Rev. 61 (3), 233–258.

MacEwan, J.P., Alston, J.M., Okrent, A.M., 2014. The consequences of obesity for the external costs of public health insurance in the United States. Appl. Econ. Perspect. Policy 36 (4), 696–716.

Maclean, J.C., Cawley, J., 2014. The effect of rising obesity on eligibility to serve in the U.S. public health service commissioned corps. Econ. Hum. Biol. 15, 213–224.

Mader, P., 2012. Attempting the production of public goods through microfinance: the case of water and sanitation. Econ. Res. 25 (Special issue 1), 190–214.

Maher, K.J., 2012. It's called fruit juice so it's good for me ... right?: an exploratory study of children's fruit content inferences made from food brand names and packaging. J. Appl. Business Res. 28 (3), 501–513.

Majumder, M.A., 2013. Does obesity matter for wages? Evidence from the United States. Econ. Pap. 32 (2), 200–217.

Malapit, H.J.L., et al., 2013. Women's Empowerment in Agriculture, Production Diversity, and Nutrition: Evidence From Nepal. IFPRI, Washington, DC, USA, IFPRI. Discussion Paper 01313.

Mamiro, P.S., Kolsteren, P., Roberfroid, D., Tatala, S., Opsomer, A.S., Van Camp, J.H., 2005. Feeding practices and factors contributing to wasting, stunting, and iron-deficiency anaemia among 3–23-month old children in Kilosa district, rural Tanzania. J. Health, Popul. Nutr. 222–230.

Mandal, B., Chern, S.W., 2011. A multilevel approach to model obesity and overweight in the U.S.. Int. J. Appl. Econ. 8 (2), 1–17.

Mandal, B., Powell, M.L., 2014. Child care choices, food intake, and children's obesity status in the United States. Econ. Hum. Biol. 14, 50–61.

Mangyo, E., 2005. Who Benefits More From Higher Household Consumption? The Intra-household Allocation of Nutrients in China (University of Michigan Doctoral theses).

Mangyo, E., 2008. Who benefits more from higher household consumption? The intra-household allocation of nutrients in China. J. Dev. Econ. 86 (2), 296–312.

Marjan, Z.M., Kandiah, M., Lin, K.G., Siong, T.E., 2002. Socioeconomic profile and nutritional status of children in rubber smallholdings. Asia Pac. J. Clin. Nutr. 11, 133–141.

Marlow, L.M., 2014. Determinants of state laws addressing obesity. Appl. Econ. Lett. 21 (1–3), 84–89.
Marlow, L.M., 2015. Big box stores and obesity. Appl. Econ. Lett. 22 (10–12), 938–944.
Marlow, L.M., Shiers, F.A., 2012. The relationship between fast food and obesity. Appl. Econ. Lett. 19 (16–18), 1633–1637.
Marlow, L.M., Shiers, F.A., 2013. The Relationship Between Fast Food and Obesity.
Martin, G.L., Schoeni, F.R., Andreski, M.P., 2010. Trends in health of older adults in the United States: past, present, future. Demography 47 (Suppl.), S17–S40.
Martinson, L.M., McLanahan, S., Brooks-Gunn, J., 2012. Race/ethnic and nativity disparities in child overweight in the United States and England. Ann. Am. Acad. Polit. Soc. Sci. 643, 219–238.
Martorell, R., Melgar, P., Maluccio, J.A., Stein, A.D., Rivera, J.A., 2010. The nutrition intervention improved adult human capital and economic productivity. J. Nutr. 140 (2), 411–414.
Mazumdar, D., 1959. The marginal productivity theory of wages and disguised unemployment. Rev. Econ. Stud. 26 (3), 190–197.
McDermott, J., Johnson, N., Kadiyala, S., Kennedy, G., Wyatt, A.J., 2015. Agricultural research for nutrition outcomes – rethinking the agenda. Food Secur. 7 (3), 593–607.
McEwan, P.J., 2013. The impact of chile's school feeding program on education outcomes. Econ. Educ. Rev. 32, 122–139.
McGranahan, G., 2015. Realizing the right to sanitation in deprived urban communities: meeting the challenges of collective action, coproduction, affordability, and housing tenure. World Dev. 68, 242–253.
Mehta, K.N., Chang, W.V., 2011. Obesity and mortality. In: The Oxford Handbook of the Social Science of Obesity, pp. 502–516, Collective Volume Article by John, Oxford Handbooks Series. Oxford University Press, Oxford and New York.
Mellor, M.J., 2011. Do cigarette taxes affect children's body mass index? The effect of household environment on health. Health Econ. 20 (4), 417–431.
Mello, M.M., 2010. Federal trade commission regulation of food advertising to children: Possibilities for a reinvigorated role. J. Health Politics Policy Law 35 (2), 227–276.
Miao, Z., Beghin, C.J., Jensen, H.H., 2012. Taxing sweets: sweetener input tax or final consumption tax? Contem. Econ. Policy 30 (3), 344–361.
Miao, Z., Beghin, C.J., Jensen, H.H., 2013. Accounting for product substitution in the analysis of food taxes targeting obesity. Health Econ. 22 (11), 1318–1343.
Miller, P.D., 2011. Maternal work and child overweight and obesity: the importance of timing. J. Family Econ. Issues 32 (2), 204–218.
Millimet, L.D., Tchernis, R., 2015. Persistence in Body Mass Index in a recent cohort of US children. Econ. Hum. Biol. 17, 157–176.
Millimet, L.D., Tchernis, R., Husain, M., 2010. School nutrition programs and the incidence of childhood obesity. J. Hum. Resour. 45 (3), 640–654.
Mojduszka, M.E., Caswell, A.J., Harris, J.M., 2001. Consumer choice of food products and the implications for price competition and government policy. Agribusiness 17 (1), 81–104.
Molini, V., Nube, M., 2007. Is the Nutritional Status of Males and Females Equally Affected by Economic Growth? Evidence From Vietnam in the 1990s. World Institute for Development Economic Research (UNU-WIDER), Working Papers: UNU-WIDER Research Paper RP2007/54.
Monteiro, C.A., Benicio, M.H.D.A., Conde, W.L., Konno, S., Lovadino, A.L., Barros, A.J., et al., 2010. Narrowing socioeconomic inequality in child stunting: the Brazilian experience, 1974–2007. Bull. W.H.O. 88 (4), 305–311.
Monteverde, M., Noronha, K., Palloni, A., Novak, B., 2010. Obesity and excess mortality among the elderly in the United States and Mexico. Demography 47 (1), 79–96.
Morris, J.R., 1999. Market constraints on child care quality. Ann. Am. Acad. 563, 130–145.
Mu, R., de Brauw, A., 2015. Migration and young child nutrition: evidence from rural china. J. Popul. Econ. 28 (3), 631–657.

Murty, K.N., Radhakrishna, R., 1982. Agricultural prices, income distribution and demand patterns in a low-income country. In: Kalman, R.E., Martinez, J. (Eds.), Computer Applications in Food Production and Agricultural Engineering. North Holland Publishing Company, Amsterdam.

Muth, M.K., 2010. Theme overview: addressing the obesity challenge. Choices, 3rd Quarter (2010) 25 (3).

Nakamuro, M., Inui, T., Senoh, W., Hiromatsu, T., 2015. Are television and video games really harmful for kids? Contem. Econ. Policy 33 (1), 29–43.

Nayga Jr., M.R., 2013. Obesity and heart disease awareness: a note on the impact of consumer characteristics using qualitative choice analysis. Appl. Econ. Weight Obes. 88–90.

Nelson, J.A., 2015. Husbandry: a (feminist) reclamation of masculine responsibility for care. Camb. J. Econ., pp. 1–15., Oxford University Press, Oxford, UK.

Nesbitt, et al., 2014. Baseline for consumer food safety knowledge and behaviour in Canada. Food Control 38, 157–173.

Ngnikam, E., Mougoué, B., Roger, F., Isidore, N., Ghislain, T., Jean, M., 2012. Water, wastes, and children's health in low-income neighbourhoods of Yaoundé. In: Charron, D.F. (Ed.), Insight and Innovation in International Development, Ecohealth Research in Practice, Innovative Applications of an Ecosystem Approach to Health. International Development Research Centre Ottawa, Ontario, Canada, pp. 215–231, Springer, Chapter 20.

Nisbett, N., Wach, E., Haddad, L.J., El Arifeen, S., 2015. What drives and constrains effective leadership in tackling child undernutrition? Findings from Bangladesh, Ethiopia, India and Kenya. Food Policy 53, 33–45. Available from: <http://dx.doi.org/10.1016/j.foodpol.2015.04.001>.

Novak, L., 2014. The impact of access to water on child health in senegal. Rev. Dev. Econ. 18 (3), 431–444.

Okali, C. 2011. Achieving transformative change for rural women's empowerment. In: Expert Group Meeting Enabling Rural Women's Economic Empowerment: Institutions, Opportunities And Participation. UN Women In cooperation with FAO, IFAD and WFP.

Okrent, M.A., Alston, M.J., 2012. The effects of farm commodity and retail food policies on obesity and economic welfare in the United States. Am. J. Agric. Econ. 94 (3), 611–646.

Oldewage-Theron, W.H., Napier, C.E., 2011. Nutrition education tools for primary school children in the Vaal Region. Dev. South. Afr. 28 (2), 283–292.

O'leary, C.J., Kline, K., 2014. Use of Supplemental Nutritional Assistance Program Benefits by Unemployment Insurance Applicants in Michigan During the Great Recession. Available at SSRN 2401120.

Oliver, J.E., Lee, T., 2005. Public opinion and the politics of obesity in America. J. Health Politics Policy Law 30 (5), 923–954.

Omamo, S.W., Gentilini, U., Sandstrom, S., 2010. Innovations in food assistance: issues, lessons and implications. In: Omamo, S.W., Gentilini, U., Sandström, S. (Eds.), Revolution: From Food Aid to Food Assistance. Innovations in Overcoming Hunger. World Food Programme, Rome, pp. 1–18.

Palmer-Jones, R., Sen, K., 2001. On India's poverty puzzles and statistics of poverty. Econ. Polit. Wkly 211–217.

Panagariya, A., 2013. Does India really suffer from worse child malnutrition than Sub-Saharan Africa? Econ. Polit. Wkly 48 (18), 98–111.

Parks, J.C., Smith, A.D., Alston, J.M., 2011. The Effects of the Food Stamp Program on Energy Balance and Obesity. University of California, Davis.

Parr, K.E., 2012. Three Essays on the Economics of Obesity (Doctoral dissertations). Paper AAI3510502. <http://digitalcommons.uconn.edu/dissertations/AAI3510502>.

Patil, S.R., Arnold, B.F., Salvatore, A., Briceno, B., Colford, J.M. Jr., Gertler, P.J., 2013. A Randomized, Controlled Study of a Rural Sanitation Behavior Change Program in Madhya Pradesh, India, The World Bank, Policy Research Working Paper Series: 6702.

Paudel, L., Adhikari, M., Houston, J., Paudel, K.P., 2013. Low carbohydrate information, consumer health preferences and market demand of fruits in the United States. Appl. Econ. Weight Obes. 102–106.

Peckham, J.G., 2013. Three Essays Evaluating Health Impacts of the National School Lunch Program. All Dissertations. Paper 1235.

Persson, T.H., 2002. Welfare calculations in models of the demand for sanitation. Appl. Econ. 34 (12), 1509–1518.

Philipson, J.T., Posner, R.A., 2011. Economic perspectives on obesity policy. In: The Oxford Handbook of the Social Science of Obesity, pp. 609–619, Collective Volume Article by John, Oxford Handbooks Series. Oxford University Press, Oxford and New York.

Pickett, E.K., Wilkinson, R.G., 2012. Income inequality and psychosocial pathways to obesity. In Insecurity, Inequality, and Obesity in Affluent Societies, pp. 179–198, Collective Volume Article by, Offer Avner, Pechey Rachel and Ulijaszek Stanley, Proceedings of the British Academy, vol. 174. Oxford University Press, Oxford and New York.

Pinstrup-Anderson, P., 2013. Nutrition-senstive food system: from rhetoric to action. vol. 382, pp. 375–376. <www.thelancet.com>.

Pinstrup-Andersen, P., Caicedo, E., 1978. The potential impact of changes in income distribution on food demand and human nutrition. Am. J. Agric. Econ. 60, 402–415.

Pitt, M.M., 1983. Food preferences and nutrition in rural Bangladesh. Rev. Econ. Stat. 65, 105–114.

Pitt, M.M., Rosenzweig, M., 1985. Health and nutrient consumption across and within farm households. Rev. Econ. Stat. 67, 212–223.

Popkin, B.M., Adair, L.S., Ng, S.W., 2012. Global nutrition transition and the pandemic of obesity in developing countries. Nutr. Rev. 70 (1), 3–21.

Porter, C., 2010. Safety nets or investment in the future: does food aid have any long-term impact on children's growth? J. Int. Dev. 22, 1134–1145.

Powell, L.M., Chriqui, F.J., 2011. Food taxes and subsidies: evidence and policies for obesity prevention. In: The Oxford Handbook of the Social Science of Obesity, pp. 639–164, Collective Volume Article by John Cawley, Oxford Handbooks Series. Oxford University Press, Oxford and New York.

Powell, L.M., Han, E., 2011. Adult obesity and the price and availability of food in the United States. Am. J. Agric. Econ. 93 (2), 378–384, Projecting the Effect of Changes in Smoking and Obesity on Future Life Expectancy in the United States Press/IFPRI, 3-21, 1993.

Price, J., 2012. The importance of parental knowledge: evidence from weight report cards in Mexico. Agric. Resour. Econ. Rev. 41 (1), 92–99.

Price, J., Swigert, J., 2012. Within-family variation in obesity. Econ. Hum. Biol. 10 (4), 333–339.

Prina, S., Royer, H., 2014. The importance of parental knowledge: evidence from weight report cards in Mexico. J. Health Econ. 37, 232–247.

Qian, Y., 2014. The effect of school and neighborhood environmental factors on childhood obesity. University of Arkansas. Q. J. Econ. 112, 729–758.

Quisumbing, A., 2012. Innovative approaches to gender and food security: changing attitudes, changing behaviours. IDS Insights issue 82.

Quisumbing, A., Meizen-Dick, R., 2012. Women in Agriculture: Closing the Gender Gap, IFPRI Global Policy Report. IFPRI, Washington, DC, www.ifpri.org/gfpr/2012/women-agriculture.

Quisumbing, A.R., Brown, L.R., Feldstein, H.S., Haddad, L., Peña, C., 1995. Women: The Key to Food Security. IFPRI, Washington, DC, USA.

Rabbani, M., Prakash, V.A., Sulaiman, M., 2006. Impact Assessment of CFPR/TUP (Challenging the Frontiers of Poverty Reduction/Targeting the Ultra Poor): A Descriptive Analysis Based on 2002–2005 Panel Data. CFPR/TUP Working Paper 12. BRAC (Bangladesh Rural Advancement Committee) Research and Evaluation Division and the Aga Khan Foundation, Dhaka and Ottawa.

Ramalingaswami, V., Jonsson, U., Rohde, J., 1996. "Commentary: The Asian Enigma," The Progress of Nations 1996. UNICEF, New York.

Ramli, Agho, K.E., Inder, K.J., Bowe, S.J., Jacobs, J., Dibley, M.J., 2009. Prevalence and risk factors for stunting and severe stunting among under-fives in North Maluku Province of Indonesia. Biomed Central (BMC) Pediatr. 9, 64.

Rashad, I., Grossman, M., Chou, S.-Y., 2006. The super size of America: an economic estimation of Body Mass Index and obesity in adults. East. Econ. J. 32 (1), 133–148.

Ravallion, M., 1990. Reaching the Poor Through Rural Public Employment: A Survey of Theory and Evidence (No. 94).

Ravi, S., Engler, M., 2015. Workfare as an effective way to fight poverty: the case of India's NREGS. World Dev. 67, 57–71.

Reardon, T., Chen, K., Minten, B., Adriano, L., 2012. The Quiet Revolution in Staple Food Value Chains. Asian Development Bank (ADB)/IFPRI, Manila/Washington, DC.

Redden, P.J., Haws, L.K., 2013. Healthy satiation: the role of decreasing desire in effective self-control. J. Consum. Res. 39 (5), 1100–1114.

Reddy, A.A., 2010. Regional disparities in food habits and nutritional intake in Andhra Pradesh, India. Reg. Sect. Econ. Stud. 10 (2), 125–134.

Reddy, B.S., Snehalatha, M., 2011. Sanitation and personal hygiene: what does it mean to poor and vulnerable women? Indian J. Gender Stud. 18 (3), 381–404.

Reifschneider, J.M., Hamrick, S.K., Lacey, N.J., 2011. Exercise, eating patterns, and obesity: evidence from the ATUS and its eating and health module. Soc. Indicators Res. 101 (2), 215–219.

Resnick, S.G., Rosenheck, R.A., 2015. Integrating peer-provided services: a quasi-experimental study of recovery orientation, confidence, and empowerment. Psychiatr. Serv. 59 (11), 1307–1314.

Reyes, H., Pérez-Cuevas, R., Sandoval, A., Castillo, R., Santos, J.I., Doubova, S.V., et al., 2004. The family as a determinant of stunting in children living in conditions of extreme poverty: a case-control study. BMC Public Health 4, 57.

Rhoe, V., Babu, S., Reidhead, W., 2008. An analysis of food security and poverty in Central Asia – case study from Kazakhstan. J. Int. Dev. 20, 452–465.

Richards, J.T., Padilla, L., 2009. Promotion and fast food demand. Am. J. Agric. Econ. 91 (1), 168–183.

Riera-Crichton, D., Tefft, N., 2014. Macronutrients and obesity: revisiting the calories in, calories out framework. Econ. Hum. Biol. 14, 33–49.

Roberto, A.C., Brownell, D.K., 2011. The imperative of changing public policy to address obesity. In: The Oxford Handbook of the Social Science of Obesity, pp. 587–608, Collective Volume Article by John, Oxford Handbooks Series. Oxford University Press, Oxford and New York.

Robinson, A.C., Zheng, X., 2011. Household food stamp program participation and childhood obesity. J. Agric. Resour. Econ. 36 (1), 1–13.

Roemling, C., Qaim, M., 2013. Dual burden households and intra-household nutritional inequality in Indonesia. Econ. Hum. Biol. 11 (4), 563–573.

Rosenbaum, P.R., Rubin, D.B., 1983. The central role of the propensity score in observational studies for causal effects. Biometrika 70 (1), 41–55.

Roy, M., Millimet, L.D., Tchernis, R., 2012. Federal nutrition programs and childhood obesity: inside the black box. Rev. Econ. Household 10 (1), 1–38.

Ruel, M.T., Alderman, H., 2013. Nutrition-sensitive interventions and programmes: how can they help accelerate progress in improving maternal and child nutrition? Lancet 382 (9891), 536–551.

Runge, C.F., 2011. Should soft drinks be taxed more heavily? Choices, 3rd Quarter 26 (3).

Sabates-Wheeler, R., Devereux, S., 2010. Cash transfers and high food prices: explaining outcomes on Ethiopia's Productive Safety Net Programme. Food Policy, 35 (4), 274–285.

Sahn, D., Alderman, H., 1997. On the determinants of nutrition in Mozambique: the importance of age-specific effects. World Dev. 25 (4), 577–588.

Sahn, D.E., 1988. The effect of price and income changes on food-energy intake in Sri Lanka. Econ. Dev. Cult. Change 36, 315–340.

Sallis, F.J., Adams, A.M., Ding D., 2011. Physical activity and the built environment. In: The Oxford Handbook of the Social Science of Obesity, pp. 433–451, Collective Volume Article by John, Oxford Handbooks Series. Oxford University Press, Oxford and New York.

Salois, J.M., 2012. Obesity and diabetes, the built environment, and the "local" food economy in the United States, 2007. Econ. Hum. Biol. 10 (1), 35–42.

Sarma, S., Zaric, G.S., Campbell, M.K., Gilliland, J., 2014. The effect of physical activity on adult obesity: evidence from the Canadian NPHS panel. Econ. Hum. Biol. 14, 1–21.

Sassi, F., 2010. The impact of interventions. In: Obesity and the Economics of Prevention: Fit Not Fat, Collective Volume Article, by Sassi, Franco, pp. 175–209, in Association with the Organisation for Economic Co-operation and Development. Elgar, Cheltenham, U.K. and Northampton, MA.

Schmeiser, D.M., 2012. The impact of long-term participation in the supplemental nutrition assistance program on child obesity. Health Econ. 21 (4), 386–404.

Schuldt, P.J., Schwarz, N., 2010. The "organic" path to obesity? Organic claims influence calorie judgments and exercise recommendations. Judgment Decis. Making 5 (3), 144–150.

Schuring, J., 2013. The Impact of Maternal Occupation and Pre-pregnancy Weight Status on Childhood Obesity: A Comparative Analysis of the United States and the United Kingdom.

Scrimshaw, N.S., SanGiovanni, J.P., 1997. Synergism of nutrition, infection, and immunity: an overview. Am. J. Clin. Nutr. 66 (2), 464S–477S.

Semba, R.D., de Pee, S., Sun, K., Sari, M., Akhter, N., Bloem, M.W., 2008. Effect of parental formal education on risk of child stunting in Indonesia and Bangladesh: a cross-sectional study. Lancet 371 (9609), 322–328.

Sen, A., 1981. Poverty and Famines: An Essay on Entitlement and Deprivation. Clarendon Press, Oxford.

Sen, A., 1985. Commodities and Capabilities. Oxford University Press, New York.

Sen, B., 2012. Is there an association between gasoline prices and physical activity? Evidence from American time use data. J. Policy Anal. Manag. 31 (2), 338–366.

Sethuraman, K., 2008. The Role of Women's Empowerment and Domestic Violence in Child Growth and Undernutrition in a Tribal and Rural Community in South India. World Institute for Development Economic Research (UNU-WIDER), Working Papers: RP2008/15, 28 p.

Shah, M., Steinberg, B.M., 2015. Workfare and Human Capital Investment: Evidence From India. National Bureau of Economic Research, Inc, NBER Working Papers: 21543.

Shah, U., 2011. Impact assessment of nutritional supplement program in urban settings: a study of under nutrition in slum community of Mumbai. J. Soc. Dev. Sci. 1 (1), 24–35.

Shekar, M., Kakietek, J., D'Alimonte, M., Walters, D., Rogers, H., Dayton, E.J., et al., 2016. Investing in Nutrition the Foundation for Development: An Investment Framework to Reach Global Nutrition Targets. World Bank.

Simmons, O.W., Zlatoper, J.T., 2010. Obesity and motor vehicle deaths in the USA: a state-level analysis. J. Econ. Stud. 37 (5-6), 544–556.

Singh, A., 2011. Inequality of opportunity in Indian children: the case of immunization and nutrition. Popul. Res. Policy Rev. 30 (6), 861–883.

Singh, A., Park, A., Dercon, S., 2014. School meals as a safety net: an evaluation of the midday meal scheme in India. Econ. Dev. Cult. Change, University of Chicago Press 62 (2), 275–306.

Skinner, C., 2012. State immigration legislation and SNAP take-up among immigrant families with children. J. Econ. Issues 46 (3), 661–682.

Skoufias, E., 2005. PROGRESA and Its Impacts on the Welfare of Rural Households in Mexico, Research Report 139. International Food Policy Research Institute, Washington, DC, pp. xiv, 84.

Skoufias, E., Di Maro, V., Gonzalez-Cossio, T., Rodriguez Ramirez, S., 2011. Food quality, calories and household income. Appl. Econ. 43 (28–30), 4331–4342.

Skoufias, E., Tiwari, S., Zaman H., 2011. Can We Rely on Cash Transfers to Protect Dietary Diversity During Food Crises? Estimates From Indonesia. The World Bank, Policy Research Working Paper Series: 5548.

Smed, S., Jensen, J.D., Denver, S., 2007. Socio-economic characteristics and the effect of taxation as a health policy instrument. Food Policy 32 (5–6), 624–639.

Smith, G.T., Tasnadi, A., 2014. The economics of information, deep capture, and the obesity debate. Am. J. Agric. Econ. 96 (2), 533–541.

Smith, L.C., Haddad, L., 2015. Reducing child undernutrition: past drivers and priorities for the post-MDG era. World Dev. 68, 180–204.

Spears, D., 2012. How Much International Variation in Child Height Can Sanitation Explain? Princeton University, Woodrow Wilson School of Public and International Affairs, Research Program in Development Studies, Working Papers: 1436.

Spears, D., Lamba, S., 2013. Effects of Early-Life Exposure to Sanitation on Childhood Cognitive Skills: Evidence from India's Total Sanitation Campaign, The World Bank, Policy Research Working Paper Series: 6659.

SPRING, 2014. Understanding the Women's Empowerment Pathway Brief #4. Improving.

Springmann, M., et al., 2016. Global and regional health effects of future food production under climate change: a modelling study. Lancet 387 (10031), 1937–1946.

Stock, J.H., Watson, M.W., 2011. Introduction to Econometrics. third ed. Addison-Wesley, New York.

Strauss, J., 1982. Determinants of food consumption in rural Sierra Leone. J. Dev. Econ. 11, 327–353.

Strauss, Thomas, 1998. Health, nutrition and economic development. J. Econ. Lit. 36 (2), 766–817.

Streletskaya, N.A., Rusmevichientong, P., Amatyakul, W., Kaiser, H.M., 2014. Taxes, subsidies, and advertising efficacy in changing eating behavior: an experimental study. Appl. Econ. Perspect. Policy 36 (1), 146–174.

Subramanian, S., Deaton, A., 1996. The demand for food and calories. J. Polit. Econ. 133–162.

Suryanarayana, M.H., Silva, D., 2007. Is targeting the poor a penalty on the food insecure? Poverty and food insecurity in India. J. Human Dev. 8 (1), 89–107.

Swain, R.B., Varghese, A., 2014. Evaluating the impact of training in self-help groups in India. Eur. J. Dev. Res. 26 (5), 870–885.

Takeuchi, L.R., 2015. Intra-Household Inequalities in Child Rights and Well-Being: A Barrier to Progress? World Institute for Development Economic Research (UNU-WIDER), Working Paper Series: UNU-WIDER Research Paper.

Teachman, J., Tedrow, L., 2013. Veteran status and body weight: a longitudinal fixed-effects approach. Popul. Res. Policy Rev. 32 (2), 199–220.

Teshome, B., Kogi-Makau, W., Getahun, Z., Taye, G., 2009. Magnitude and determinants of stunting in children underfive years of age in food surplus region of Ethiopia: the case of west gojam zone. Ethiop. J. Health Dev. 23.

Thapa, R.J., Lyford, P.C., 2014. Behavioral economics in the school lunchroom: can it affect food supplier decisions? A systematic review. Int. Food Agribusiness Manag. Rev. 17 (Special issue A), 187–208.

Thomas, D., 1990. Intra-household resource allocation: an inferential approach. J. Hum. Resour. 25 (4), 635–664.

Thorpe, E.K., Allen, L., Joski, P., 2015. The role of chronic disease, obesity, and improved treatment and detection in accounting for the rise in healthcare spending between 1987 and 2011. Appl. Health Econ. Health Policy 13 (4), 381–387.

Tian, X., Yu, X., 2013. The demand for nutrients in China. Front. Econ. China 8 (2), 186–206.

Tiffin, R., Dawson, P.J., 2002. The demand for calories: some further estimates from Zimbabwe. J. Agricu. Econ. 53 (2), 221–232.

Timmer, C.P., 1981. Is there "curvature" in the Slutsky matrix? Rev. Econ. Stat. 63, 395–402.

Timmer, C.P., 2013. Coping with climate change: a food policy approach. In: 2013 Conference (57th), February 5–8, 2013, Sydney, Australia (No. 152188). Australian Agricultural and Resource Economics Society.

Timmer, C.P., 2015. Food Security and Scarcity: Why Ending Hunger Is So Hard. University of Pennsylvania Press, Philadelphia.

Timmer, C.P., Alderman, H., 1979. Estimating consumption parameters for food policy analysis. Am. J. Agric. Econ. 61, 982–994.

Todd, E.J., Winters, P., 2011. The effect of early interventions in health and nutrition on on-time school enrollment: evidence from the oportunidades program in rural Mexico. Econ. Dev. Cult. Change 59 (3), 549–581.

Todd, E.J., Zhen, C., 2010. Can taxes on calorically sweetened beverages reduce obesity? Choices, 3rd Quarter 25 (3).

Tomer, J., 2011. What causes obesity? And why has it grown so much? Challenge 54 (4), 22–49.

Tsegai, W.D., Kormawa, P., 2009. The determinants of urban households' demand for cassava and cassava products in Kaduna, Northern Nigeria: an application of the AIDS model. Eur. J. Dev. Res. 21 (3), 435–447.

Ukwuani, F.A., Suchindran, C.M., 2003. Implications of women's work for child nutritional status in Sub-Saharan Africa: a case study of Nigeria. Soc. Sci. Med. 56 (10), 2109–2121.

Ulijaszek, S., Schwekendiek, D., 2013. Intercontinental differences in overweight of adopted Koreans in the United States and Europe. Econ. Hum. Biol. 11 (3), 345–350.

UNICEF, 2013. <http://www.unicef.org/publications/index_73682.html>.

UNICEF, World Health Organization, and World Bank, 2014. Joint Child Malnutrition Estimates: Levels and Trends (July 2015 update).

USAID, 2014. DHS Program. US Agency for International Development. <http://dhsprogram.com/>.

Van Hanswijck de Jonge, Waller, G., Stettler, N., 2003. Ethnicity modifies seasonal variations in birth weight and weight gain of infants. J. Nutr. 133 (5), 1415–1418.

Vander, W., Jillon, S., 2012. The relationship between Body Mass Index and unhealthy weight control behaviors among adolescents: the role of family and peer social support. Econ. Hum. Biol. 10 (4), 395–404.

Vandewater, E.A., Wartella, E.A., 2011. Food marketing, television, and video games. In: The Oxford Handbook of the Social Science of Obesity, pp. 350–66, Collective Volume Article by John, Oxford Handbooks Series. Oxford University Press, Oxford and New York.

Varela-Silva, M.I., Azcorra, H., Dickinson, F., Bogin, B., Frisancho, A., 2009. Influence of maternal stature, pregnancy age, and infant birth weight on growth during childhood in Yucatan, Mexico: a test of the intergenerational effects hypothesis. Am. J. Hum. Biol. 21, 657–663.

Variyam, J.N., 1999b. Mother's nutrition knowledge and children's dietary intakes. Am. J. Agric. Econ. 81 (2), 373–384.

Variyam, J.N., Blaylock, J., Smallwood, D., 1998. Information effects of nutrient intake determinants on cholesterol consumption. J. Agric. Resour. Econ. 23 (1), 110–125.

Variyam, J.N., Blaylock, J., Smallwood, D., 1999a. Information, endogeneity, and consumer health behavior: application to dietary intakes. Appl. Econ. 31 (2), 217–226.

Variyam, J.N., Blaylock, J., Smallwood, D., 2002. Characterizing the distribution of macronutrient intake among U.S. adults: a quantile regression approach. Am. J. Agric. Econ. 84 (2), 454–466.

Vecchi, G., Coppola, M., 2006. Nutrition and growth in Italy, 1861–1911: what macroeconomic data hide explorations. Econ. History 43 (3), 438–464.

Ver Ploeg, M., 2010. Food environment, food store access, consumer behavior, and diet. Choices, 3rd Quarter 25 (3).

Ver Ploeg, M., 2011. Food assistance and obesity. In: The Oxford Handbook of the Social Science of Obesity, pp. 415–432, Collective Volume Article by Cawley, John, Oxford Handbooks Series. Oxford University Press, Oxford and New York.

Vermeersch, C., Kremer, M., 2005. School Meals, Educational Achievement, and School Competition: Evidence From a Randomized Evaluation. World Bank.

Viego, V.N., Temporelli, L.K., 2011. Sobrepeso y obesidad en Argentina. Un analisis basado en tecnicas de econometria espacial. (Overweight and obesity in Argentina: a spatial approach. With English summary). Estudios de Economia Aplicada 29 (3).

Villa, K.M., Barrett, C.B., Just, D.R., 2011a. Differential nutritional responses across various income sources among East African Pastoralists: intrahousehold effects, missing markets and mental accounting. J. Afr. Econ. 20 (2), 341–375.

Villa, K.M., Barrett, C.B., Just, D.R., 2011b. Whose fast and whose feast? Intrahousehold asymmetries in dietary diversity response among East African Pastoralists. Am. J. Agric. Econ. 93 (4), 1062–1081.

Waage, J., Yap, C., Bell, S., Levy, C., Mace, G., Pegram, T., et al., 2015. Governing the UN sustainable development goals: interactions, infrastructures, and institutions. Lancet Glob. Health 3 (5), e251–e252.

Wada, R., Tekin, E., 2010. Body composition and wages. Econ. Hum. Biol. 8 (2), 242–254.

Walker, E.R., Kawachi, I., 2011. Race, ethnicity, and obesity. In: The Oxford Handbook of the Social Science of Obesity, pp. 257–275, Collective Volume Article by Cawley, John, Oxford Handbooks Series. Oxford University Press, Oxford and New York.

Wamani, H., Tylleskär, T., Åstrøm, A.N., Tumwine, J.K., Peterson, S., 2004. Mothers' education but not fathers' education, household assets or land ownership is the best predictor of child health inequalities in rural Uganda. Int. J. Equity Health 3, 9.

Wansink, B., 2011. Mindless eating: environmental contributors to obesity. In: The Oxford Handbook of the Social Science of Obesity, pp. 385–414, Collective Volume Article by Cawley, John, Oxford Handbooks Series. Oxford University Press, Oxford and New York.

Ward, J.O., Sanders, J.H., 1980. Nutritional determinants and migration in the Brazilian Northeast: a case study of rural and Urban Ceara. Econ. Dev. Cult. Change 29, 141–163, Washington.

Webb, P., Block, S., 2012. Support for agriculture during economic transformation: impacts on poverty and undernutrition. Proc. Natl. Acad. Sci. U.S.A. 109 (31), 12309–12314.

Wehby, L., George, L., Yang, M., 2012. Depression, antidepressant use and weight gain. Int. J. Appl. Econ. 9 (2), 1–38.

Wen, M., Maloney, N.T., 2014. Neighborhood socioeconomic status and BMI differences by immigrant and legal status: evidence from Utah. Econ. Hum. Biol. 12, 120–131.

Whittington, D., Jeuland, M., Barker, K., Yuen, Y., 2012. Setting priorities, targeting subsidies among water, sanitation, and preventive health interventions in developing countries. World Dev. 40 (8), 1546–1568.

WHO, 2005 and 2006. Data tables on deaths by age, sex and cause for the year 2002; 20 leading causes of deaths and burden of disease at all ages; 20 leading causes of DALYs due to selected risk factors for each OECD countries, BRIICS countries and the world. Data provided by MHI/EIP/WHO between November 2005 and April 2006.

WHO, UNICEF, 2005. World Malaria Report 2005. Geneva (Switzerland) and New York (USA).

Willey, B.A., Cameron, N., Norris, S.A., Pettifor, J.M., Griffiths, P.L., 2009. Socio-economic predictors of stunting in preschool children: a population-based study from Johannesburg and Soweto. S. Afr. Med. J. 99, 450–456.

Williamson-Gray, C., 1982. Food Consumption Parameters for Brazil and Their Application to Food Policy. IFPRI Research Report 32. International Food Policy Research Institute, Washington, DC.

Wilson, E.S., 2012. Marriage, gender and obesity in later life. Econ. Hum. Biol. 10 (4), 431–453.

Wisman, D.J., Capehart, W.K., 2010. Creative destruction, economic insecurity, stress, and epidemic obesity. Am. J. Econ. Sociol. 69 (3), 936–982.

Wolfe, B.L., Behrman, J.R., 1983. Is income overrated in determining adequate nutrition? Econ. Dev. Cult. Change 31 (3), 525–550.

Wooldridge, J.M., 2002. Econometric analysis of cross section and panel data. MIT Press, Boston MA and nutrition: rice in a West African setting. World Dev. 16 (9), 1083–1098.

World Bank, 2007. Global Monitoring Report 2007, Ch. 3. Washington, DC.

World Bank, 2008. World Development Report (WDR) 2008: Agriculture for Development, Ch. 7. Washington, DC.

World Bank, 2015a. World Development Indicators. <http://data.worldbank.org/indicator> (accessed 07.05.15.).

World Bank, 2015b. Global Monitoring Report 2015. <http://www.worldbank.org/en/publication/global-monitoring-report> (accessed 29.04.16.).

World Health Organization, 2014. Global Nutrition Targets 2025: Policy Brief Series.

Yang, M., Huang, R., 2014. Asymmetric association between exposure to obesity and weight gain among adolescents. East. Econ. J. 40 (1), 96–118.

Yang, Y., Goldhaber-Fiebert, J.D., Wein, L.M., 2013. Analyzing screening policies for childhood obesity. Manag. Sci. 59 (4), 782–795.

Yin, H., Wallace, H., Abebayehu, T., 2011. Impacts of economic and psychological factors on adult obesity and food program participation: evidence from the NLSY panel. Am. J. Agric. Econ. 94 (2), 331–337.

You, J., Imai, K., Gaiha, R., 2016. Declining nutrient intake in a growing China: does household heterogeneity matter? World Dev. 77, 171–191.

Yu, H.J., 2011. Parental communication style's impact on children's attitudes toward obesity and food advertising. J. Consum. Aff. 45 (1), 87–107.

Zagorsky, J.L., Smith, P.K., 2011. The freshman 15: a critical time for obesity intervention or media myth? Soc. Sci. Q. 92 (Special issue), 1389–1407.

Zavodny, M., 2013. Does Weight Affect Children's Test Scores and Teacher Assessments differently? Econ. Educ. Rev. 34, 135–145.

Zeng, W., 2013. Adult obesity: panel study from native Amazonians. Econ. Hum. Biol. 11 (2), 227–235.

Zewdie, T., Abebaw, D., 2013. Determinants of child malnutrition: empirical evidence from Kombolcha District of Eastern Hararghe Zone, Ethiopia. Q. J. Int. Agric. 52 (4), 357–372.

Zhao, Z., Kaestner, R., 2010. Effects of urban sprawl on obesity. J. Health Econ. 29 (6), 779–787.

Zhen, C., Brissette, I.F., Ruff, R.R., 2014. By ounce or by calorie: the differential effects of altenative sugar-sweetened beverage tax strategies. Am. J. Agric. Econ. 96 (4), 1070–1083.

Zheng, Y., McLaughlin, W.E., Kaiser, M.H., 2013. Taxing food and beverages: theory, evidence, and policy. Am. J. Agric. Econ. 95 (3), 705–723.

Index

Note: Page numbers followed by "*f*," "*t*," and "*b*" refer to figures, tables, and boxes, respectively.

G

H

I

S

T

Printed in the United States
By Bookmasters